ACKNOWLEDGMENTS

The SFI Press would not exist without the support of William H. Miller and the Miller Omega Program, Andrew Feldstein and the Feldstein Program on History, Law, and Regulation, and Alana Levinson-Labrosse.

THE ENERGETICS OF
COMPUTING IN LIFE & MACHINES

DAVID H. WOLPERT, CHRIS KEMPES,
PETER F. STADLER & JOSHUA A. GROCHOW

editors

© 2019 Santa Fe Institute
All rights reserved.

THE SANTA FE INSTITUTE PRESS

1399 Hyde Park Road
Santa Fe, New Mexico 87501

The Energetics of Computing in Life & Machines
ISBN (PAPERBACK): 978-1-947864-07-8
Library of Congress Control Number: 2019942205

The SFI Press is supported by the
Feldstein Program on History, Regulation, & Law,
the Miller Omega Program, and Alana Levinson-LaBrosse.

This volume emerged from workshops supported by the
National Science Foundation under Grant Nos. 1741021
and 1749019. Any opinions, findings, and conclusions or
recommendations expressed in this material are those of
the author(s) and do not necessarily reflect the views
of the National Science Foundation.

WHETHER WE ARE BASED on carbon or
on silicon makes no fundamental difference;
we should each be treated with appropriate respect.

ARTHUR C. CLARKE
2010: Odyssey Two (1982)

CONTRIBUTORS

Lee Altenberg, *University of Hawai'i*
Jakob L. Andersen, *University of Southern Denmark*
Rory A. Brittain, *Imperial College London*
Anne Condon, *University of British Columbia*
Bernat Corominas-Murtra, *Institute of Science & Technology Austria*
Jonathan Dorn, *GrammaTech, Inc.*
Marta Dueñas-Díez, *Repsol Technology Center*
Massimiliano Esposito, *University of Luxembourg*
Harold Fellermann, *Newcastle University*
Christoph Flamm, *University of Vienna*
Stephanie Forrest, *Arizona State University*
Joshua A. Grochow, *University of Colorado Boulder*
Chris Kempes, *Santa Fe Institute*
Jeremy Lacomis, *Carnegie Mellon University*
Manfred Laubichler, *Arizona State University*
GW C. McElfresh, *University of Kansas*
Daniel Merkle, *University of Southern Denmark*
Hildegard Meyer-Ortmanns, *Jacobs University Bremen*
Thomas E. Ouldridge, *Imperial College London*
Juan Pérez-Mercader, *Harvard University*
Blake S. Pollard, *Carnegie Mellon University*
Sonja J. Prohaska, *University of Leipzig*
Riccardo Rao, *University of Luxembourg*
J. Christian J. Ray, *University of Kansas*
Paul M. Riechers, *University of California, Davis*
Ricard Solé, *Pompeu Fabra University*
Peter F. Stadler, *University of Leipzig*
Susanne Still, *University of Hawai'i*
Elan Stopnitzky, *University of Hawai'i*
Pieter Rein ten Wolde, *FOM Institute AMOLF*
Chris Thachuk, *California Institute of Technology*
Westley Weimer, *University of Michigan*
David H. Wolpert, *Santa Fe Institute*

TABLE OF CONTENTS

Preface
*Chris Kempes, David H. Wolpert, Peter F. Stadler,
and Joshua A. Grochow*... xiii

1: Overview of Information Theory, Computer Science
Theory, and Stochastic Thermodynamics of Computation
David H. Wolpert.. 3

2: A Compositional Chemical Architecture
for Asynchronous Computation
Blake S. Pollard.. 63

3: Information Processing in Chemical Systems
*Jakob L. Andersen, Christoph Flamm, Daniel Merkle,
and Peter F. Stadler*... 83

4: Native Chemical Automata and the Thermodynamic
Interpretation of Their Experimental Accept/Reject Responses
Marta Dueñas-Díez and Juan Pérez-Mercader...................... 105

5: Intergenerational Cellular Signal Transfer and Erasure
GW C. McElfresh and J. Christian J. Ray........................ 127

6: Protocell Cycles as Thermodynamic Cycles
Bernat Corominas-Murtra, Harold Fellermann, and Ricard Solé.. 149

7: How and What Does a Biological System Compute?
Sonja J. Prohaska, Peter F. Stadler, and Manfred Laubichler..... 169

8: Toward Space- and Energy-Efficient Computations
Anne Condon and Chris Thachuk.................................. 191

9: Beyond Number of Bit Erasures: Computer Science
Theory of the Thermodynamics of Computation
Joshua A. Grochow and David H. Wolpert......................... 215

10: Automatically Reducing Energy Consumption of Software
*Jeremy Lacomis, Jonathan Dorn, Westley Weimer,
and Stephanie Forrest*... 263

11: Trade-Offs between Cost and Precision
and Their Possible Impact on Aging
Hildegard Meyer-Ortmanns 285

12: The Power of Being Explicit: Demystifying Work, Heat,
and Free Energy in the Physics of Computation
Thomas E. Ouldridge, Rory A. Brittain, and Pieter Rein ten Wolde.. 307

13: Transforming Metastable Memories: The Nonequilibrium
Thermodynamics of Computation
Paul M. Riechers 353

14: Physical Limitations of Work Extraction
from Temporal Correlations
*Elan Stopnitzky, Susanne Still, Thomas E. Ouldridge,
and Lee Altenberg* 383

15: Detailed Fluctuation Theorems: A Unifying Perspective
Riccardo Rao and Massimiliano Esposito 405

Index 457

PREFACE

Chris Kempes, Santa Fe Institute
David H. Wolpert, Santa Fe Institute
Peter F. Stadler, University of Leipzig
Joshua A. Grochow, University of Colorado Boulder

Currently about 5% of all energy consumption in the US goes to running computers (Desroches *et al.*, n.d.),[1] and similar figures are reported in Europe (Avgerinou, Bertoldi, and Castellazzi 2017). This is a huge monetary cost to the US economy, and the associated burning of carbon is a huge cost to the worldwide environment. Such costs of computation also arise at the much smaller scale of individual computers. For example, a large fraction of the lifetime budget of a modern high-performance computing center goes to pay its energy bill.

Despite these energetic costs of computing, the amount of computation will continue to grow at a prodigious rate. Accordingly, improving the energy efficiency of current and near-future computers is a crucial challenge facing humanity. Indeed, the benefits of improved energy efficiency would extend beyond reducing economic and environmental costs. In particular, reducing the rate of heat production is crucial for the development of next-generation high-performance computers, as dumping heat (i.e., cooling the components) is a major engineering challenge.

---

[1] The percentage of US energy consumption dedicated to computers is surprisingly hard to ascertain with precision, with reported estimates ranging from 1.5% to 10%. See Mills (2013), Clark (2013), Walsh (2013), Fanara *et al.* (2007), and Bramfitt *et al.* (2012). Seth Blumsack (personal communication, 2015) said a figure of 5% was about right for the modern system.

These issues do not arise only in artificial, silicon-based digital computers. There are many naturally occurring computers, and they, too, require huge amounts of energy. To give a rather pointed example, the human brain is a computer. This particular computer uses some 10–20% of all the calories that a human consumes. Ultimately, the cost of this computer, which uses energy that could be applied towards reproduction, must be balanced by the gains provided by the computations it performs, such as a greater ability to find food and avoid becoming prey. This implies that natural selection acts on the overall combined energetic and computational efficiency of the brain. Thus, analyzing the evolutionary biology of the thermodynamic costs of biological intelligence should be an important aspect of future research.

There are other biological computers besides brains, and they too consume large amounts of energy. In particular, many cellular systems can be viewed as computers. Moreover, the comparison of thermodynamic costs of our artificial computers with cellular computers can be extremely humbling for modern computer engineers. For example, a large fraction of the energy budget of a cell goes to translating mRNA into sequences of amino acids (i.e., proteins), through the cell's ribosome. But the thermodynamic efficiency of this computation—the amount of energy required by a ribosome per elementary operation—is many orders of magnitude superior to the thermodynamic efficiency of our current artificial computers (e.g., Kempes *et al.* (2017)).

More generally, there has been an explosion of recent research focused on both natural and artificial biological and chemical computers (Qian and Winfree 2011; Soloveichik *et al.* 2008; Murphy *et al.* 2018). These include designing simple genetic or chemical circuits, constructing computational structures and functions out of DNA and RNA, and trying to understand the entire information-processing dynamics of metabolic networks. These investigations necessarily encompass some thermodynamic aspects of the systems they involve. However, this research remains disorganized, as a set of one-off techniques and specialized applications, rather than a comprehensive analysis of the scientific issues at the intersection of biology, thermodynamics, and computation. In particular, despite several interesting and useful efforts (Benenson 2012; Arnold, Stadler, and Prohaska 2013; Varghese *et al.* 2015; Krakauer *et al.* 2016; Mehta and Schwab 2012; Mehta, Lang, and Schwab 2016; Sartori *et al.* 2014; Sartori and Pigolotti 2015), this diverse research has not been systematically connected with a combination of computer science theory and nonequilibrium statistical physics.

At a fundamental level though, why do computers—be they artificial, natural, silicon-based, or chemistry-based—use so much energy in the first place? What are the fundamental physical laws governing the relationship between the precise computation a system runs and how much energy it requires? Can we make our computers more energy-efficient by redesigning how they implement their algorithms? Can we learn how to do so by figuring out how biological computers manage to be so efficient?

These issues have been investigated for over a century and a half. The early work resulted in many important insights, in particular the analyses in the mid-to-late-twentieth century by Landauer (1961, 1996), Bennett (1973, 1982), Levine and Sherman (1990), Zurek (1989), and Bennett *et al.* (1998). However, this early research was limited by the fact that, given the state of statistical physics at the time, it was forced to try to analyze the thermodynamics of computers using equilibrium statistical physics. The problem is that, by definition, an equilibrium system is one whose state never changes. So, whatever else they are, computers are definitely *non*equilibrium systems. In fact, they are often very-far-from-equilibrium systems.

Fortunately, completely independent of this early work on the thermodynamics of computation, there have been some major breakthroughs in the past few decades in the field of nonequilibrium statistical physics (closely related to a field called "stochastic thermodynamics"). These breakthroughs allow us to analyze a variety of issues concerning how heat, energy, and information get transformed in nonequilibrium systems (Esposito and Van den Broeck 2010; Horowitz and Esposito 2014; Van den Broeck and Esposito 2015; Parrondo, Horowitz, and Sagawa 2015; Seifert 2012; Jarzynski 1997; Sagawa 2014; Ito and Sagawa 2013; Crooks 1998; Gingrich *et al.* 2016). These analyses have provided some astonishing predictions. For example, we can now calculate the (nonzero) probability that a given nanoscale system will violate the second law, reducing its entropy, in a given time interval. Moreover, laboratory experiments have confirmed these predictions. We now understand that the second law does not say that the entropy of a closed system cannot decrease, only that its expected entropy cannot decrease.

In addition, these analyses have uncovered some quite elaborate connections between information theory and the thermodynamic behavior of out-of-equilibrium physical systems. As a result, arguably modern nonequilibrium statistical physics is the only body of work that satisfies, even partially, the famous desire of John Wheeler to derive "It from Bit," and to show formally ". . . that every item of the physical world has at bottom . . . an immaterial source and explanation; that which we call reality arises in the last analysis from the posing of yes–no questions and the registering of equipment-evoked responses" (Wheeler 1990, 5).

These new theoretical tools of nonequilibrium statistical physics allow us to revisit the entire topic of the thermodynamics of computation in a fully formal manner. This has already been done for bit erasure—the topic of concern to Landauer and others—and we now have a fully formal understanding of the nonequilibrium thermodynamic costs of erasing a bit.[2]

Computer science extends far beyond counting the number of bit erasures in a given computation, in essence ignoring that issue entirely (Arora and Barak 2009; Sipser 2006; Li and Vitányi 2008). Nevertheless, one would be hard-pressed to find any mention of the topic of the thermodynamics of computation in most computer science theory textbooks. Thanks to the breakthroughs of nonequilibrium statistical physics, we can now begin to investigate the rich entirety of computer science from a thermodynamic perspective. In particular, much of conventional computer science concerns unavoidable trade-offs between the memory resources and number of timesteps needed to perform a given computation. It now seems that there might also be unavoidable thermodynamic trade-offs in performing a computation, to add to the trade-offs typically considered. Such multifaceted trade-offs apply in both artificial and biological computers, and these may be extremely important both for designing artificial computers and for understanding biological ones.

Clearly there is a huge amount of research to be done to develop this modern, nonequilibrium thermodynamics of computation. This book is an attempt to start to cultivate this research, by drawing connections among the many fields that consider different aspects of the thermodynamics of computation. The genesis of this book is a pair of workshops held at the

---

[2] The formal physics of these operations turns out to be surprisingly subtle (Sagawa 2014; Parrondo, Horowitz, and Sagawa 2015).

Santa Fe Institute during the week of August 14–18, 2017, which had the explicit aim of bringing together researchers in theoretical computer science, thermodynamics, our growing understanding of the computational processes in natural and engineered biological systems, and energy-efficient computational science and engineering.[3]

This book integrates new research from many authors that begins to address these issues. These contributions span both pure and applied research, and cross many disciplinary boundaries. Contributions that focus on nonequilibrium statistical physics and computer science are central and include work on the nonequilibrium statistical physics of the "building blocks" of computers, such as information storage and error correction. But there are also contributions that focus on the thermodynamics of computation in biological and chemical systems. These include contributions on the thermodynamics of protocell cycles, and on extracting work from temporal correlations. Other contributions in this vein focus on information processing in chemical systems, including illustrative examples of chemical computers. Yet other contributions concern computation using explicitly biological components, such as DNA and RNA, and discussions of what exactly it is that biological systems compute. Below we have provided a summary of the individual chapters.

Chapter Summaries

Chapter 1 introduces the basic theoretical concepts of modern nonequilibrium statistical physics that are relevant for a thermodynamical view on computation. It focuses on the decomposition of the entropy flow out of a physical system that

---

[3] Interestingly, Wheeler first presented the idea of It from Bit at another Santa Fe Institute conference, in the spring of 1989, before it was published (Wheeler 1989).

implements a given map into a sum of three terms: the Landauer cost measuring the drop in Shannon entropy as the system evolves; the mismatch cost arising from implementing π in a system with given priors; and the residual entropy production, which is the only term that depends on the details of the physical system.

Chapter 2 argues that the striking parallel between the processing of information by computing systems and the processing of free energy by biochemical systems suggests that biochemical motifs could be interpreted as implementing some algorithm. The question is, *what* algorithm?[4] As part of a preliminary consideration of this issue, this chapter highlights three connected lines of work: chemical reactions as a platform for asynchronous computation, a compositional framework for open chemical reaction networks based on the mathematics of category theory, and the thermodynamics of open chemical reaction networks.

Chemical structure formulae treat molecules as labeled graphs that represent atoms and chemical bonds while neglecting geometric details. Chemical reactions then correspond to graph transformations that preserve atom types and bond orders. From a thermodynamic perspective, each reaction is associated with a well-defined change in the internal free energy. Chapter 3 outlines this formal model and describes a toolkit to explore such systems.

Chapter 4 presents recent nonbiological chemical realizations of the operation of a finite automaton, a push-down au-

---

[4] There are many papers that make broad claims about biological systems performing "computation." However, outside of some specific examples of very simple "computations," like translation, transcription, replication, and genetic circuits, there is an ongoing debate regarding how much formal substance those claims have in terms of computer science theory (Benenson 2012, see chapter 7).

tomaton, and a full Turing machine. For each of the three realizations, authors define an associated thermodynamic "metric" based on enthalpy for the finite automaton and push-down automaton, and on the Gibbs free energy for the Turing machine. They use these metrics both to assess the results of the computation they investigate and as a first step towards quantifying the energetic cost of such computations.

Cells respond to information about modulation of the physicochemical environment in and around themselves. Chapter 5 analyzes the effect of cellular memory of such information in a broad class of information transfer systems taking one generation of bacteria to the next. This analysis places nongenetic intergenerational information transfer in a computational context and raises the question of the appropriate scales for analyzing the thermodynamics of information in living systems.

Chapter 6 considers the thermodynamic efficiency of photocell cycles. The goal is to demonstrate that a realistic, feasible chemical system consisting of a population of protocells can be tuned such that it undergoes successive growth and replication cycles, and that within this system, life cycles can be characterized as thermodynamic cycles.

Chapter 7 emphasizes that, while living systems are often viewed as "computing," it is much less obvious what, exactly, is computed. The chapter goes on to argue that different time scales and levels of description imply different decompositions between the "computational process" and the underlying (molecular) machinery on which it runs. They argue that this structure is important because it has allowed evolution to construct molecular machinery that drastically reduces the necessary computational efforts for the control of processes such as replication, transcription, translation, or development by effectively limiting the system to a very small number of degrees of freedom.

Chapter 8 starts with the observation that, for current chemical-based computers, an external and energy-efficient metabolism is often simply assumed. This chapter goes on to review how chemical reaction networks in the abstract, and DNA strand displacement systems (DSDs) specifically, can realize space- and energy-efficient computation without making any assumption concerning metabolism. The authors argue that this makes DSDs a promising platform to pursue space- and energy-efficient computation.

As mentioned elsewhere, modern nonequilibrium statistical physics has shown that, while logically reversible computing may have some thermodynamic benefits in certain specific contexts, it is far from the end of the story. Chapter 9 first presents a review of the salient aspects of nonequilibrium statistical physics for investigating the thermodynamics of computers. It then presents some of the many new open questions in computational complexity concerning the thermodynamic costs of different computational architectures that have been revealed by nonequilibrium statistical physics. These new questions touch on randomized algorithms, average-case complexity, the thermodynamic cost of error-correcting codes, and noisy, inexact, or approximate computation.

The investigation in Chapter 10 starts by noting that energy consumption in modern, artificial computers is typically addressed by optimizing hardware or scheduling algorithms. However, this leaves open the question of how to write energy-efficient application programs. This chapter concludes with a description of emerging techniques and experimental results that use evolutionary computation to automatically reduce the energy consumption of such programs.

Nature makes use of a whole repertoire of measures to prevent an immediate accumulation of errors. However, such mechanisms like kinetic proofreading are subject themselves to a trade-off between invested efforts and successful repair. This suggests that errors may accumulate in the course of time, leading to a basic feature of aging, if insufficient resources for the simultaneous processing of information and its correction are available. Chapter 11 reports on results on the time evolution of such errors that are obtained from a simple model, and relates them to various aspects of aging.

Chapter 12 provides an explicit treatment of erase and copy operations with molecular bits, along with the full operational cycle of a Szilard engine, in order to show that measurement and copy operations can only be performed at a cost. More generally, the chapter argues for the importance of reifying abstract nonequilibrium statistical physics in terms of concrete systems, in order to gain a clearer understanding of the important issues. That chapter also argues against the general emphasis placed on the significance of erasure operations that is found even in the recent nonequilibrium statistical physics literature.

Chapter 13 explores the fundamental thermodynamic limits of computation as the transformation of metastable memories. To do so it derives a novel decomposition of nonequilibrium free energy, which provides a rigorous thermodynamic description of coarse-grained memory systems. This shows that beyond the reversible work that Landauer's bound requires be expended in any computation, dissipation must be incurred both for modular computation and for neglected statistical structure among memory elements used in a computation.

Chapter 14 revisits some recently proposed information-exploiting systems designed to extract work from a single heat bath utilizing temporal correlations on an input tape. This chapter shows that enforcing time-continuous dynamics—which is necessary to ensure that the device is physically realizable—constrains possible designs and can drastically diminish thermodynamic efficiency. This chapter goes on to show that these problems can be circumvented by means of applying an external, time-varying protocol.

Chapter 15 presents a general method to identify an arbitrary number of fluctuating quantities which satisfy a detailed fluctuation theorem for all times within the framework of time-inhomogeneous Markovian jump processes. In doing so the authors provide a unified perspective on many fluctuation theorems derived in nonequilibrium statistical physics and related fields.

Outlook

Despite all the advances recounted in these chapters, we are nowhere close to having a theory for unifying computer science and thermodynamics. Nor do we have a broad-ranging theory for addressing detailed problems concerning the thermodynamics of computation within biological and chemical contexts. We do not even have a commonly agreed definition of what it means for biological systems to "compute," only individual case studies. Part of the challenge is that any such synthesis would need to have connections to such a wide range of fields. Nonetheless, making progress on such a synthesis would result in novel computer science theory, reduction in computational energy requirements of our artificial computers, and the opportunity for us to both understand the types of computation that have evolved in life and to extract novel computational and engineering concepts from living computation. Such a synthesis would also deepen our understanding of some of the most profound issues concerning the relation between physics and computational systems at the most fundamental levels.

As a final comment, there is now a wiki[5] combining lists of papers, researcher websites, events pages, job openings, and the like. We highly encourage people to visit it, sign up, and start improving it; the more scientists get involved, from the more fields, the better the resultant science will be.

Acknowledgments

We would like to thank the Santa Fe Institute for support. The editors would also like to acknowledge the thoughts, guidance, and assistance of Sienna Latham, Laura Egley Taylor, Lucy Fleming, and Katherine Mast, whose broad-scale and detailed contributions were invaluable. This work was supported in part by the National Science Foundation INSPIRE 1648973 and is based upon workshops supported by the National Science Foundation under Grant Nos. 1741021 and 1749019. Any opinions, findings, and conclusions or recommendations expressed in this material are those of the author(s) and do not necessarily reflect the views of the National Science Foundation. Some of the material in this preface appeared originally in Scientific American (Wolpert 2018).

REFERENCES

Arnold, C., P. F. Stadler, and S. J. Prohaska. 2013. "Chromatin Computation: Epigenetic Inheritance as a Pattern Reconstruction Problem." *J. Theor. Biol.* 336:61–74.

---

[5] http://centre.santafe.edu/thermocomp

Arora, S., and B. Barak. 2009. *Computational Complexity: A Modern Approach*. Cambridge University Press.

Avgerinou, M., P. Bertoldi, and L. Castellazzi. 2017. "Trends in Data Centre Energy Consumption under the European Code of Conduct for Data Centre Energy Efficiency." *Energies* 10:1470.

Benenson, Y. 2012. "Biomolecular Computing Systems: Principles, Progress and Potential." *Nat Rev Genet.* 13:455–468.

Bennett, C. H. 1973. "Logical Reversibility of Computation." *IBM Journal of Research and Development* 17 (6): 525–532.

———. 1982. "The Thermodynamics of Computation—A Review." *International Journal of Theoretical Physics* 21 (12): 905–940.

Bennett, C. H., P. Gacs, M. Li, P. M. B. Vitanyi, and W. H. Zurek. 1998. "Information Distance." *IEEE Transactions on Information Theory* 44 (4): 1407–1423.

Bramfitt, M., A. Bard, R. Huang, and M. McNamara. 2012. *Understanding and Designing Energy Efficiency Programs for Data Centers*. https://www.energystar.gov/buildings/tools-and-resources/understanding-and-designing-energy-efficiency-programs-data-centers.

Clark, J. 2013. "IT Now 10 Percent of World's Electricity Consumption, Report Finds." *The Register* (August). https://www.theregister.co.uk/2013/08/16/it_electricity_use_worse_than_you_thought/.

Crooks, G. E. 1998. "Nonequilibrium Measurements of Free Energy Differences for Microscopically Reversible Markovian Systems." *Journal of Statistical Physics* 90 (5–6): 1481–1487.

Desroches, L.-B., H. Fuchs, J. B. Greenblatt, S. Pratt, H. Willem, E. Claybaugh, B. Beraki, M. Nagaraju, S. K. Price, and S. J. Young. n.d. *Computer Usage and National Energy Consumption: Results from a Field-Metering Study*. Technical report LBNL # 6876E. Lawrence Berkeley National Laboratory.

Esposito, M., and C. Van den Broeck. 2010. "Three Faces of the Second Law: 1. Master Equation Formulation." *Physical Review E* 82 (1): 011143.

Fanara, A., J. Abelson, A. Bailey, K. Crossman, R. Shudak, A. Sullivan, M. Vargas, and M. Zatz. 2007. *Report to Congress on Server and Data Center Energy Efficiency Public Law 109-431,* August. https://www.energystar.gov/buildings/tools-and-resources/report-congress-server-and-data-center-energy-efficiency-opportunities.

Gingrich, T. R., G. M. Rotskoff, G. E. Crooks, and P. L. Geissler. 2016. "Near-Optimal Protocols in Complex Nonequilibrium Transformations." *Proceedings of the National Academy of Sciences* 113 (37): 10263–10268.

Horowitz, J. M., and M. Esposito. 2014. "Thermodynamics with Continuous Information Flow." *Physical Review X* 4 (3): 031015.

Ito, S., and T. Sagawa. 2013. "Information Thermodynamics on Causal Networks." *Physical Review Letters* 111 (18): 180603.

Jarzynski, C. 1997. "Nonequilibrium Equality for Free Energy Differences." *Physical Review Letters* 78 (14): 2690–2693.

Kempes, C. P., D. Wolpert, Z. Cohen, and J. Pérez-Mercader. 2017. "The Thermodynamic Efficiency of Computations Made in Cells Across the Range of Life." *Phil. Trans. R. Soc. A* 375:20160343.

Krakauer, D. C., L. Müller, S. J. Prohaska, and P. F. Stadler. 2016. "Design Specifications for Cellular Regulation." *Th. Biosci.* 231–240.

Landauer, R. 1961. "Irreversibility and Heat Generation in the Computing Process." *IBM Journal of Research and Development* 5 (3): 183–191.

———. 1996. "The Physical Nature of Information." *Physics Letters A* 217 (4–5): 188–193.

Levine, R. Y., and A. T. Sherman. 1990. "A Note on Bennett's Time Space Tradeoff for Reversible Computation." *SIAM Journal on Computing* 19 (4): 673–677.

Li, M., and P. Vitányi. 2008. *An Introduction to Kolmogorov Complexity and Its Applications.* 3rd ed. Springer.

Mehta, P., A. H. Lang, and D. J. Schwab. 2016. "Landauer in the Age of Synthetic Biology: Energy Consumption and Information Processing in Biochemical Networks." *Journal of Statistical Physics* 162 (5): 1153–1166.

Mehta, P., and D. J. Schwab. 2012. "Energetic Costs of Cellular Computation." *Proceedings of the National Academy of Sciences* 109 (44): 17978–17982.

Mills, M. P. 2013. *The Cloud Begins with Coal: Big Data, Big Networks, Big Infrastructure, and Big Power—An Overview of the Electricity Used by the Global Digital Ecosystem,* August. https://www.tech-pundit.com/wp-content/uploads/2013/07/Cloud_Begins_With_Coal.pdf.

Murphy, N., R. Petersen, A. Phillips, B. Yordanov, and N. Dalchau. 2018. "Synthesizing and Tuning Stochastic Chemical Reaction Networks with Specified Behaviours." *Journal of the Royal Society Interface* 15.

Parrondo, J. M. R., J. M. Horowitz, and T. Sagawa. 2015. "Thermodynamics of Information." *Nature Physics* 11:131–139.

Qian, L., and E. Winfree. 2011. "Scaling Up Digital Circuit Computation with DNA Strand Displacement Cascades." *Science* 332 (6034): 1196–1201.

Sagawa, T. 2014. "Thermodynamic and Logical Reversibilities Revisited." *Journal of Statistical Mechanics: Theory and Experiment* 2014 (3): P03025.

Sartori, P., L. Granger, C. F. Lee, and J. M. Horowitz. 2014. "Thermodynamic Costs of Information Processing in Sensory Adaptation." *PLOS Computational Biology* 10 (12): e1003974.

Sartori, P., and S. Pigolotti. 2015. "Thermodynamics of Error Correction." *Physical Review X* 5 (4): 041039.

Seifert, U. 2012. "Stochastic Thermodynamics, Fluctuation Theorems, and Molecular Machines." *Reports on Progress in Physics* 75 (12): 131–139.

Sipser, M. 2006. *Introduction to the Theory of Computation*. 2nd ed. Thomson Course Technology.

Soloveichik, D., M. Cook, E. Winfree, and J. Bruck. 2008. "Computation with Finite Stochastic Chemical Reaction Networks." *Natural Computing* 7 (4): 615–633.

Van den Broeck, C., and M. Esposito. 2015. "Ensemble and Trajectory Thermodynamics: A Brief Introduction." *Physica A: Statistical Mechanics and its Applications* 418:6–16.

Varghese, S., J. A. A. W. Elemans, A. E. Rowan, and R. J. M. Nolte. 2015. "Molecular Computing: Paths to Chemical Turing Machines." *Chem Sci.* 6:6050–6058.

Walsh, B. 2013. "The Surprisingly Large Energy Footprint of the Digital Economy." *Time* (August). http://science.time.com/2013/08/14/power-drain-the-digital-cloud-is-using-more-energy-than-you-think/#.

Wheeler, J. A. 1989. "Information, Physics, Quantum: The Search for Links." In *Proc. 3rd Int. Symp. Foundations of Quantum Mechanics,* edited by H. Ezawa, S. I. Kobayashi, and Y. Murayama, 354–368. Tokyo: Phys. Soc. Japan.

———. 1990. "Information, Physics, Quantum: The Search for Links." Chap. 1 in *Complexity, Entropy, and the Physics of Information,* edited by W. H. Zurek, 3–28. Addison-Wesley Publishing Company.

Wolpert, D. H. 2018. "Why Do Computers Use So Much Energy?" https://blogs.scientificamerican.com/observations/why-do-computers-use-so-much-energy.

Zurek, W. H. 1989. "Thermodynamic Cost of Computation, Algorithmic Complexity, and the Information Metric." *Nature* 341 (6238): 119–124.

OVERVIEW OF INFORMATION THEORY, COMPUTER SCIENCE THEORY, AND STOCHASTIC THERMODYNAMICS OF COMPUTATION

David H. Wolpert, Santa Fe Institute

Introduction

In this chapter, I give a quick overview of some of the theoretical background necessary for using modern nonequilibrium statistical physics to investigate the thermodynamics of computation. I begin by presenting some general terminology, and then I review some of the most directly relevant concepts from information theory. Next I introduce several of the most important kinds of computational machines studied in computer science theory. After this I summarize the parts of nonequilibrium statistical physics (more precisely, stochastic thermodynamics; see Seifert (2012), Van den Broeck and Esposito (2015), and Esposito and Van den Broeck (2010b)) that are necessary for analyzing the thermodynamics of such computational machines. The interested reader is directed to Wolpert (2019) for a more detailed overview.

The key tool provided by stochastic thermodynamics is a decomposition of the entropy flow out of any open physical system that implements some desired map π on some initial distribution p_0, into the sum of three terms (Wolpert and

Kolchinsky 2018):

1. The *Landauer cost* of π. This is an information-theoretic quantity: the drop in Shannon entropy of the actual distribution over the states of the system as the system evolves. The Landauer cost depends only on π and p_0, being independent of the precise details of the physical system implementing π.

2. The *mismatch cost* of implementing π with a particular physical system. This is also an information-theoretic quantity: a drop in Kullback–Leibler distance, between p_0 and a counterfactual "prior" distribution q_0, as those distributions both get transformed by π. The mismatch cost depends only on π, p_0, and q_0, being independent of any details of the physical system implementing π that are not captured in q_0.

3. The *residual entropy production* due to using a particular physical system. This is a *linear* term that depends on the graph-theoretic nature of π, on p_0, and on the precise details of the physical system implementing π.

This decomposition allows us to analyze how the thermodynamic costs of a fixed computational device vary if we change the physical environment of that device, that is, change the distribution of inputs to the device. This decomposition also plays a key role in analyzing the thermodynamics of systems with multiple components that are interconnected, since the physical environments of each of those components will depend on the logical maps performed by the "upstream" components (Wolpert and Kolchinsky 2018; Wolpert 2019; Grochow and Wolpert 2018).

To illustrate this decomposition of entropy flow, I end by using it to analyze the relative thermodynamic advantages and disadvantages of performing a computation with a logically reversible circuit (e.g. constructed out of Fredkin gates) rather than with a conventional circuit that uses logically irreversible gates.

Terminology and General Notation

As usual, for any set A, A^* is the set of all finite strings of elements from A. I write the Kronecker delta function as

$$\delta(a,b) = \begin{cases} 1 \text{ if } a = b \\ 0 \text{ otherwise.} \end{cases} \quad (1.1)$$

I write the indicator function for any Boolean function $f(z)$ as

$$\boldsymbol{I}(f) = \begin{cases} 1 \text{ if } f(z) = 1 \\ 0 \text{ otherwise.} \end{cases} \quad (1.2)$$

In general, random variables are written with uppercase letters, and instances of those random variables are written with the corresponding lowercase letters. When the context makes the meaning clear, I will often also use the uppercase letter indicating a random variable, for example, X, to indicate the set of possible outcomes of that random variable. For any distribution $p(x)$ defined over the set X, and any $X' \subseteq X$, I write $p(X') = \sum_{x \in X'} p(x)$. Finally, given any conditional distribution $\pi(y \mid x)$ and a distribution p over X, I write πp for the distribution over Y induced by π and p:

$$(\pi p)(y) := \sum_{x \in X} \pi(y \mid x) p(x). \quad (1.3)$$

Computational machines are most often defined in terms of conditional distributions π that are (very good approximations of) single-valued state-update functions. That means that there are transitions in the state of the system that cannot occur in a single step; that is, there are restrictions on which entries of the transition matrix can be nonzero. To analyze the entropy production in a physical system with this character, it helps to "decompose" the dynamics of the system into the entropy production associated with a set of "submaps" defined by zero entries in the transition matrix.

Let f be a single-valued function from a set X into itself. An *island* of f is a preimage $f^{-1}(x)$ for some $x \in f(X)$ (Wolpert and Kolchinsky 2018; Wolpert 2019). I will write the set of all islands of a function f as $L(f)$.

As an example, the logical AND operation

$$\pi(c|a,b) = \delta(c, a \wedge b)$$

has two islands, corresponding to $(a,b) \in \{\{0,0\}, \{0,1\}, \{1,0\}\}$ and $(a,b) \in \{\{1,1\}\}$, respectively. I write the set of all distributions over an island $c \in L(f)$ as Δ_c. I make the obvious definitions that, for any distribution $p(x)$ and any $c \in L(f)$, the associated distribution over islands is $p(c) = \sum_{x \in c} p(x)$. As shorthand, I also write

$$p^c(x) = p(x|X \in c) = p(x)\mathbf{I}(x \in c)/p(c).$$

Intuitively, the islands of a function are different systems that are isolated from one another for the duration of any process that implements that function. As this suggests, the islands of a dynamic process depend on how long the process runs. For example, suppose $X = \{a, b, c\}$, and $f(a) = a$, $f(b) = a$, while $f(c) = b$. Then f has two islands, $\{a, b\}$ and $\{c\}$. However, if we iterate f, we have just a single island, because all three states get mapped under f^2 to the state a.

In the more general case where the physical process implements an arbitrary stochastic matrix π, the islands of π are the transitive closure of the equivalence relation

$$x \sim x' \Leftrightarrow \exists x'' : \pi(x''|x) > 0, \pi(x'' \mid x') > 0. \qquad (1.4)$$

(Note that x and x' may be in the same island even if there is no x'' such that both $P(x'' \mid x) > 0$ and $P(x' \mid x) > 0$, due to the transitive closure requirement.) Equivalently, the islands of π are a partition $\{X^i\}$ of X such that for all X^i, $x \notin X^i$,

$$\mathrm{supp}\pi(x' \mid x) \cap \bigcup_{x'' \in X^i} \mathrm{supp}\pi(x' \mid x'') = \varnothing. \tag{1.5}$$

Although the focus of this chapter is computers, which are typically viewed as implementing single-valued functions, all of the results herein also hold for physical processes that implement general stochastic matrices, if one uses this more general definition of islands. Note that for either single-valued or non-single-valued stochastic matrices π, the islands of π will in general be a refinement of the islands of π^2, because π may map two of its islands to (separate) regions that are mapped on top of one another by π^2. This will not be the case, though, if π permutes its islands.

Information Theory

The Shannon entropy of a distribution over a set X, the Kullback–Leibler (KL) divergence between two distributions both defined over X (sometimes called the *relative entropy* of those two distributions), and the cross-entropy between two such distributions, respectively, are defined as

$$S(p(X)) = -\sum_{x \in X} p(x) \ln p(x),$$

$$D(p(X) \parallel r(X)) = \sum_{x \in X} p(x) \ln \frac{p(x)}{r(x)}, \tag{1.6}$$

$$S(p(X) \parallel r(X)) = S(p(X)) + D(p(X) \parallel r(X))$$
$$= -\sum_{x \in X} p(x) \ln r(x).$$

I adopt the convention of using natural logarithms rather than logarithms base 2 for most of this chapter. I sometimes refer to the second arguments of KL divergence and of cross-entropy as a *reference distribution*. Note that the entropy of a distribution p is just the negative of the KL divergence from p to the uniform reference distribution, up to an overall (negative) constant.

The conditional entropy of a random variable X, conditioned on a variable Y under joint distribution p, is defined as

$$S(p(X \mid Y)) = \sum_{y \in Y} p(y) S(p(X \mid y))$$
$$= - \sum_{x \in X, y \in Y} p(y) p(x \mid y) \ln p(x \mid y), \quad (1.7)$$

and similarly for conditional KL divergence and conditional cross-entropy. The *chain rule* for entropy (Cover and Thomas 2012) says that

$$S(p(X \mid Y)) + S(p(Y)) = S(p(X, Y)). \quad (1.8)$$

Similarly, given any two distributions p and r, both defined over $X \times Y$, the conditional cross-entropy between them equals the associated conditional entropy plus the associated conditional KL divergence:

$$\begin{aligned} & S(p(X \mid Y) \parallel r(X \mid Y)) \\ &= S(p(X \mid Y)) + D(p(X \mid Y) \parallel r(X \mid Y)) \\ &= - \sum_{x \in X, y \in Y} p(x, y) \ln r(x \mid y). \end{aligned} \quad (1.9)$$

The mutual information between two random variables X and Y jointly distributed according to p is defined as

$$\begin{aligned} I_p(X; Y) &\equiv S(p(X)) + S(p(Y)) - S(p(X, Y)) \\ &= S(p(X)) - S(p(X \mid Y)). \end{aligned} \quad (1.10)$$

I drop the subscript p where the distribution is clear from context. The *data processing inequality* for mutual information (Cover and Thomas 2012) states that if we have random variables X, Y, and Z, and Z is a stochastic function of Y, then $I(X;Z) \leq I(X;Y)$.

Where the random variable is clear from context, I sometimes simply write $S(p)$, $D(p \parallel r)$, and $S(p \parallel r)$. I also sometimes abuse notation and (for example), if a and b are specified, write $S(A = a \mid B = b)$ to mean the conditional entropy of the random variable $\delta(A, a)$ conditioned on the event that the random variable B has the value b. When considering a set of random variables, I usually index them and their outcomes with subscripts, as in $X_1, X_2, \ldots$ and $x_1, x_2 \ldots$ I also use notation like $X_{1,2}$ to indicate the joint random variable (X_1, X_2).

I write $S(\pi p)$ to refer to the entropy of distributions over Y induced by $p(x)$ and the conditional distribution π, as defined in equation (1.6). I use similar shorthand for the other information-theoretic quantities, $D(\cdot \parallel \cdot)$, $S(\cdot \parallel \cdot)$, and $I(\cdot)$. In particular, the *chain rule* for KL divergence and the *data-processing inequality* for KL divergence (Cover and Thomas 2012), respectively, are as follows:

1. For all distributions p, r over the space $X^a \times X^b$,

$$D(p(X^a, X^b) \parallel r(X^a, X^b))$$
$$= D(p(X^b) \parallel r(X^b)) + D(p(X^a \mid X^b) \parallel r(X^a \mid X^b)).$$
$$(1.11)$$

2. For all distributions p, r over the space X and conditional distributions $\pi(y \mid x)$,

$$D(p \parallel r) \geq D(\pi p \parallel \pi r). \qquad (1.12)$$

Note that by combining the chain rule for KL divergence with the chain rule for entropy, we get a chain rule for cross-entropy.

Computational Machines

STRAIGHT-LINE CIRCUITS

A *circuit* C is a tuple (V, E, F, X) that can be viewed as a special type of Bayes net (Koller and Friedman 2009; Ito and Sagawa 2013, 2015). The pair (V, E) specifies the vertices and edges of a directed acyclic graph (DAG). Intuitively, this DAG is the wiring diagram of the circuit. In the circuit engineering industry, this is sometimes called the *netlist* of the circuit. X is a Cartesian product $\prod_v X_v$, where each X_v is the space of the variable associated with node v.[1] F is a set of conditional distributions, one for each nonroot node v of the DAG, mapping the joint value of the (variables at the) parents of v, $x_{\mathrm{pa}(v)}$ to the value of (the variable at) v, x_v. In conventional circuit theory, F is a set of single-valued functions. Note that, following the convention in the Bayes nets literature, with this definition of a circuit, we orient edges in the direction of logical implication, that is, in the direction of information flow. So the inputs to the circuit are the roots of the associated DAG, and the outputs are the leaves. The reader should be warned that this is the *opposite* of the convention in computer science theory. When there is no risk of confusion, I simply refer to a circuit C, with all references to V, E, F, or X implicitly assumed as the associated elements defining C.[2]

Straight-line circuits are an example of *nonuniform* computers. These are computers that can only work with

---

[1] In real electronic circuits, x_v will typically specify an element in a coarse graining of the microstate space of the system, but as mentioned earlier, that is not relevant here.

[2] A more general type of circuit than the one considered here allows branching conditions at the nodes and allows loops. Such circuits cannot be represented as a Bayes net. To make clear what kind of circuit is being considered, sometimes the branch-free, loop-free type of circuit is called a straight-line circuit (Savage 1998).

inputs of some fixed length. In the case of a circuit, that length is specified by the number of root nodes in the circuit's DAG. One can use a single circuit to compute the output for any input in a set of inputs, so long as all those inputs have the same length. If, on the other hand, one wishes to consider using circuits to compute the output for any input in a set that contains inputs of all possible lengths, then one must use a *circuit family*, that is, an infinite set of circuits $\{C_i : i \in \mathbb{Z}^+\}$, where each circuit C_i has i root nodes.

In contrast to nonuniform computers, *uniform* computers are machines that can work with arbitrary length inputs.[3] In general, the number of iterations a particular uniform computational machine requires to produce an output is not prefixed, in contrast to the case of any particular nonuniform computational machine. Indeed, for some inputs, a uniform computational machine may never finish computing. The rest of this section introduces some of the more prominent uniform computational machines.

FINITE AUTOMATA

One important class of uniform computational machines comprises the finite automata. There are several different, very similar definitions of finite automata, some of which overlap with common definitions of finite state machines. To fix the discussion, here I adopt the following definition:

Definition 1. A finite automaton *(FA) is a 5-tuple (R, Λ, $r^\varnothing$, r^A, ρ) where*

   1. R is a finite set of computational states;

   2. Λ is a finite (input) alphabet;

---

[3] The reader should be warned that computer scientists also consider "uniform circuit families," which are related but different.

3. $r^\varnothing \in R$ *is the* start state;

4. $r^A \in R$ *is the* accept state; *and*

5. $\rho : R \times \Lambda \to R$ *is the* update function, *mapping a current input symbol and the current computational state to a next computational state.*

A finite string of successive input symbols, that is, an element of Λ^*, is sometimes called an *input word*, written as $\vec{\lambda}$. To operate a finite automaton on a particular input word, one begins with the automaton in its start state and feeds that state, together with the first symbol in the input word, into the update function to produce a new computational state. Then, one feeds in the next symbol in the input word, if any, to produce a next computational state, and so on. I will sometimes say that the *head* is in some state $r \in R$ rather than that the computational state of the automaton is r.

Often one is interested in whether, after the last symbol from the input word is processed, the head is in state r^A. If so, one says that the automaton *accepts* that input word. In this way, any given automaton uniquely specifies a *language* of all input words that that automaton accepts, which is called a *regular* language. As an example, any finite language (consisting of a finite set of words) is a regular language. On the other hand, the set of all palindromes over Λ, to give a simple example, is *not* a regular language.

Importantly, any particular FA can process input words of arbitrary length. This means that one cannot model a given FA as some specific and therefore fixed-width circuit, in general. The FA will have properties that are not captured by that circuit. In this sense, individual FAs are computationally more powerful than individual circuits. This does not mean that individual FAs are more powerful than entire circuit *families*, however; see the discussion in the section "Turing Machines" of how circuit families can be even more powerful than Turing machines.

Finite automata play an important role in many different fields, including electrical engineering, linguistics, computer science, philosophy, biology, mathematics, and logic. In computer science specifically, they are widely used in the design of hardware digital systems, compilers, and network protocols and in the study of computation and languages more broadly.

For a given physical system to qualify as a FA, we must be able to map the dynamics of a subset of the (physical) variables in that system and its surrounding environment to the (logical) variables R, Λ specified in definition 1. In general, that physical system will include other variables as well. For example, the physical system may include a variable representing the computational state of the FA and a buffer holding the current input symbol, but also various counters, etc.

In addition, though, to analyze the thermodynamic efficiency of a physical system we wish to identify as a computational machine, we have to specify which of the physical variables in that system will be viewed as "inputs" of the machine and which will be viewed as "outputs." In general, this specification depends on the precise way the machine will be used rather than being part of the more general definition of the computational machine. In particular, in all applications of FAs mentioned just previously, the automaton is viewed as a system that maps an input *word* to an output *computational state*. This motivates the following alternative definition of a FA.

Definition 2. *A* word-based *finite automaton is a 6-tuple* $(R, \Lambda, r^\emptyset, r^A, \rho^*, \tau)$ *where*

1. *R is a finite set of* computational states;

2. Λ *is a finite (input)* alphabet;

3. $r^\emptyset \in R$ *is the* start state;

4. $r^A \in R$ *is the* accept state;

5. $\tau \in \mathbb{Z}^+$ *is the* counter; *and*

6. $\hat{\rho} : R \times \Lambda^* \times \mathbb{Z}^+ \to R$ *is the* (computational state) update function, *mapping a current input word, current counter pointing to one of that word's symbols, and current computational state to a next computational state.*

When I need to distinguish FAs as defined in definition 1 from word-based FAs, I will refer to the former as *symbol-based* FAs.

To map a symbol-based FA (def. 1) with update function ρ into an equivalent word-based update function $\hat{\rho}$, we set

$$\hat{\rho}(r, \lambda^*, \tau) = \rho(r, \lambda^*_\tau). \qquad (1.13)$$

At the first iteration of a word-based FA, not only is the computational state set to the start state but the counter has the value 0. In addition, at the end of each iteration m of the FA, after its computational state is updated by $\hat{\rho}$, the counter is incremented; that is, $\tau_m \to \tau_{m+1}$. From now on, I will implicitly move back and forth between the two definitions of a FA as is convenient.

To allow us to analyze a physical system that implements the running of a FA on many successive input words, we need a way to signal to the system when one input word ends and another one begins. Accordingly, without loss of generality, we assume that Λ contains a special *blank* state, written b, that delimits the end of a word. I write Λ_- for $\Lambda \setminus \{b\}$, so that words are elements of Λ_-^*.

In a *stochastic* finite automaton (sometimes called a *probabilistic automaton*), the single-valued function ρ is replaced by a conditional distribution. To use notation that covers all iterations i of a stochastic finite automaton, I write this *update distribution* as $\pi(r_{i+1}, |\ r_i, \lambda_i)$. The analogous extension of the word-based definition of FA into a word-based definition of a stochastic FA is immediate. For simplicity, from now on, I will simply refer to "finite automaton," using the acronym "FA," to refer to a finite automaton that is either stochastic or deterministic.

Typically in the literature, there is a set of multiple accept states—called *terminal states* or *accepting states*—not just one state. Sometimes there are also multiple start states.

TRANSDUCERS: MOORE MACHINES AND MEALY MACHINES

In the literature, the definition of FA is sometimes extended so that in each transition from one computational state to the next, an output symbol is generated. Such systems are also called *transducers* in the computer science community.

Definition 3. *A transducer is a 6-tuple* $(R, \Lambda, \Gamma, r^\varnothing, x^A, \rho)$ *such that*

1. *R is a finite set, the set of* computational states;

2. *Λ is a finite set, called the* input alphabet;

3. *Γ is a finite set, called the* output alphabet;

4. *$r^\varnothing \in R$ is the* start *state*;

5. $r^A \in R$ *is the* accept state;

6. $\rho : R \times \Lambda \to R \times \Gamma$ *is the* update rule.

Sometimes the computational states of a transducer are referred to as the states of its *head*. I refer to the (semifinite) string of symbols that have yet to be processed by a transducer at some moment as the *input (data) stream* at that moment. I refer to the string of symbols that have already been produced by the information ratchet at some moment as the *output (data) stream* at that moment.

To operate a transducer on a particular input data stream, one begins with the machine in its start state and feeds that state together with the first symbol in the input stream into the update function to produce a new computational state and a new output symbol. Then one feeds in the next symbol in the input stream, if any, to produce a next computational state and associated output symbol, and so on.

In a *stochastic* transducer, the single-valued function ρ is replaced by a conditional distribution. To use notation that covers all iterations i of the transducer, I write this *update distribution* as $\pi(\gamma_{i+1}, r_{i+1} \mid r_i, \lambda_i)$. Stochastic transducers are used in fields ranging from linguistics to natural language processing (in particular, machine translation) to machine learning more broadly. From now on, I implicitly mean "stochastic transducer" when I use the term *transducer*. [4]

As with FAs, typically in the literature, transducers are defined to have a set of multiple accept states, not just one state. Sometimes there are also multiple start states. Similarly, in some of the literature, the transition function allows the

---

[4] The reader should be warned that some of the literature refers to both FAs and transducers as "finite state machines," using the term *acceptor* or *recognizer* to refer to the system defined in definition 1.

transducer to receive the empty string as an input and/or produce the empty string as an output.

A *Moore machine* is a transducer in which the output γ is determined purely by the current state of the transducer, r. In contrast, a transducer in which the output depends on both the current state x and the current input λ is called a *Mealy machine*.

As a final comment, an interesting variant of the transducers defined in definition 3 arises if we remove the requirement that there be accept states, and maybe even the requirement of start states. In this variant, rather than feeding a (perhaps infinite) sequence of input words into the system, each of which results in its own output word, one feeds in a single input word that is infinitely long, producing a single (infinitely long) output word, in the infinite time limit. This variant is used to define so-called automata groups or self-similar groups (Nekrashevych 2005).

Somewhat confusingly, although the computational properties of such systems differ in crucial ways from those defined in definition 3, this variant is also called a "transducer" in the literature on "computational mechanics," a branch of hidden Markov model theory (Barnett and Crutchfield 2015). Fortunately, these same systems have been given a different name, *information ratchets*, in other work analyzing their statistical physics properties (Mandal and Jarzynski 2012). Accordingly, here I adopt that term for this variant of computer science transducers.

TURING MACHINES

Perhaps the most famous class of computational machines are the Turing machines (Hopcroft, Motwani, and Rotwani 2000; Arora and Barak 2009; Savage 1998). One reason for their fame is that it seems one can model any computational machine that is constructable by humans as a Turing machine. A bit more formally, the *Church-Turing hypothesis* states that "a function on the natural numbers is computable by a human being following an algorithm, ignoring resource limitations, if and only if it is computable by a Turing machine." The *physical Church-Turing thesis* modifies this to say that the set of Turing machines includes all mechanical algorithmic procedures admissible by the laws of physics.

In part due to this thesis, Turing machines form one of the keystones of the entire field of computer science theory, and in particular of computational complexity (Moore and Mertens 2011). For example, the famous Clay prize question of whether P = NP—widely considered one of the deepest and most profound open questions in mathematics—concerns the properties of Turing machines. As another example, the theory of Turing machines is intimately related to deep results on the limitations of mathematics, such as Gödel's incompleteness theorems, and has broader, philosophical implications (Aaronson 2013). As a result, it seems that the foundations of physics may be restricted by some of the properties of Turing machines (Barrow 2011; Aaronson 2005).

Along these lines, some authors have suggested that the foundations of statistical physics should be modified to account for the properties of Turing machines, for example, by adding terms to the definition of entropy. After all, given the Church-Turing hypothesis, one might argue that the probability distributions at the heart of statistical physics are distributions "stored in the mind" of the human being analyzing a given statistical physical system—that is, running a particular algorithm to compute a property of a given system (Caves 1990, 1993; Zurek 1989).

Many different definitions of Turing machines are "computationally equivalent" to one another. This means that any computation that can be done with one type of Turing machine can be done with another. It also means that the "scaling function" of one type of Turing machine, mapping the size of a computation to the minimal amount of resources needed to perform that computation by that type of Turing machine, is at most a polynomial function of the scaling function of any other type of Turing machine. (See, for example, the relation between the scaling functions of single-tape and multitape Turing machines (Arora and Barak 2009).) The following definition will be useful for our purposes, even though it is more complicated than is strictly needed

Definition 4. *A* Turing machine *(TM) is a 7-tuple ($R, \Lambda, b, v, r^{\varnothing}, r^A, \rho$) where*

1. *R is a finite set of* computational states;

2. *Λ is a finite* alphabet;

3. *$b \in \Lambda$ is a special* blank *symbol;*

4. *$v \in \mathbb{Z}$ is a* pointer;

5. *$r^{\varnothing} \in R$ is the* start state;

6. *$r^A \in R$ is the* accept state; *and*

7. *$\rho : R \times \mathbb{Z} \times \Lambda^{\infty} \to R \times \mathbb{Z} \times \Lambda^{\infty}$ is the* update function. *It is required that for all triples (r, v, T), if we write $(r', v', T') = \rho(r, v, T)$, then v' does not differ by more than 1 from v, and the vector T' is identical to the*

vectors T for all components, with the possible exception of the component with index v.[5]

State r^A is often called the *halt state* of the TM rather than the accept state. In addition, in some alternative (computationally equivalent) definitions, there is a set of multiple accept states rather than a single accept state. Function ρ is sometimes called the *transition function* of the TM. We sometimes refer to R as the states of the *head* of the TM, and we refer to the third argument of ρ as a *tape*, writing a value of the tape (i.e., a semi-infinite string of elements of the alphabet) as T. The set of triples that are possible arguments to the update function of a given TM is sometimes called the set of *instantaneous descriptions* (IDs) of the TM. These are sometimes instead referred to as "configurations." Note that, as an alternative to definition 4, we could define any TM as a map over an associated space of IDs.

Any TM $(R, \Sigma, b, v, \rho, r^\varnothing, \rho)$ starts with $r = r^\varnothing$, with the counter set to a specific initial value (e.g, 0), and with T consisting of a finite contiguous set of nonblank symbols, with all other symbols equal to b. The TM operates by iteratively applying ρ, until the computational state falls in r^A, at which time it stops. Note that the definition of ρ for $r = r^A$ is arbitrary and irrelevant.

If running a TM on a given initial state of the tape results in the TM eventually halting, the state of T when it halts is called the TM's *output*. The initial state of T (excluding the blanks) is sometimes called the associated *input* or *program*. (However, the reader should be warned that the term *program* has been

---

[5] Technically, the update function only needs to be defined on the "finitary" subset of $R \times \mathbb{Z} \times \Lambda^\infty$, namely, those elements of $R \times \mathbb{Z} \times \Lambda^\infty$ for which the tape contents have a nonblank value in only finitely many positions.

used by some physicists to mean specifically the shortest input to a TM that results in it computing a given output.) We also say that the TM *computes* an output from an input. In general, there will be inputs for which the TM never halts. The set of all those inputs to a TM that cause it to halt eventually is called its *halting set*.

As mentioned, there are many variants of the definition of TMs provided above. In one particularly popular variant, the single tape in definition 4 is replaced by multiple tapes. Typically one of those tapes contains the input, one contains the TM's output (if and) when the TM halts, and there are one or more intermediate "work tapes" that are in essence used as scratch pads. The advantage of using this more complicated variant of TMs is that it is often easier to prove theorems for such machines than for single-tape TMs. However, there is no difference in their computational power.

Returning to the TM variant defined in definition 4, a *universal Turing machine* (UTM) M is one that can be used to emulate any other TM. More precisely, a UTM M has the property that for any other TM M', there is an invertible map f from the set of possible states of the tape of M' into the set of possible states of the tape of M, such that if we

1. apply f to an input string σ' of M' to fix an input string σ of M;

2. run M on σ until it halts;

3. apply f^{-1} to the resultant output of M;

then we get exactly the output computed by M' if it is run directly on σ'.

An important theorem of computer science is that universal TMs exist. Intuitively, this just means that there exist programming languages that are "universal," in that we can use them to implement any desired program in any other language, after appropriate translation of that program from that other language. This universality leads to a very important concept:

Definition 5. *The* Kolmogorov complexity *of a UTM* M *computing a string* $\sigma \in \Lambda^*$ *is the length of the shortest input string* s *such that* M *computes* σ *from* s.

Intuitively, (output) strings that have low Kolmogorov complexity for some specific UTM M are those with short, simple programs in the language of M. For example, in all common (universal) programming languages (e.g., C, Python, Java), the first m digits of π have low Kolmogorov complexity because those digits can be generated using a relatively short program. Strings that have high (Kolmogorov) complexity are sometimes referred to as "incompressible." These strings have no patterns in them that can be generated by a simple program. As a result, it is often argued that the expression "random string" should only be used for strings that are incompressible.

A *prefix-free TM* is such that no one input in its halting set is a proper prefix of another string in its halting set when both are viewed as symbol strings. The *coin-flipping prior* of a prefix TM M is the probability distribution over the strings in M's halting set generated by IID "tossing a coin" to generate those strings, in a Bernoulli process, and then normalizing.[6] So any string σ in the halting set has probability $2^{-|\sigma|}/\Omega$ under the coin-flipping prior, where Ω is the normalization constant for the TM in question.

---

[6] Kraft's inequality guarantees that because the set of strings in the halting set is prefix-free, the sum over all its elements of their probabilities cannot exceed 1 and so can be normalized (Li and Vitanyi 2008).

The coin-flipping prior provides a simple Bayesian interpretation of Kolmogorov complexity: under that prior, the Kolmogorov complexity of any string σ for any prefix TM M is just (the log of) the maximum *a posteriori* (MAP) probability that any string σ' in the halting set of M was the *input* to M, conditioned on σ being the *output* of that TM.[7]

The normalization constant Ω for any fixed-prefix UTM, sometimes called *Chaitin's Omega*, has some extraordinary properties. For example, the successive digits of Ω provide the answers to *all* well-posed mathematical problems. So if we knew Chaitin's Omega for some particular prefix UTM, we could answer every problem in mathematics. Alas, the value of Ω for any prefix UTM M cannot be computed by any TM (either M or some other one). So under the Church–Turing hypothesis, we cannot calculate Ω.[8]

It is now conventional to analyze Kolmogorov complexity using prefix UTMs, with the coin-flipping prior, since this removes some undesirable technical properties that Kolmogorov complexity has for more general TMs and priors. Reflecting this, all analyses in the physics community that concern TMs assume prefix UTMs.[9]

Interestingly, for all their computational power, there are some surprising ways in which TMs are *weaker* than the other computational machines introduced earlier. For example, an infinite number of TMs are more powerful than any given

---

[7] Strictly speaking, this result is only true up to an additive constant, given by the log of the normalization constant of the coin-flipping prior for M.

[8] See also Baez and Stay (2012) for a discussion of a "statistical physics" interpretation of Ω that results if we view the coin-flipping prior as a Boltzmann distribution for an appropriate Hamiltonian, so that Ω plays the role of a partition function.

[9] See Li and Vitanyi (2008) for a discussion of related concepts like conditional Kolmogorov complexity.

circuit; that is, given any circuit C, an infinite number of TMs compute the same function as C. Indeed, any single UTM is more powerful than *every* circuit in this sense. On the other hand, it turns out that there are circuit *families* that are more powerful than any single TM. In particular, there are circuit families that can solve the halting problem (Arora and Barak 2009).

Entropy Dynamics

This section reviews those aspects of stochastic thermodynamics that are necessary to analyze the dynamics of various types of entropy during the evolution of computational machines. As illustrated with examples, the familiar quantities at the heart of thermodynamics (e.g., heat, dissipation, thermodynamic entropy, work) arise in special cases of this analysis.

In the first subsection, I review the conventional decomposition of the entropy flow (EF) out of a physical system into the change in entropy of that system plus the (irreversible) entropy creation (EP) produced as that system evolves (Van den Broeck and Esposito 2015; Seifert 2012). To fix details, I will concentrate on the total amounts of EF, EP, and entropy change that arise over a time interval $[0, 1]$.[10]

In the second subsection, I review recent results (Wolpert and Kolchinsky 2018) that specify how the EP generated by the evolution of some system depends on the initial distribution of

---

[10] In this chapter, I will not specify units of time and often implicitly change them. For example, when analyzing the entropy dynamics of a given circuit, sometimes the time interval $[0, 1]$ will refer to the time to run the entire circuit and the attendant entropic costs. However, at other times, $[0, 1]$ will refer to the time to run a single gate within that circuit and the entropic costs of running just that gate. In addition, for computational machines that take more than one iteration to run, I will usually just refer to a "time interval $[0, 1]$" without specifying which iteration of the machine that interval corresponds to. The context will always make the meaning clear.

states of the system. These recent results allow us to evaluate how the EF of an arbitrary system, whose dynamics implements some conditional distribution π of final states given initial states, depends on the initial distribution of states of the system that evolves according to π. As elaborated in subsequent sections, this dependence is one of the central features determining the entropic costs of running any computational machine.

I end this section with some general cautions about translating a computer science definition of a computational machine into a physics definition of a system that implements that machine.

ENTROPY FLOW, ENTROPY PRODUCTION, AND LANDAUER COST

To make an explicit connection with thermodynamics, consider a physical system with countable state space X that evolves over time interval $t \in [0, 1]$ while in contact with one or more thermal reservoirs, while possibly also undergoing driving by one or more work reservoirs.[11] In this chapter, I focus on the scenario where the system dynamics over the time interval is governed by a continuous-time Markov chain (CTMC). However, many of the results presented in what follows are more general.

Let $W_{x,x'}(t)$ be the rate matrix of the CTMC. The probability that the system is in state x at time t evolves according to the linear, time-dependent equation

$$\frac{d}{dt}p_x(t) = \sum_{x'} W_{x,x'}(t)p_{x'}(t), \qquad (1.14)$$

---

[11] In statistical physics, a *reservoir* R in contact with a system S is loosely taken to mean an infinitely large system that interacts with S on time scales infinitely faster than the explicitly modeled dynamical evolution of the state of S. For example, a *particle reservoir* exchanges particles with the system, a *thermal reservoir* exchanges heat, and a *work reservoir* is an external system that changes the energy spectrum of the system S.

which I can write in vector form as $\dot{p}(t) = W(t)p(t)$. I just write W to refer to the entire time history of the rate matrix. W and $p(0)$ jointly fix the conditional distribution of the system's state at $t = 1$ given its state at $t = 0$, which I write as π. As shorthand, I sometimes abbreviate $x(0)$ as x and sometimes abbreviate the initial distribution $p(0)$ as p. So, for example, πp is the ending distribution over states. I will also sometimes abbreviate $p(1)$ as p' and $x(1)$ as x'; the context should always make the meaning clear.

Next, define the *entropy flow (rate)* at time t as

$$\sum_{x',x''} W_{x',x''}(t) p_{x''}(t) \ln \left[\frac{W_{x',x''}}{W_{x'',x'}} \right]. \qquad (1.15)$$

Physically, this corresponds to an entropy flow rate out of the system into reservoirs to which it is coupled.

In order to define an associated total amount of EF during a noninfinitesimal time interval, define $\boldsymbol{x} = (N, \vec{x}, \vec{\tau})$ as a trajectory of $N+1$ successive states $\vec{x} = (x(0), x(1), \ldots, x(N))$, along with times $\vec{\tau} = (\tau_0 = 0, \tau_1, \tau_2, \ldots, \tau_{N-1})$ of the associated state transitions, where $\tau_{N-1} \leq 1$ is the time of the end of the process, $x(0)$ is the beginning, $t = 0$ is the state of the system, and $x(N)$ is the ending, $t = 1$ state of the system. Then, under the dynamics of equation (1.14), the probability of that $\boldsymbol{x}$ given the initial state x_0 is (Esposito and Van den Broeck 2010a; Seifert 2012) is:

$$p(\boldsymbol{x}|x(0)) = \left(\prod_{i=1}^{N-1} S^{\tau_i}_{\tau_{i-1}}(x(i-1)) W_{x(i),x(i-1)}(\tau_i) \right) S^1_{\tau_{N-1}}(x_N), \qquad (1.16)$$

where $S^{\tau'}_{\tau}(x) = e^{\int_\tau^{\tau'} W_{x,x}(t)dt}$ is the "survival probability" of remaining in state x throughout the interval $t \in [\tau, \tau']$. The total EF out of the system during the interval can be written as an integral weighted by these probabilities:

$$\mathcal{Q}_{(p_0)} = \int p_0(x_0) p(\boldsymbol{x} \mid x(0)) \sum_{i=1}^{N-1} W_{x(i),x(i-1)}(\tau_i) \ln \frac{W_{x(i),x(i-1)}(\tau_i)}{W_{x(i-1),x(i)}(\tau_i)} D\boldsymbol{x}. \tag{1.17}$$

Note that I use the convention that EF reflects total entropy flow *out* of the system, whereas much of the literature defines EF as the entropy flow *into* the system.

EF will be the central concern in the following analysis. By plugging in the evolution equation for a CTMC, we can decompose EF as the sum of two terms. The first is just the change in entropy of the system during the time interval. The second is the *(total) entropy production* (EP) in the system during the process (Esposito and Van den Broeck 2011; Seifert 2012; Van den Broeck and Toral 2015). We write EP as $\sigma(p)$. It is the integral over the interval of the instantaneous EP rate,

$$\sum_{x',x''} W_{x',x''}(t) p_{x''}(t) \ln \left[\frac{W_{x',x''} p_{x''}(t)}{W_{x'',x'} p'_x(t)} \right]. \tag{1.18}$$

I will use the expressions "EF incurred by running a process," "EF to run a process," or "EF generated by a process" interchangeably, and similarly for EP.[12] EF can be positive or negative. However, for any CTMC, the EP rate given in equation (1.18) is nonnegative (Esposito and Van den Broeck 2011; Seifert 2012), and therefore so is the EP generated by the process. So,

---

[12] Confusingly, sometimes in the literature the term "dissipation" is used to refer to EP, and sometimes it is used to refer to EF. Similarly, sometimes EP is instead referred to as "irreversible EP," to contrast it with any change in the entropy of the system that arises due to entropy flow.

$$\mathcal{Q}(p_0) = \sigma(p_0) + S(p_0) - S(\pi p_0) \qquad (1.19)$$

$$\geq S(p_0) - S(\pi p_0), \qquad (1.20)$$

where, throughout this section, π refers to the conditional distribution of the state of the system at $t = 1$ given its state at $t = 0$, which is implicitly fixed by $W(t)$.

Total entropy flow across a time interval can be written as a linear function of the initial distribution:

$$\mathcal{Q}_{(p_0)} = \sum_{x_0} \mathcal{F}(x_0) p_0(x_0) \qquad (1.21)$$

for a function $\mathcal{F}(x)$ that depends on the entire function $W_{x;x'}(t)$ for all $t \in [0, 1)$ and so is related to the discrete time dynamics of the entire process, $\pi(x_1 \mid x_0)$ (see eq. 1.17). However, the *minimal* entropy flow for a fixed transition matrix π is the drop in entropy from $S(p_0)$ to $S(\pi P_0)$. This is not a linear function of the initial distribution p_0. In addition, the entropy production—the difference between actual entropy flow and minimal entropy flow—is not a linear function of p_0. These nonlinearities are the basis of much of the richness of statistical physics, in particular, of its relation with information theory.

There are no temperatures in any of this analysis. Indeed, in this very general setting, temperatures need not even be defined. However, often the system is coupled to a heat bath with a well-defined temperature T.[13] If in addition the Hamiltonian of the system obeys *detailed balance* (DB) with

---

[13] Sometimes in the literature, a *heat bath* is defined to be a thermal reservoir at (canonical ensemble) equilibrium, which is sometimes also presumed to be infinite. The context will make it clear whenever the discussion requires this presumption.

that heat bath, EF can be written as (Esposito and Van den Broeck 2010b)

$$\mathcal{Q} = k_B T^{-1} Q, \quad (1.22)$$

where k_B is the Boltzmann constant and Q is the expected amount of heat transferred from the system into the bath ν during the course of the process.

Example 1. *Consider the special case of an* isothermal *process, meaning there is a single heat bath at temperature T (although possibly one or more work reservoirs and particle reservoirs). Suppose that the process transforms an initial distribution p and Hamiltonian H into a final distribution p' and Hamiltonian H'. There are no a priori requirements that either p or p' is at equilibrium for the associated Hamiltonian.*

As mentioned, in this scenario EF equals $(k_B T)^{-1}$ *times the total heat flow into the bath. EP, on the other hand, equals* $(k_B T)^{-1}$ *times the* dissipated work *of the process, which is the work done on the system over and above the minimal work required by any isothermal process that performs the transformation* $(p, H) \mapsto (p', H')$ *(Parrondo, Horowitz, and Sagawa 2015). So by equation (1.20) and energy conservation, the minimal work is the change in the expected energy of the system plus ($k_B T$ times) the drop in Shannon entropy of the system. This is just the change in the* nonequilibrium free energy *of the system from the beginning to the end of the process (Deffner and Jarzynski 2013; Parrondo, Horowitz, and Sagawa 2015; Hasegawa et al. 2010).*

Many different physical phenomena can result in nonzero EP. One broad class of such phenomena arises if we take an "inclusive" approach, modeling the dynamics of the system and bath together.

Example 2. *Continuing with the special case of an isothermal process, suppose that the heat bath never produces any entropy by itself, that is, that the change in the entropy of the bath equals the EF from the system into the bath. Then the change in the sum, {marginal entropy of the system} + {marginal entropy of the heat bath}, must equal the EF from the system to the bath plus the change in the marginal entropy of the system by itself. By equation (1.19), though, this is just the EP of the system.*

On the other hand, Liouville's theorem tells us that the joint entropy of the system and the bath is constant. Combining establishes that EP of the system equals the change in the difference between the joint entropy and the sum of the marginal entropies; that is, EP equals the change in the mutual information between the system and the bath.

To illustrate this, suppose we start with system and bath statistically independent. So the mutual information between them originally equals zero. Since mutual information cannot be negative, the change of that mutual information during the process is nonnegative. This confirms that EP is nonnegative for this particular case where we start with no statistical dependence between the system and the bath (see Esposito, Lindenberg, and Van den Broeck 2010).

Variants of equation (1.20) are sometimes referred to in the literature as the *generalized Landauer's bound*. To motivate this name, suppose that there is a single heat bath at temperature T and that the system has two possible states $X = \{0,1\}$. Suppose further that the initial distribution $p(x)$ is uniform over these two states and that the conditional distribution π implements the function $\{0,1\} \mapsto 0$, that is, a 2-to-1 "bit-erasure" map. So by equation (1.22) and the nonnegativity of EP, the minimal heat flow *out* of the system accompanying any process that performs that bit erasure is $k_B T(\ln 2 - \ln 1) = k_B T \ln 2$, in accord with the bound proposed by Landauer (1961).

Note, though, that in contrast to the bound proposed by Landauer, the generalized Landauer's bound holds for systems with an arbitrary number of states, an arbitrary initial distribution over their states, and an arbitrary conditional distribution π. Most strikingly, the generalized Landauer bound holds even if the system is coupled to multiple thermal reservoirs, all at different temperatures, for example, in a steady-state heat engine (Esposito, Lindenberg, and Van den Broeck 2009; Pietzonka and Seifert 2018) (see ex. 4). In such a case, $k_B T \ln 2$ is not defined. Indeed, the generalized Landauer bound holds even if the system does not obey detailed balance with any of the one or more heat reservoirs to which it's coupled.

Motivated by the generalized Landauer's bound, we define the *(unconstrained) Landauer cost* as the minimal EF required to compute π on initial distribution p using *any* process, with no constraints:

$$\mathcal{L}(p, \pi) := S(p) - S(\pi p). \qquad (1.23)$$

With this definition, we can write

$$\mathcal{Q}(p) = \mathcal{L}(p, \pi) + \sigma(p). \qquad (1.24)$$

Example 3. *Landauer's bound is often stated in terms of the minimal amount of* work *that must be done to perform a given computation, rather than the* heat *that must be generated. This is appropriate for physical processes that both begin and end with a constant, state-independent energy function. For such* *processes, there cannot be any change in expected energy between the beginning and end of the process. Moreover, by the first law of thermodynamics,*

$$\Delta E = W - \mathcal{Q}(p)$$

where ΔE is the change in expected energy from the beginning and end of the process; W is work incurred by the process; and, as before, $\mathcal{Q}(p)$ is the expected amount of heat that leaves the system and enters the bath. Since $\Delta E = 0$, $W = Q$. So the bounds in example 1 on the minimal heat that must flow out of the system also give the minimal work that must be done on the system.

Any process that achieves $\sigma = 0$ (i.e., the generalized Landauer's bound) for some particular initial distribution p is said to be *thermodynamically reversible* when run on that distribution. A necessary condition for a process to be thermodynamically reversible is that if we run it forward on an initial distribution p to produce p', and then "run the process backward" by changing the signs of all momenta and reversing the time sequence of any driving, we return to p (Jarzynski 2011; Van den Broeck and Esposito 2015; Sagawa 2014; Ouldridge, Brittain, and ten Wolde 2019).

A process being "logically reversible" means it implements an invertible map over its state space. However, a process being "thermodynamically reversible" does *not* mean it implements an invertible map over the space of all distributions. Indeed, unless a process implements a logically reversible map over the state space, in general, it will map multiple initial distributions to the same final distribution, up to any desired accuracy (Owen, Kolchinsky, and Wolpert 2018).

As an example, bit erasure is a noninvertible map over the associated unit simplex. However, for any initial distribution q_0, there is a process that implements it thermodynamically reversibly (Esposito *et al.* 2010).

Note, though, that if we run the bit erasure process backward from the ending (delta function) distribution, we have to arrive back at the initial distribution q_0 to satisfy the necessary condition for thermodynamic reversibility when run on q_0. So if we were to run that bit-erasure process on any initial distribution $p_0 \neq q_0$ and then run it backward, we would not arrive back at p_0 (we would arrive at q_0 instead). This proves that the bit-erasure process cannot be thermodynamically reversible when run on any such $p_0 \neq q_0$.

This bit-erasure example underscores that thermodynamic reversibility is a joint property of a process W and the initial distribution p_0; if we run the same process on a different initial distribution, in general the amount of EP it generates will change. This dependence of EP on the initial distribution is a central issue in analyzing the entropic costs of computation, and is addressed in the next subsection.

MISMATCH COST AND RESIDUAL EP

Computational machines are built of multiple interconnected computational devices. A crucial concern in calculating the entropic costs of running such a computational machine is how the costs incurred by running any of its component devices, implementing some distribution π, depend on the distribution over the inputs to that device, p_0. It is crucial how the entropic cost of running an AND gate depends on its inputs, how the cost of running an OR gate depends on its inputs, and so on.

For a fixed π, we can write the Landauer cost of any process that implements π as a single-valued function of the initial distribution p_0; no properties of the rate matrix W matter for calculating the Landauer cost beyond the fact that that matrix implements π. However, even if we fix π, we cannot write EP as a single-valued function of p_0, because EP *does* depend on the details of how W implements π. Intuitively, it is the EP, not the Landauer cost, that reflects the "nitty-gritty" details of the dynamics of the rate matrix implementing the computation. In this subsection, I review recent results establishing precisely how W determines the dependence of EP on p_0.

It has long been known how the entropy production *rate*, at a single moment t, jointly depends on the rate matrix $W(t)$ and on the distribution over states p_t. In fact, those dependencies are given by the expression in equation (1.18), which defines entropy production. On the other hand, until recently, nothing was known about how the EP of a discrete time process, evolving over an extended time interval, depends on the initial distribution over states. Initial progress was made in Kolchinsky and Wolpert (2017), in which the dependence of EP on the initial distribution was derived for the special case where $\pi(x_1 \mid x_0)$ is nonzero for all x_0, x_1. However, this restriction on the form of π is violated in deterministic computations.

Motivated by this difficulty, Wolpert and Kolchinsky (2018) extended the earlier work in Kolchinsky and Wolpert (2017) to give the full dependence of EP on the initial distribution for arbitrary π. That extended analysis shows that EP can always be written as a sum of two terms. Each of those terms depends on p_0 as well as on the nitty-gritty details of the process, embodied in $W(t)$.

The first of those EP terms depends on p_0 linearly. By appropriately constructing the nitty-gritty details of the system (e.g., by having the system implementing π run a quasistatic process), it is possible to have this first term equal zero identically, for all p_0. The second of the EP terms instead is given by a drop in the KL divergence between p and a distribution q that is specified by the nitty-gritty details, during the time interval $t \in [0, 1]$. For nontrivial distributions π, this term *cannot* be made to equal zero for any distribution p_0 that differs from q_0. This is unavoidable EP incurred in running the system, which arises whenever one changes the input distribution to a device away from the optimal distribution without modifying the device itself.

To review these recent results, recall the definition of islands c and associated distributions Δ_c from the section "Terminology and General Notation." I begin with the following definition:

Definition 6. *For any conditional distribution π implemented by a CTMC and any island $c \in L(\pi)$, the associated prior is*

$$q^c \in \underset{r:\text{supp}(r)\in\Delta_c}{\arg\min} \; \sigma(r).$$

We write the associated lower bound on EP as

$$\sigma^{min}(c) := \min_{r:supp(r)\in\Delta_c} \sigma(r).$$

It will simplify the exposition to introduce an arbitrary distribution over islands, $q(c)$, and define

$$q(x) := \sum_{c \in L(\pi)} q(c) q^c(x).$$

In Wolpert and Kolchinsky (2018), it is shown that[14]

$$\sigma(p) = D(p \,||\, q) - D(\pi p \,||\, \pi q) + \sum_{c \in L(\pi)} p(c) \sigma^{min}(c). \quad (1.25)$$

The drop of KL divergences on the RHS of equation (1.25) is called the the *mismatch cost* of running the CTMC on the

---

[14] Owing to the definition of islands, while the choice of distribution $q(c)$ affects the precise distribution q inside the two KL divergences, it has no effect on their difference, and so has no effect on EP (see Wolpert and Kolchinsky 2018).

initial distribution p and is written as $\mathcal{E}(p)$.[15] Given the priors q^c, both of these KL divergences in the mismatch cost depend only on p and on π. By the data-processing inequality for KL divergence, mismatch cost is nonnegative. It equals zero if $p^c = q^c$ for all c, or if π is a measure-preserving map, that is, a permutation of the elements of X.

The remaining sum on the RHS of equation (1.25) is called the *residual EP* of the CTMC. It is a linear function of $p(c)$, without any information-theoretic character. In addition, it has no explicit dependence on π. It is (the $p(c)$-weighted average of) the minimal EP within each island. $\sigma^{min}(c) \geq 0$ for all c, and residual EP equals zero if and only if the process is thermodynamically reversible. I will refer to $\sigma^{min}(c)$ as the *residual EP (parameter) vector* of the process. The nitty-gritty physics details of how the process operates are captured by the residual EP vector together with the priors.

Combining equation (1.25) with the definition of EF and of cross-entropy establishes the following set of equivalent ways of expressing the EF.

Proposition 1. The total EF incurred in running a process that applies map π to an initial distribution p is

$$\mathcal{Q}(p) = \mathcal{L}(p, \pi) + \mathcal{E}(p) + \sum_{c \in L(\pi)} p(c)\sigma^{min}(c)$$
$$= [S(p \parallel q) - S(\pi p \parallel \pi q)] + \sum_{c} p(c)\sigma^{min}(c).$$

Unlike the generalized Landauer's bound, which is an inequality, proposition 1 is exact. It holds for both macroscopic and

---

[15] In Kolchinsky and Wolpert (2017), owing to a Bayesian interpretation of q, the mismatch cost is instead called the "dissipation due to incorrect priors."

microscopic systems, whether or not they are computational devices.

I will use the term *entropic cost* to refer broadly to entropy flow, entropy production, mismatch cost, residual entropy, or Landauer cost. Note that the entropic cost of any computational device is only properly defined if we have fixed the distribution over possible inputs of the device (or strings of inputs, as the case might be).

It is important to realize that we *cannot* ignore the residual EP when calculating EF of real-world computational devices. In particular, in real-world computers—even real-world quantum computers, presuming they are coupled to input/output devices—a sizable portion of the heat generation occurs in the wires connecting the devices inside the computer (often a majority of the heat generation, in fact). However, wires are designed simply to copy their inputs to their outputs, which is a logically invertible map. As a result, the Landauer cost of running a wire is zero (to within the accuracy of the wire's implementing the copy operation with zero error), no matter the initial distribution over states of the wire p_0. For the same reason, the mismatch cost of any wire is zero. This means that the entire EF incurred by running any wire is just the residual EP incurred by running that wire. So in real-world wires, in which $\sigma^{min}(c)$ invariably varies with c (i.e., in which the heat generated by using the wire depends on whether it transmits a 0 or a 1), the dependence of EF on the initial distribution p_0 must be linear. In contrast, for the other devices in a computer (e.g., the digital gates in the computer), both Landauer cost and mismatch cost can be quite large, resulting in nonlinear dependencies on the initial distribution.

Example 4. *It is common in the literature to decompose the rate matrix into a sum of rate matrices of one or more mechanisms* v:

$$W_{x,x'}(t) = \sum_\nu W^v_{x,x'}(t). \tag{1.26}$$

In such cases, one replaces the definitions of the EF rate and EP rate in equations (1.15) and (1.18) with the similar definitions

$$\sum_{x',x'',\nu} W^\nu_{x',x''}(t) p_{x''}(t) \ln\left[\frac{W^\nu_{x',x''}}{W^\nu_{x'',x'}}\right] \tag{1.27}$$

and

$$\sum_{x',x'',\nu} W^\nu_{x',x''}(t) p_{x''}(t) \ln\left[\frac{W^\nu_{x',x''} p_{x''}(t)}{W^\nu_{x'',x'} p'_x(t)}\right], \tag{1.28}$$

respectively.

When there is more than one mechanism, since the log of a sum is not the same as the sum of a log, these redefined EF and EP rates differ from the analogous quantities given by plugging $\sum_\nu W^v_{x,x'}(t)$ into equations (1.15) and (1.18). For example, if we were to evaluate equation (1.15) for this multiple-mechanism $W(t)$, we would get

$$\sum_{x',x'',\nu} W^\nu_{x',x''}(t) p_{x''}(t) \ln\left[\frac{\sum_{\nu'} W^{\nu'}_{x',x''}}{\sum_{\nu''} W^{\nu''}_{x'',x'}}\right], \tag{1.29}$$

which differs from the expression in equation (1.27).

Nonetheless, all the results presented above apply just as well with these redefinitions of EF and EP. In particular, under these redefinitions, the time derivative of the entropy still equals the difference between the EP rate and the EF rate, total EP is still nonnegative, and total EF is still a linear function of the initial distribution. Moreover, that linearity of EF means that with this redefinition, we can still write (total) EP as a sum of the

Chapter 1: Overview of Theory

mismatch cost, defined in terms of a prior and a residual EP that is a linear function of the initial distribution.

By themselves, neither the pair of definitions in equations (1.15) and (1.18) nor the pair in equations (1.27) and (1.28) is "right" or "wrong." Rather, the primary basis for choosing between them arises when we try to apply the resulting mathematics to analyze specific thermodynamic scenarios. The development starting from equations (1.15) and (1.18), for a single mechanism, can be interpreted as giving heat flow rates and work rates for the thermodynamic scenario of a single heat bath coupled to the system (see examples 2 and 3). However, in many thermodynamic scenarios, there are multiple heat baths coupled to the system. The standard approach for analyzing these scenarios is to identify each heat bath with a separate mechanism so that there is a separate temperature for each mechanism, T^ν. Typically, one then assumes local detailed balance (LDB), meaning that separately for each mechanism ν, the associated matrix $W^\nu(t)$ obeys detailed balance for the (shared) Hamiltonian $H(t)$ and resultant (ν-specific) Boltzmann distribution defined in terms of the temperature T^ν, that is, for all ν, x, x', t:

$$\frac{W^\nu_{x,x'}(t)}{W^\nu_{x',x}(t)} = e^{[H_{x'}(t) - H_x(t)]/T^\nu}. \qquad (1.30)$$

This allows us to identify the EF rate in equation (1.27) as the rate of heat flow to all of the baths. So the EP rate in equation (1.28) is the rate of irreversible gain in entropy that remains after accounting for that EF rate and for the change in entropy of the system (Van den Broeck and Esposito 2015; Esposito and Van den Broeck 2010a; Seifert 2012).

As a final comment, it is important to emphasize that all of the foregoing analysis assumes that there are no constraints on how the physical system can implement π. For example, the Landauer cost given in equation (1.23) and proposition 1 is the unconstrained minimal amount of EF necessary to implement the conditional distribution π on any physical system when there are no restrictions on the rate matrix underlying the dynamics of that system. However, in practice, there will always be *some* constraints on what rate matrices the engineer of a system can use to implement a desired logical state dynamics. In particular, the architectures of the computational machines defined in the section "Computational Machines" constrain which variables in a system implementing those machines are allowed to be directly coupled with one another by the rate matrix.

The minimal amount of EF needed to implement a desired distribution π if one is constrained to use a particular computational machine to implement that dynamics is called the *machine Landauer cost*. Trivially, since it is the solution to the same optimization problem that defines Landauer cost, only with extra constraints imposed, the machine Landauer cost is never less than the (unconstrained) Landauer cost; that is, the machine Landauer cost of some particular computational machine is never less than the value given in equation (1.23). In fact, the machine Landauer cost can be substantially greater than the unconstrained Landauer cost, as illustrated by the following simple example.

Example 5. *Suppose our computational machine's state space is two bits, x^1 and x^2, and that the function $f(x)$ erases both of those bits. Let $p_0(x)$ be the initial distribution over joint states of the two bits. As a practical matter, p_0 would be determined by the preferences of the users of the system, for example, as given by the frequency counts over a long time interval in which they repeatedly use the system. In this scenario, the unconstrained Landauer cost is:*

$$\begin{aligned}S(p_0(X)) - S(p_1(X)) &= S(p_0(X)) \\ &= S(p_0(X^1)) + S(p_0(X^2 \mid X^1)).\end{aligned} \quad (1.31)$$

Now modify this scenario by supposing that we are constrained to implement the parallel bit erasure with two subsystems acting independently of one another—one subsystem acting on the first bit and one subsystem acting on the second bit. This changes the Landauer cost to

$$\begin{aligned}S(p_0(X^1)) - S(p_1(X^1)) &+ S(p_0(X^2)) - S(p_1(X^2)) \\ &= S(p_0(X^1)) + S(p_0(X^2)).\end{aligned} \quad (1.32)$$

The gain in Landauer cost due to the constraint—the machine Landauer cost—is $S(p_0(X^2)) - S(p_0(X^2 \parallel X^1))$. This is just the mutual information between the two bits under the initial distribution p_0, which in general is nonzero.

To understand the implications of this phenomenon, suppose that the parallel bit erasing subsystems are thermodynamically reversible when considered by themselves. It is still the case that if they are run in parallel as two subsystems of an overall system, and if their initial states are statistically correlated, then that overall system is not thermodynamically reversible. Indeed, if we start with p_0, implement the parallel bit erasure using two thermodynamically reversible bit erasers, and then run that process in reverse, we end up with the distribution $p_0(x^1)p_0(x^2)$ rather than $p_0(x^1, x^2)$.

A general analysis of how the architecture of a computational machine affects the associated machine Landauer cost can be found in Wolpert and Kolchinsky (2018). The special case when the open system in question is an information ratchet, there is only a single heat bath, and local detailed balance holds is considered in Boyd (2018). See also Riechers (2019) and Grochow and Wolpert (2018).

Entropy Dynamics of Logically Reversible Circuits

Our modern understanding of nonequilibrium statistical physics makes clear that there is no *a priori* relation between the logical reversibility of a function taking inputs to outputs, $f : X^{IN} \to X^{OUT}$, and the thermodynamic reversibility of a system that implements f (Sagawa 2014; Wolpert 2015, 2016b, 2016a; Esposito *et al.* 2010; Maroney 2005). This should not be surprising; logical reversibility is all about maps from the *state* of a system at one time to its state at another, whereas thermodynamic reversibility is all about trajectories of *marginal probability distributions* over the states of the system as the system evolves from one time to another (together with, e.g., properties of any external reservoirs to which the system is coupled). These are completely different kinds of mathematical structures, with the result that thermodynamic reversibility need not imply anything about logical reversibility, or vice versa.

However, the pioneering work of Landauer and Bennett (Landauer 1961; Bennett 1973, 1982) on the thermodynamics of computation led to a common misperception that logical and thermodynamic reversibility are in fact identical. This has in turn motivated research on "reversible circuits." In this research, one is presented with a conventional circuit C made of logically *irreversible* gates that implements some logically irreversible function f and tries to construct a logically *reversible* circuit, C', that emulates C. The underlying insight is that we can always do this by appropriately wiring together a set of logically reversible gates (e.g., Fredkin gates) to create a circuit C' that maps any input bits $x^{IN} \in X^{IN}$ to a set of output bits that contains both $f(x^{IN})$ and a copy of x^{IN} (Fredkin and Toffoli 1982; Drechsler and Wille 2012;

Perumalla 2013; Frank 2005).[16] Tautologically, the entropy of the distribution over the states of this circuit after the map has completed is identical to the entropy of the initial distribution over states. So the Landauer cost is zero, it would appear. This has led to claims in the literature suggesting that, by replacing a conventional, logically irreversible circuit with an equivalent logically reversible circuit, we can reduce the "thermodynamic cost" of computing $f(x^{IN})$ to zero. More precisely, it would appear that we can reduce the total EF expended to zero.

This general line of reasoning should be worrisome. As mentioned, we now know that we can directly implement any map $x^{IN} \rightarrow f(x^{IN})$—even logically irreversible maps—in a thermodynamically reversible manner. So, by running such a direct implementation of a logically reversible f in reverse (which can be done thermodynamically reversibly), we would extract heat from a heat bath. If we do that, and then implement f forward using a logically reversible circuit, we will return the system to its starting distribution, seemingly having extracted heat from the heat bath—and thereby violating the second law.

As it turns out, there *are* some thermodynamic advantages to using a logically reversible circuit rather than an equivalent logically irreversible circuit. However, there are also some disadvantages to using logically reversible circuits. Moreover, the advantages of logically reversible circuits cannot be calculated simply by counting the "number of bit erasures" performed by the equivalent logically irreversible circuit. In the following two subsections, I elaborate these relative advantages and disadvantages of using reversible circuits, to illustrate the results presented in the preceding sections.

---

[16] Or at least contains $f(x^{IN})$ and some additional data to make the function logically reversible.

Before doing that, though, in the remainder of this subsection, I present some needed details concerning logically reversible circuits that are constructed out of logically reversible gates. One of the properties of logically reversible gates that initially caused problems in designing circuits out of them is that running those gates typically produces "garbage" bits, to go with the bits that provide the output of the conventional gate that they emulate. The problem is that these garbage bits need to be reinitialized after the gate is used so that the gate can be used again. Recognizing this problem, Fredkin and Toffoli (1982) shows how to avoid the costs of reinitializing any garbage bits produced by using a reversible gate in a reversible circuit C', by extending C' with yet more reversible gates (e.g., Fredkin gates). The result is an *extended circuit* that takes as input a binary string of input data x, along with a binary string of "control signals" $m \in M$, whose role is to control the operation of the reversible gates in the circuit. The output of the extended circuit is a binary string of the desired output for input x^{IN}, $x^{OUT} = f(x^{IN})$, together with a copy of m and a copy of x^{IN}, which I will write as x^{IN}_{copy}. So, in particular, none of the output garbage bits produced by the individual gates in the original, unextended circuit of reversible gates still exists by the time we get to the output bits of the extended circuit.[17]

---

[17] More precisely, in one popular form of reversible circuits, a map $f : X^{IN} \to X^{OUT}$ is implemented in several steps. First, in a "forward pass," the circuit made out of reversible gates sends $(x^{IN}, m, \vec{0}^{GARBAGE}, \vec{0}^{OUT}) \to (x^{IN}, m, m', f(x^{IN}))$, where m' is the set of garbage bits, $\vec{0}^{OUT}$ is defined as the initialized state of the output bits, and similarly for $\vec{0}^{GARBAGE}$. After completing this forward pass, an offboard copy is made of x^{OUT}, that is, of $f(x^{IN})$. Then the original circuit is run "in reverse," sending $(x^{IN}, m, m', f(x^{IN})) \to (x^{IN}, m, \vec{0}^{GARBAGE}, \vec{0}^{OUT})$. The end result is a process that transforms the input bit string x^{IN} into the offboard copy of $f(x^{IN})$, together with a copy of x^{IN} (conventionally stored in the same physical variables that contained the original version of x^{IN}), all while leaving the control bit string m unchanged.

Chapter 1: Overview of Theory

While it removes the problem of erasing the garbage bits, this extension of the original circuit with more gates does not come for free. In general, it requires doubling the total number of gates (i.e., the circuit's size), doubling the running time of the circuit (i.e., the circuit's depth), and increasing the number of edges coming out of each gate, by up to a factor of three. In special cases, though, these extra costs can be reduced, sometimes substantially.

REVERSIBLE CIRCUITS COMPARED TO ALL-AT-ONCE DEVICES

In practice, typically we want to reuse a given circuit many times, with different inputs each time. Most simply, we can assume that those inputs are generated by IID sampling a fixed distribution (which is ultimately determined by the user of the circuit). To ensure that the (average) entropic costs are independent of the number of times the circuit has been used, we need to reinitialize the output variables of the circuit after each use. I will refer to a process that does this as the *answer-reinitialization* of that circuit.

In general, there are many different "basis sets" of allowed gates we can use to construct a conventional (logically irreversible) circuit that computes any given logically irreversible function f. Indeed, even once we fix a set of allowed gates, an infinite number of logically irreversible circuits implement f using that set of gates. Consequently, we need to clarify precisely what "logically irreversible circuit" we wish to compare to any given extended circuit implementing the same function f as that circuit.

One extreme possibility is to compare the extended circuit to a single, monolithic gate that computes the same function, that is, to a physical system that directly maps $(x^{IN}, \vec{0}^{OUT}) \to (x^{IN}, f(x^{IN}))$. However, this map is logically reversible, just like the extended circuit, and so not of interest for the comparison.

A second possibility is to compare the extended circuit to a system with a state space X that directly maps $x \in X \to$

$f(x) \in X$, without distinguishing input and output variables. Such a map is *not* logically reversible, but, as mentioned earlier, it can be implemented with a thermodynamically reversible system, whatever the initial distribution over X. I will refer to such a physical system that maps $x \to f(x)$ in a manner that is thermodynamically reversible for some initial distribution $q(x)$ as an *all-at-once* (AO) device, or AO circuit. The reason for this terminology is that the underlying Hamiltonian may simultaneously need to couple all components of x to achieve thermodynamic reversibility. If we implement f with an AO device, then the minimal EF we must expend to calculate f is the drop in entropy of the distribution over X as that distribution evolves according to f. This drop is nonzero, assuming f is not logically invertible. This would seem to mean that there is an advantage to using the equivalent extended circuit rather than the AO device, since the minimal EF with the extended circuit is zero.

However, we must be careful to compare apples to apples. The number of information-carrying bits in the extended circuit after it completes computing f is $\log|X^{IN}| + \log|X^{OUT}| + \log|M|$. The number of information-carrying bits in the AO device when it completes is just $\log|X^{OUT}|$. Therefore the minimal number of bits there must be in the AO device is $\max(\log|X^{OUT}|, \log|X^{IN}|)$. So it may be that the extended circuit and the AO circuit implement functions over different spaces, of different sizes.

This means that the entropic costs of answer-reinitializing the two circuits (i.e., reinitializing their variables in preparation for the next inputs) will differ. In general, the Landauer cost and mismatch cost of answer-reinitialization of an extended circuit will be greater than the corresponding answer-reinitialization costs of an equivalent AO device. This is for the simple reason that the answer-reinitialization of the extended circuit must reinitialize the bits containing copies of x and m, which do not even exist in the AO device.

To take a more quantitative approach, first, for simplicity, assume that the initial distribution over the bits in the extended circuit that encodes m is a delta function. This would be the case if we did not want the physical circuit to implement a different computation from one run to the next, so only one vector of control signals m is allowed. This means that the ending distribution over those bits is also a delta function. The Landauer cost of reinitializing those bits is zero, and assuming that we perform the reinitialization using a prior that equals the delta function over m, the mismatch cost is also zero. So assuming that the residual EP of reinitialization of those bits containing a copy of m is zero, we can ignore those bits from now on.

To proceed further in our comparison of the entropic costs of the answer-reinitialization of an AO device with those of an equivalent extended circuit, we need to specify the detailed dynamics of the answer-reinitialization process that is applied to the two devices. Both the AO device and the equivalent extended circuit have a set of output bits that contains $f(x^{\text{IN}})$, which need to be reinitialized, with some associated entropic costs. In addition, though, the extended circuit needs to reinitialize its ending copy of x^{IN}, whereas there is no such requirement of the equivalent AO device. To explore the consequences of this, I now consider several natural models of the answer-reinitialization:

1. In one model, we require that the answer-reinitialization of the circuit be performed within each output bit g itself, separately from all other variables. Define $Fr(C)$ to mean an extended circuit that computes the same input–output function

f^C as a conventional circuit C, and define $AO(C)$ similarly. Assuming for simplicity that the residual entropy of reinitializing all output bits is zero, the EF for the answer-reinitialization of $Fr(C)$ using such a bit-by-bit process is

$$\mathcal{Q}_{X'}(p,q) = \sum_{g \in V^{\text{Cout}}} S(p_g \parallel q_g), \quad (1.33)$$

where V^{OUT} indicates the set of all bits containing the final values of x^{OUT} and $x^{\text{IN}}_{\text{copy}}$.

Using gate-by-gate answer-reinitialization, the EF needed to erase the output bits containing $f^C(x^{\text{IN}})$ is the same for both $AO(C)$ and $Fr(C)$. Therefore the additional Landauer cost incurred in answer-reinitialization due to using $Fr(C)$ rather than $AO(C)$ is the Landauer cost of erasing the output bits in $Fr(C)$ that store x^{IN}:

$$\Delta S_{Fr(C),C}(p) := \sum_{v \in V_{\text{IN}}} S(p_v), \quad (1.34)$$

where I write $v \in V_{\text{IN}}$ to mean the output bits that contain $x^{\text{IN}}_{\text{copy}}$ and p_v to mean the ending marginal distributions over those bits. Similarly, the difference in mismatch cost is

$$\Delta D_{Fr(C),C}(p,q) := \sum_{v \in V_{\text{IN}}} D_v(p^v \parallel q^v), \quad (1.35)$$

where q_v refers to a prior used to reinitialize the output bits in $v \in V_{\text{IN}}$.

However, independent of issues of answer-reinitialization, the Landauer cost of implementing a function using an AO device that is optimized for an initial distribution $p_{/Cin}$ can be bounded as follows:

$$S(p_{\text{IN}}) - S(f^C p_{\text{IN}}) \leq S(p_{\text{IN}})$$
$$\leq \sum_{v \in V_{\text{IN}}} S_v(p_v) \quad (1.36)$$
$$= \Delta S_{Fr(C),C}(p).$$

Combining this with equation (1.34) shows that under gate-by-gate answer-reinitialization, the *total* Landauer cost of implementing a function using an AO device—including the costs of reinitializing the gates containing the value $f^C(x^{\text{IN}})$—is upper bounded by the *extra* Landauer cost of implementing that same function with an equivalent extended circuit, that is, just that portion of the cost that occurs in answer-reinitializing the extra output bits of the extended circuit. This disadvantage of using the extended Fredkin circuit holds even if the equivalent AO device is logically irreversible. So as far as Landauer cost is concerned, there is no reason to consider using an extended circuit to implement a logically irreversible computation with this first type of answer-reinitialization.

On the other hand, in some situations, the mismatch cost of the AO device will be *greater* than the mismatch cost of the answer-reinitialization of $x_{\text{copy}}^{\text{IN}}$ in the equivalent extended circuit. This is illustrated in the following example:

Example 6. *Suppose that the input to the circuit consists of two bits, a and b, where the actual distribution over those bits, p, and prior distribution over those bits, q, are*

$$p(x_b) = \delta(x_b, 0),$$
$$p(x_a \mid x_b = 0) = 1/2, \quad \forall x_a,$$
$$q(x_b) = \delta(x_b, 1),$$
$$q(x_a \mid x_b = 0) = 1/2, \quad \forall x_a,$$
$$q(x_a \mid x_b = 1) = \delta(x_a, 0).$$

Suppose as well that f^C is a many-to-one map. Then plugging in gives

$$D(p_{\text{IN}} \parallel q_{\text{IN}}) - D(f^C p_{\text{IN}} \parallel f^C q_{\text{IN}}) = D(p_{\text{IN}} \parallel q_{\text{IN}})$$
$$> \sum_{v \in V_{\text{IN}}} D(p_v \parallel q_v). \quad (1.37)$$

This sum equals the mismatch cost of the answer-reinitialization of $x_{\text{copy}}^{\text{IN}}$, which establishes the claim.

However, care should be taken in interpreting this result, because there are subtleties in comparing mismatch costs between circuits and AO devices owing to the need to compare apples to apples (Wolpert and Kolchinsky 2018).

2. A second way we could answer-reinitialize an extended circuit involves using a system that simultaneously accesses all of the output bits to reinitialize $x_{\text{copy}}^{\text{IN}}$, including the bits storing $f(x^{\text{IN}})$.

To analyze this approach, for simplicity, assume there are no restrictions on how this reinitializing system operates, in other words, that it is an AO device. The Landauer cost of this type of answer-reinitialization of $x_{\text{copy}}^{\text{IN}}$ is just $S(p(X^{\text{IN}} \mid X^{\text{OUT}})) - \ln[1]$, since this answer-reinitialization process is a many-to-one map over the state of $X_{\text{copy}}^{\text{IN}}$. Assuming f^C is a deterministic map, though, by Bayes's theorem,

$$S(p(X^{\text{IN}} \mid X^{\text{OUT}})) = S(p(X^{\text{IN}})) - S(p(X^{\text{OUT}})). \quad (1.38)$$

So in this type of answer-reinitialization, the extra Landauer cost of the answer-reinitialization in the extended circuit that computes f^C is identical to the total Landauer cost of the AO device that computes the same function f^C. On the other hand, in this type of answer-reinitialization process, the mismatch cost of the extended circuit may be either greater or smaller than that of the AO device, depending on the associated priors.

3. A third way we could answer-reinitialize $x_{\text{copy}}^{\text{IN}}$ in an extended circuit arises if, after running the circuit, we happen upon a set of initialized external bits, just lying around, as it were, ready to be exploited. In this case, after running the circuit, we could simply swap those external bits with the output bits of the

circuit that contains a copy of x^{IN}, thereby answer-reinitializing the output bits at zero cost.

Arguably, this is more sleight of hand than a real proposal for how to reinitialize the output bits. Even so, it's worth pointing out that, rather than using those initialized external bits to contain a copy of x^{IN}, we could have used them as an information battery, extracting up to a maximum of $k_B T \ln 2$ from each one by thermalizing it. So the opportunity cost in using those external bits to reinitialize the output bits of the Fredkin circuit rather than using them as a conventional battery is $|V_{\text{IN}}| k_B T \ln 2$. This is an upper bound on the Landauer cost of implementing the desired computation using an AO device. So, again, as far as Landauer cost is concerned, there is no advantage to using an extended circuit to implement a logically irreversible computation with this third type of answer-reinitialization.

Summarizing, it is not clear that there is a way to implement a logically irreversible function with an extended circuit built out of logically reversible gates that reduces the Landauer cost below the Landauer cost of an equivalent AO device. The effect on the mismatch cost of using such a circuit rather than an AO device is more nuanced, varying with the priors, the actual distribution, and so on.

However, it is important to emphasize that all of the foregoing analysis compares the entropic costs of an extended circuit with the costs of a computationally equivalent *AO device*, not with the costs of a computationally equivalent conventional circuit built with logically irreversible gates. Almost by the definition of an AO device, the comparison of an extended circuit's cost with that of a conventional circuit can only be *worse* than its comparison with an AO device. It is instructive to see where some of this increase in the cost of a conventional circuit compared to an AO device arises.

REVERSIBLE CIRCUITS VERSUS IRREVERSIBLE CIRCUITS

I now extend the analysis from comparing the entropic costs of an extended circuit and those of a computationally equivalent AO device to consider the costs of a computationally equivalent conventional circuit, built with multiple logically irreversible gates. As illustrated in what follows, the entropic costs of the answer-reinitialization of a conventional circuit (appropriately modeled) are the same as the entropic costs of the answer-reinitialization of a computationally equivalent AO device. The analysis of the preceding subsection carries over, giving the relationship between the entropic costs of answer-reinitialization of conventional circuits and the entropic costs of answer-reinitialization of computationally equivalent extended circuits. In particular, the minimal EF required to answer-reinitialize a conventional circuit is in general lower than the minimal EF required to answer-reinitialize a computationally equivalent extended circuit.

Accordingly, in this subsection, I focus instead on comparing the entropic costs of running conventional circuits, before they undergo any answer-reinitialization, with the entropic costs of running computationally equivalent extended circuits, before *they* undergo any answer-reinitialization. While the full analysis of the entropic costs of running conventional circuits is rather elaborate (Wolpert and Kolchinsky 2018), some of the essential points can be illustrated with the following simple example.

Suppose we have a system that comprises two input bits and two output bits, with state space written as $X = X_1^{IN} \times X_2^{IN} \times X_1^{OUT} \times X_2^{OUT}$. Consider mapping the input bits to the output bits by running the "parallel bit erasure" function. Suppose that while doing that, we simultaneously reinitialize the input bits x_1^{IN} and x_2^{IN}, in preparation for the next run of the system on

Chapter 1: Overview of Theory

a new set of inputs. Assuming that both of the output bits are initialized before the process begins to the erased value 0, the state space evolves according to the function $f : (x_1^{\text{IN}}, x_2^{\text{IN}}, 0, 0) \to (0, 0, 0, 0)$.

Consider the following three systems that implement this f:

1. an AO device operating over $X_1^{\text{IN}} \times X_2^{\text{IN}} \times X_1^{\text{OUT}} \times X_2^{\text{OUT}}$ that directly implements f;

2. a system that implements f using two bit-erasure gates that are physically isolated from one another, as briefly described in example 5. Under this model, the system first uses one bit-erasure gate to send $(X_1^{\text{IN}}, X_1^{\text{OUT}}) = (x_1^{\text{IN}}, 0) \to (0, 0)$ and then uses a second bit-erasure gate to apply the same map to the second pair of bits, $(X_2^{\text{IN}}, X_2^{\text{OUT}})$. The requirement that the gates be physically isolated means that the rate matrix of the first gate is only allowed to involve the pair of bits $(X_1^{\text{IN}}, X_1^{\text{OUT}})$; that is, it is of the form $W_{x_1^{\text{IN}}, x_1^{\text{OUT}}; (x_1^{\text{IN}})', (x_1^{\text{OUT}})'}(t)$. So the dynamics of $(x_1^{\text{IN}}, x_1^{\text{OUT}})$ is independent of the values of the other variables, $(x_2^{\text{IN}}, x_2^{\text{OUT}})$. Similar restrictions apply to the rate matrix of the second gate. So, in the language of Wolpert and Kolchinsky (2018), each of the two gates runs a "solitary process;"

3. a system that uses two bit-erasure gates to implement f, just as in model 2, but does *not* require that those gates run in sequence and that they be physically isolated. In other words, the rate matrix that drives the first bit-erasure gate as it updates the variables $(x_1^{\text{IN}}, x_1^{\text{OUT}})$ *is* allowed to do so based on the values $(x_2^{\text{IN}}, x_2^{\text{OUT}})$, and vice versa. Formally, this means that both of those rate matrices are of the form $W_{x_1^{\text{IN}}, x_1^{\text{OUT}}, x_2^{\text{IN}}, x_2^{\text{OUT}}; (x_1^{\text{IN}})', (x_1^{\text{OUT}})'; (x_2^{\text{IN}})', (x_2^{\text{OUT}})'}(t)$.

Models 2 and 3 are both physical models of a conventional circuit made out of two gates that each implements logically irreversible functions. However, they differ in whether they only allow physical coupling among the variables in the circuit that are logically needed for the circuit to compute the desired function (model 2) or instead allow arbitrary coupling, for example, to reduce entropic costs (model 3).

The Landauer cost of the first model is the minimal EF needed to implement the parallel bit erasure with a single monolithic gate,

$$\begin{aligned}
& S(p_0(X_1^{\text{IN}}, X_2^{\text{IN}}, X_1^{\text{OUT}}, X_2^{\text{OUT}})) \\
& \quad - S(\hat{f}_{1,2}\, p_0(X_1^{\text{IN}}, X_2^{\text{IN}}, X_1^{\text{OUT}}, X_2^{\text{OUT}})) \\
& = S(p_0(X_1^{\text{IN}}, X_2^{\text{IN}}, X_1^{\text{OUT}}, X_2^{\text{OUT}})) \\
& = S(p_0(X_1^{\text{IN}}, X_2^{\text{IN}})),
\end{aligned} \qquad (1.39)$$

where $\hat{f}_{1,2}$ is the conditional distribution implementing the parallel bit erasure so that $\hat{f}_{1,2}\, p_0(x)$ is the ending distribution, which is a delta function centered at $(0, 0, 0, 0)$.

Next, assume that both of the gates in model 2 are thermodynamically reversible if they are considered in isolation, separately from any other systems (and if they are run on the appropriate initial distributions). In other words, *considered in isolation from any other systems*, they are both AO devices. The minimal EF needed to run the first of those gates is

$$S(p_0(X_1^{\text{IN}})) - S(\hat{f}_1\, p_0(X_1^{\text{IN}}, X_1^{\text{OUT}})) = S(p_0(X_1^{\text{IN}})). \qquad (1.40)$$

Similarly, the minimal EF needed to run the second gate is $S(p_0(X_2^{\text{IN}}))$.

The difference between {the minimal EF needed to run a conventional circuit constructed as in model 2} and {the minimal EF needed to run a computationally equivalent AO device (model 1)} is

$$S(p_0(X_1^{\text{IN}})) + S(p_0(X_2^{\text{IN}})) - S(p_0(X_1^{\text{IN}}, X_2^{\text{IN}})). \qquad (1.41)$$

This is just the initial mutual information between X_1^{IN} and X_2^{IN}. So the minimal EF needed to run model 2 will exceed the minimal EF needed to run model 1 whenever X_1^{IN} and X_2^{IN} are statistically coupled under the initial distribution p_0 (see ex. 5).

On the other hand, because of the increased flexibility in their rate matrices, we can assume that the bit-erasure gates in the circuit defined in model 3 each achieves zero EP *even when considered as a system operating over the full set of four bits.* So each of those bit-erasure gates is thermodynamically reversible even when considered in the context of the full system. As a result, running the circuit defined in model 3 requires the same minimal EF as running an AO device. So, in general, the minimal EF needed to run the conventional circuit defined in model 3 is less than the minimal EF needed to run the conventional circuit defined in model 2. This increase in the minimal EF of running a circuit as modeled in 2 compared to a circuit as modeled in 3 is precisely the difference between machine Landauer cost and (unconstrained) Landauer cost, discussed at the end of the section "Entropy Dynamics." A detailed analysis of this difference in costs, involving explicit rate matrices, can be found in Wolpert (2019).

Next, note that we only need to reinitialize the two output bits after running any of the three models. In contrast, the extended circuit that runs f needs to reinitialize its input bits as well in preparation for receiving a new set of inputs for a next run of the system. So models 1, 2, and 3 have lower minimal EF for answer-reinitialization than the computationally equivalent extended circuit, in general.

Summarizing, the minimal total EF (including the EF needed both to run the system and to answer-reinitialize it) that is needed by model 2 exceeds the minimal total EF needed by either the equivalent AO device (model 1) or the equivalent conventional circuit as defined by model 3, with the difference equal to the mutual information of the input bits under p_0. In turn, the minimal EF to run (only) either model 1 or model 3 exceeds the minimal EF needed to run an equivalent extended circuit, by $S(p_0(X_1^{IN}, X_2^{IN}))$. However, the minimal *total* EF of models 1 and 3 will in general be no greater than the minimal total EF of the extended circuit, and may be smaller, depending on the details of the answer-reinitialization process in the extended circuit.

On the other hand, as a purely practical matter, constructing a conventional circuit as in model 3 for circuits substantially larger than parallel bit erasures may be quite challenging; to do so requires that we identify all sets of variables that are *statistically* coupled, at any stage of running the circuit, and that we make sure that our gates are designed to *physically* couple those variables. There are no such difficulties with constructing an extended circuit. Another advantage of an extended circuit is that, no matter what the true distribution p_0 is, an extended circuit has zero mismatch cost because there is no drop of KL divergence between p_0 and *any* q_0 under a logically reversible dynamics. In contrast, all three models can have nonzero mismatch cost, in general.

As yet another point of comparison, an extended circuit will often have far more wires than an equivalent conventional circuit. And, as mentioned earlier, the residual EP generated in wires is one of the major sources of EF in modern digital gates. Thus, even in a situation where a conventional circuit has nonzero mismatch cost, when the EF generated in the wires is taken into account, there may be no disadvantage to using that conventional circuit rather than a computationally equivalent extended circuit.

Clearly there is a rich relationship between the detailed wiring diagram of a conventional, logically irreversible circuit; the procedure for answer-reinitializing the outputs of a computationally equivalent extended circuit; the distribution over the input bits of those circuits; and how the aggregate entropic costs of those two circuits compare. Precisely delineating this relationship is a topic for future research. 🌿

Acknowledgments

I would like to thank Josh Grochow and Peter Stadler for helpful discussion, and thank the Santa Fe Institute for helping to support this research. This paper was made possible through Grant No. CHE-1648973 from the US National Science Foundation and Grant No. FQXi-RFP-1622 from the FQXi foundation. The opinions expressed in this paper are those of the author and do not necessarily reflect the view of the SFI, the National Science Foundation, or FQXi.

REFERENCES

Aaronson, S. 2005. "NP-Complete Problems and Physical Reality." Quant-ph/0502072.

———. 2013. "Why Philosophers Should Care About Computational Complexity." *Computability: Turing, Gödel, Church, and Beyond:* 261–327.

Arora, S., and B. Barak. 2009. *Computational Complexity: A Modern Approach.* Cambridge University Press.

Baez, J., and M. Stay. 2012. "Algorithmic Thermodynamics." *Mathematical Structures in Computer Science* 22 (05): 771–787.

Barnett, N., and J. P. Crutchfield. 2015. "Computational Mechanics of Input–Output Processes: Structured Transformations and the ϵ-Transducer." *Journal of Statistical Physics* 161 (2): 404–451.

Barrow, J. D. 2011. "Gödel and Physics." In *Kurt Gödel and the Foundations of Mathematics: Horizons of Truth,* edited by M. Baaz, C. H. Papadimitriou, H. W. Putnam, D. S. Scott, and C. L. Harper Jr., 255. Cambridge University Press.

Bennett, C. H. 1973. *IBM Journal of Research and Development* 17:525–532.

———. 1982. "The Thermodynamics of Computation—A Review." *International Journal of Theoretical Physics* 21 (12): 905–940.

Boyd, A. B. 2018. "Thermodynamics of Modularity: Structural Costs Beyond the Landauer Bound." *Physical Review X* 8 (3).

Caves, C. M. 1990. "Entropy and Information: How Much Information is Needed to Assign a Probability." *Complexity, Entropy and the Physics of Information:* 91–115.

———. 1993. "Information and Entropy." *Physical Review E* 47 (6): 4010.

Cover, T. M., and J. A. Thomas. 2012. *Elements of Information Theory*. John Wiley & Sons. Accessed January 11, 2014.

Deffner, S., and C. Jarzynski. 2013. "Information Processing and the Second Law of Thermodynamics: An Inclusive, Hamiltonian Approach." *Physical Review X* 3 (4): 041003.

Drechsler, R., and R. Wille. 2012. "Reversible Circuits: Recent Accomplishments and Future Challenges for an Emerging Technology." In *Progress in VLSI Design and Test,* 383–392. Springer.

Esposito, M., R. Kawai, K. Lindenberg, and C. Van den Broeck. 2010. "Finite-Time Thermodynamics for a Single-Level Quantum Dot." *EPL (Europhysics Letters)* 89 (2): 20003.

Esposito, M., K. Lindenberg, and C. Van den Broeck. 2009. "Thermoelectric Efficiency at Maximum Power in a Quantum Dot." *EPL (Europhysics Letters)* 85 (6): 60010.

———. 2010. "Entropy Production as Correlation between System and Reservoir." *New Journal of Physics* 12 (1): 013013.

Esposito, M., and C. Van den Broeck. 2010a. "Three Detailed Fluctuation Theorems." ArXiv: 0911.2666, *Physical Review Letters* 104, no. 9 (March). Accessed January 23, 2018. http://arxiv.org/abs/0911.2666.

———. 2010b. "Three Faces of the Second Law. I. Master Equation Formulation." *Physical Review E* 82 (1): 011143.

———. 2011. "Second Law and Landauer Principle Far from Equilibrium." *EPL (Europhysics Letters)* 95 (4): 40004.

Frank, M. P. 2005. "Introduction to Reversible Computing: Motivation, Progress, and Challenges." In *Proceedings of the 2nd Conference on Computing Frontiers,* 385–390. ACM.

Fredkin, E., and T. Toffoli. 1982. "Conservative Logic." *Internat. J. Theoret. Phys.* 21 (3): 219–253.

Grochow, J., and D. H. Wolpert. 2018. "Beyond Number of Bit Erasures: New Complexity Questions Raised by Recently Discovered Thermodynamic Costs of Computation." *ACM SIGACT News.*

Hasegawa, H.-H., J. Ishikawa, K. Takara, and D. J. Driebe. 2010. "Generalization of the Second Law for a Nonequilibrium Initial State." *Physics Letters A* 374 (8): 1001–1004.

Hopcroft, J. E., R. Motwani, and U. Rotwani. 2000. *JD: Introduction to Automata Theory, Languages and Computability.*

Ito, S., and T. Sagawa. 2013. "Information Thermodynamics on Causal Networks." *Physical Review Letters* 111 (18): 180603.

———. 2015. "Information Flow and Entropy Production on Bayesian Networks." ArXiv: 1506.08519, *arXiv:1506.08519 [cond-mat]* (June). Accessed July 19, 2017.

Jarzynski, C. 2011. "Equalities and Inequalities: Irreversibility and the Second Law of Thermodynamics at the Nanoscale." *Annu. Rev. Condens. Matter Phys.* 2 (1): 329–351. Accessed January 13, 2016.

Kolchinsky, A., and D. H. Wolpert. 2017. "Dependence of Dissipation on the Initial Distribution over States." *Journal of Statistical Mechanics: Theory and Experiment:* 083202.

Koller, D., and N. Friedman. 2009. *Probabilistic Graphical Models.* MIT Press.

Landauer, R. 1961. "Irreversibility and Heat Generation in the Computing Process." *IBM Journal of Research and Development* 5 (3): 183–191.

Li, M., and P. Vitanyi. 2008. *An Introduction to Kolmogorov Complexity and Its Applications.* Springer.

Mandal, D., and C. Jarzynski. 2012. "Work and Information Processing in a Solvable Model of Maxwell's Demon." *Proceedings of the National Academy of Sciences* 109 (29): 11641–11645.

Maroney, O. J. E. 2005. "The (Absence of a) Relationship between Thermodynamic and Logical Reversibility." *Studies in History and Philosophy of Science Part B: Studies in History and Philosophy of Modern Physics* 36, no. 2 (June): 355–374.

Moore, C., and S. Mertens. 2011. *The Nature of Computation.* Oxford University Press.

Nekrashevych, V. 2005. *Self-Similar Groups.* Vol. 117. Mathematical Surveys and Monographs. American Mathematical Society.

Ouldridge, T., R. Brittain, and P. R. ten Wolde. 2019. "The Power of Being Explicit: Demystifying Work, Heat, and Free Energy in the Physics of Computation." In *The Energetics of Computing in Life and Machines,* edited by D. H. Wolpert, C. P. Kempes, P. Stadler, and J. Grochow. SFI Press.

Owen, J. A., A. Kolchinsky, and D. H. Wolpert. 2018. "Number of Hidden States Needed to Physically Implement a Given Conditional Distribution." *New Journal of Modern Physics.*

Parrondo, J. M. R., J. M. Horowitz, and T. Sagawa. 2015. "Thermodynamics of Information." *Nature Physics* 11 (2): 131–139.

Perumalla, K. S. 2013. *Introduction to Reversible Computing.* CRC Press.

Pietzonka, P., and U. Seifert. 2018. "Universal Trade-Off between Power, Efficiency, and Constancy in Steady-State Heat Engines." *Physical Review Letters* 120 (19): 190602.

Riechers, P. 2019. "Transforming Metastable Memories: The Nonequilibrium Thermodynamics of Computation." In *The Energetics of Computing in Life and Machines,* edited by D. H. Wolpert, C. P. Kempes, P. Stadler, and J. Grochow. SFI Press.

Sagawa, T. 2014. "Thermodynamic and Logical Reversibilities Revisited." *Journal of Statistical Mechanics: Theory and Experiment* 2014 (3): P03025.

Savage, J. E. 1998. *Models of Computation.* Vol. 136. Addison-Wesley.

Seifert, U. 2012. "Stochastic Thermodynamics, Fluctuation Theorems and Molecular Machines." *Rep. Progress Phys.* 75 (12): 126001.

Van den Broeck, C., and M. Esposito. 2015. "Ensemble and Trajectory Thermodynamics: A Brief Introduction." *Physica A: Statistical Mechanics and its Applications* 418:6–16.

Van den Broeck, C., and R. Toral. 2015. "Stochastic Thermodynamics for Linear Kinetic Equations." *Physical Review E* 92 (1): 012127.

Wolpert, D. H. 2015. "Extending Landauer's Bound from Bit Erasure to Arbitrary Computation." ArXiv:1508.05319 [cond-mat.stat-mech].

———. 2016a. "Correction: Wolpert, D. H. The Free Energy Requirements of Biological Organisms; Implications for Evolution. Entropy 2016, 18, 138." *Entropy* 18 (6): 219.

———. 2016b. "The Free Energy Requirements of Biological Organisms: Implications for Evolution." *Entropy* 18 (4): 138.

———. 2019. "The Stochastic Thermodynamics of Computation." *Journal of Physics A: Mathematical and Theoretical.*

Wolpert, D. H., and A. Kolchinsky. 2018. "The Entropic Costs of Straight-Line Circuits." *arXiv preprint arXiv:1806.04103.*

Zurek, W. H. 1989. "Algorithmic Randomness and Physical Entropy." *Phys. Rev. A* 40 (8): 4731–4751.

6

A COMPOSITIONAL CHEMICAL ARCHITECTURE FOR ASYNCHRONOUS COMPUTATION

Blake S. Pollard, Carnegie Mellon University

Introduction

There is a sense in which chemical reactions perform computation, transducing energy as various molecules interact and alter one another and their environment. This idea of molecular computation provides multiple avenues of exploration as we as a species investigate various alternative notions of computation and their accompanying architectures: understanding the computational class of various reaction paradigms (Doty and Zhu 2018), implementing digital computation using chemical reactions (Soloveichik *et al.* 2008), mimicking chemical reactions using designer DNA sequences and their interactions (Soloveichik, Seelig, and Winfree 2010), and understanding the thermodynamic efficiency of computations performed by cells (Kempes *et al.* 2017).

We can interpret a number of biochemical reaction motifs as implementing an algorithm, often in an approximate or obfuscated way. For example, the process whereby a cell transitions from inactive to active mitosis is implemented by a robust biochemical switching mechanism: cyclin-dependent kinase (Cdk). At a structural level, this mechanism can be interpreted as a robust and biologically feasible implementation of an algorithm known as *approximate majority* (AM), which provides the asymptotically fastest way to reach a consensus between two possible outcomes across all members of a population (Cardelli and Csikász-Nagy 2012). As the name suggests, the final outcome approximately matches the opinion of the initial majority. This is an example of an algorithmic interpretation of a biochemical motif providing insight into the relationship between the structure of the underlying reactions and the overall function of the motif.

In recent work, Cardelli, Kwiatkowska, and Whitby (2018) go on to show how certain simple reaction motifs structurally related to AM provide a basis for performing asynchronous computation using chemical reaction networks (CRNs). The reliance of synchronous logic on a global clock makes molecular implementation difficult. This difficulty is bypassed in asynchronous logic using a *Muller C-element*, which can be utilized to synchronize various components.

A Muller C-element is a two-input x_1, x_2, single-output y gate whose behavior differs slightly from either AND or OR:

AND		
x_1	x_2	y
0	0	1
0	1	0
1	0	0
1	1	1

OR		
x_1	x_2	y
0	0	0
0	1	1
1	0	1
1	1	1

C-element		
x_1	x_2	y
0	0	0
0	1	unchanged
1	0	unchanged
1	1	1

If both inputs are 0, a 0 is returned. If both inputs are 1, a 1 is returned. This aligns with the behavior of a simple OR gate. However, when one input is a 1 and the other is a 0, rather than outputting a 1, as an OR gate would, a Muller C-element leaves the output state unchanged (Spars and Furber 2002). The utility of such a gate is best understood from the point of view of an observer monitoring only the output. For a C-element, a change in the output from 1 to 0 indicates that *both* inputs are now zero, and vice versa for a change from 0 to 1. For an OR gate, observing the output change from 1 to 0 gives the observer knowledge of *both* inputs. A change from 0 to 1 only tells the observer that at least one of the inputs has changed. It does not allow one to deduce the actual values of the new inputs.

A key contribution of Cardelli, Kwiatkowska, and Whitby (2018) is a robust CRN implementation of a Muller C-element based on the AM CRN. In this chapter, we highlight how that motif itself is generated via certain basic compositions and how those compositions provide the basic operations to build up complex asynchronous logic circuits. Combined with the work of the author together with John Baez (Baez and Pollard 2017), in which we formalize a compositional framework for chemical reaction networks using the mathematics of *category theory*, this provides a compositional framework for the design and analysis of asynchronous circuits implemented via CRNs.

In the next section, we define a graphical syntax for chemical reaction networks using bipartite graphs, commonly known as Petri nets. One can assign various semantics to this syntax, including a linear stochastic dynamics on a discrete state space of molecule counts or a nonlinear deterministic dynamics on a continuous state space of molecular concentrations. Next, we describe *open* chemical reaction networks, in which certain chemicals can flow in and out of the system, and how such open reaction networks fit into a *compositional* framework that allows one to build up complicated networks from simpler ones. We then show how the chemical circuits that form a basis for performing asynchronous computation can be understood as composites of a few simple building blocks and operations. We end by aligning our compositional framework for open chemical reaction networks with a framework for analyzing the thermodynamics of such systems.

Reaction Networks

Petri nets or reaction networks were invented in 1939 by the then thirteen-year-old Carl Adam Petri as a notation for describing chemical processes. First formally defined by Aris (1965), they are bipartite graphs representing interactions among various entities. In the context of chemistry, one type of node represents the various chemical "species" that interact via intermediate node types called *transitions* or *reactions* in chemistry. Petri nets have numerous applications in the description and modeling of distributed systems and concurrency (Peterson 1977). One can think of chemical reactions as a distributed computing system, where the overall computations result from many localized reactions. Various terminologies and definitions exist: Petri nets, place/transition nets, reaction networks, and so on. We stick to the term *reaction network* for the remainder of the chapter.

Definition 1. *A reaction network* (S, T, s, t, r) *consists of a set S, a set T, functions $s, t: T \to \mathbb{N}^S$, and a function $r: T \to (0, \infty)$.*

We call the elements of S *species*, those of T *transitions*, and those of $\mathbb{N}^S$ *complexes*. Any transition $\tau \in T$ has a *source* $s(\tau)$, a *target* $t(\tau)$, and a *rate constant* $r(\tau)$. Note that here we are including 0 in the natural numbers $\mathbb{N}$.

This amounts to specifying the structure of a *hypergraph*, a graph whose edges connect arbitrary numbers of nodes, in which the multiedges are labeled by nonnegative numbers. For example, we can draw the simple reaction $2A + C \xrightarrow{\alpha} B + C$ in the following way:

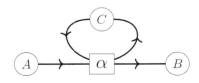

The circles represent species, the squares transitions. The directed edges specify the source and target complexes, and $\alpha \in (0, \infty)$ denotes the rate constant of the transition. The number of edges from (to) a species to (from) a transition is the coefficient of that species in the source (target) complex of that transition, hence the pair of edges heading from A to α. When rate constants are omitted, they can be assumed to be equal to 1.

Reaction networks serve as a graphical syntax admitting at least two interesting notions of semantics or dynamics: the nonlinear *rate equation* or the stochastic *chemical master equation*.

The dynamical variables for the rate equation are the *concentrations* of each chemical species, which can be interpreted as the "number per unit volume." A set of coupled, nonlinear differential equations, called the *rate equation*, describes the time evolution of these concentrations under the assumptions that the reactor vessel is well mixed and that the reactions proceed according to the law of mass action.

Introducing the notation

$$c^{s(\tau)} = \prod_{\sigma \in S} c_\sigma^{s_\sigma(\tau)},$$

we can write the *rate equation* for any reaction network (S, T, s, t, r) as

$$\frac{dc}{dt} = \sum_{\tau \in T} r(\tau) \left(t(\tau) - s(\tau) \right) c^{s(\tau)},$$

where $c \in [0, \infty)^S$ is a vector of nonnegative concentrations.

The rate equation for our previous example reaction network is given by

$$2A + C \xrightarrow{\alpha} B + C \quad \mapsto \quad \begin{aligned} \dot{A} &= -\alpha 2A^2 C \\ \dot{B} &= \alpha A^2 C \\ \dot{C} &= 0. \end{aligned} \quad (2.1)$$

Slightly abusing notation, A, B, and $C \in [0, \infty)^{\mathbb{R}}$ denote the time-dependent concentrations of their respective species and $\dot{A} = \frac{dA}{dt}$ their derivatives with respect to time.

When large numbers of molecules of each species are present, the rate equation provides an accurate description of the dynamics. When particle numbers become small, that is, tens or fewer, the fact that the number of particles fluctuates as reactions proceed becomes relevant, and one should use the chemical master equation to describe the system.

The state space for the chemical master equation consists of integer-valued molecular counts of each species. The master equation describes the dynamics of probability distributions on this space. Traditional thermodynamics of chemical reaction networks require large numbers of molecules and hence concern the rate equation, while the emerging field of stochastic thermodynamics provides an approach to thermodynamic reasoning when particle number fluctuations are of prime importance (Rao and Esposito 2016).

In addition to the Muller C-element, for asynchronous computation to be Turing complete, an *isochronous fork* is required, which copies a signal and broadcasts it so as to reach its recipient at essentially the same time (Spars and Furber 2002). The behavior of the CRN implementation of an isochronous fork can be verified in the deterministic semantics but not the stochastic semantics (Cardelli, Kwiatkowska, and Whitby 2018).

Open Reaction Networks

We now turn our attention to *open* chemical reaction networks, open in the sense that they interact with their environment or with other networks. Such open reaction networks can be composed to build more complicated networks. We formalize this using the mathematics of *category theory*. The deterministic dynamics of such open reaction networks also fits within such a compositional framework, allowing one to build up complicated sets of nonlinear differential equations by combining simpler ones. This provides a compositional framework for reasoning about asynchronous circuits as implemented by chemical reaction networks.

Definition 2. *Given finite sets X and Y, an* **open reaction network** *from X to Y is a pair of functions into the same set S*

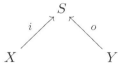

together with a reaction network with rates on $R = (S, T, s, t, r)$ on S. We often abbreviate this datum as $R\colon X \to Y$.

We can think of the sets X and Y as "input" and "output" ports through which chemicals can flow. The functions $i\colon X \to S$ and $o\colon Y \to S$ specify which species can pass through which ports. These functions need not be one-to-one or onto, nor do their images need be disjoint. We refer to the union of their images, $i(X) \cup o(Y) \subseteq S$, as the *boundary* of the open reaction network.

We can draw open reaction networks in the following way:

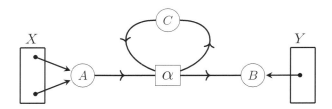

with X a two-element input set and Y a single-element output set. If the outputs of one reaction network $R\colon X \to Y$ match the inputs of another $R'\colon Y \to Z$,

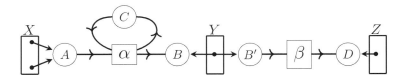

the two can be "composed" to form an open reaction network $R' \circ R\colon X \to Z$:

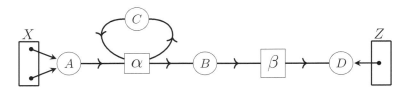

Composition here corresponds to gluing two open reaction networks together along those states that share a common element of Y. This notion of composition endows the collection of open reaction networks with the mathematical structure of a *category*.

Definition 3. *A category C consists of a collection of objects together with a collection of morphisms that is closed under a unital, associative composition operation.*

We can draw objects $X, Y, \ldots$ in a category as dots and morphisms $f\colon X \to Y$ and $g\colon Y \to Z$ as arrows:

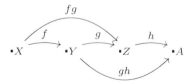

For example, there is a category Set where objects are sets and morphisms are functions. Function composition is associative, and every set has an identity map sending each element to itself. There is a category RxNet whose objects are finite sets $X, Y, \ldots$ and whose morphisms are isomorphism classes of open reaction networks $R\colon X \to Y, R'\colon Y \to Z, \ldots$. Composition in RxNet corresponds to gluing together networks along their common boundary. The detailed construction of this category can be found in Baez and Pollard (2017). Composition is the fundamental structure that makes a category a category. Functors are maps between categories that preserve this structure.

Definition 4. *A functor $F\colon \mathcal{C} \to \mathcal{D}$ is a mapping between categories $\mathcal{C}$ and $\mathcal{D}$ sending objects to objects and morphisms to morphisms while respecting composition and preserving identities,*

$$F(f \cdot g) = F(f) \circ F(g)$$

and

$$F(1_X) = 1_{F(X)},$$

for all morphisms $f, g \ldots$ in $\mathcal{C}$ and all objects $X, \ldots$ in $\mathcal{C}$.

Note that here the operation $- \cdot -$ represents composition in $\mathcal{C}$, while $- \circ -$ denotes composition in $\mathcal{D}$.

We can use functors to formalize the fact that reaction networks serve as a notation for a set of coupled nonlinear differential equations. Because functors preserve composition, this also gives a way to build up complicated chemical circuits from a few simple motifs, much in the way that complicated electrical circuits are composed.

In the aforementioned paper, we show that there is a functor ▨ : RxNet → Dynam sending an open reaction network to its corresponding rate equation, considered as an open dynamical system. Here Dynam is a category whose morphisms are certain types of open dynamical systems. The fact that this is accomplished via a functor means that the open rate equation of a composite open reaction network is the composite of the open rate equations of its constituents. Different reaction networks can give rise to the same rate equation.

We go on to "black-box" these open dynamical systems by characterizing them in terms of the relations imposed among the concentrations and flows along the *boundary* of the reaction network in steady state, effectively throwing out information about the internal species. This is also accomplished via a functor, namely,

$$■ : \text{Dynam} \to \text{Rel},$$

where Rel is the category whose objects are pairs of vector spaces (concentrations and flows) and whose morphisms are relations, or subspaces, of those.

The composition of two functors is again a functor. In fact, there is a category Cat whose objects are categories and whose morphisms are functors. The composite of the functors,

$$\text{RxNet} \xrightarrow{▨} \text{Dynam} \xrightarrow{■} \text{Rel},$$

provides a compositional characterization of open reaction networks in terms of the relations satisfied among the concentrations and flows of their boundary species in steady state.

One should note that both the categories RxNet and Dynam are examples of *decorated cospan categories* or *hypergraph categories*. To learn more about these ubiquitous structures, see Fong (2015) and Fong and Spivak (2018).

Mappings between CRNs

Mappings between reaction networks prove useful on a number of levels. As is formalized using category theory in Cardelli *et al.* (2017) and Cardelli (2014), the existence of certain mappings *between* reaction networks implies certain notions of behavioral equivalence between the source and target reaction networks. Such mappings provide a natural approach to simplifying complicated biochemical networks down to certain forms or motifs that elucidate their function. In another vein, the operation of composition in RxNet is accomplished via a mapping between reaction networks that identifies or glues together certain species to give a new composite reaction network.

At the heart of the construction of the category RxNet is a (lax monoidal) functor

$$F: \text{FinSet} \to \text{Set}$$

assigning to a finite set S the set of all reaction networks $F(S)$ with S as its set of species (Fong 2015). Because F is a functor, it acts not only on objects (finite sets) but also on morphisms (functions). Thus, given a function $f: S \to S'$, we need a function $F(f): F(S) \to F(S')$ transmuting a reaction network on S into one on S'.

Recall that a reaction network on S consists of the data (S, T, s, t, r), where S and T are finite sets, while $s, t: T \to \mathbb{N}^S$ and $r: T \to (0, \infty)$ are functions. Given such a reaction network and a function $f: S \to S'$, we can define functions $f_*(s), f_*(t): T \to \mathbb{N}^{S'}$ as

$$f_*(s(\tau))(\sigma') = \sum_{\{\sigma | f(\sigma) = \sigma'\}} s(\tau)(\sigma),$$

and similarly for $f_*(t(\tau))$. With these in hand, we can define a new reaction network on S' as $(S', T, f_*(s), f_*(t), r)$, leaving the transitions untouched and pushing forward the functions $s(\tau)$, $t(\tau) \in \mathbb{N}^S$ along $f\colon S \to S'$ to get functions in $\mathbb{N}^{S'}$. This gives a recipe for turning a reaction network on S together with a function $f\colon S \to S'$ into a reaction network on S'. We saw an example of composition in which certain states were identified. This is accomplished by applying $F\colon \mathtt{FinSet} \to \mathtt{Set}$ to a function specifying which states get identified.

We now turn our attention to the building blocks from which CRN models of asynchronous computation can be built up using this notion composition.

CRN Motifs for Asynchronous Computation

Cardelli, Kwiatkowska, and Whitby (2018) utilize the *dual-rail methodology* (Spars and Furber 2002) to encode Boolean variables using CRNs. Each Boolean is represented by two species: X_{hi}, X_{lo}. The authors build up their circuits from the following motif:

$$Y_{hi} \underset{X_{hi}}{\overset{X_{lo}}{\rightleftarrows}} Y_{lo}$$

Such CRNs have the property that either X_{hi} or X_{lo} is more abundant once the CRN has stabilized and all reactions have ceased, indicating either a 1 or a 0. Thus one can view this circuit as taking a Boolean input in the form of the relative concentration of Y_{hi} and Y_{lo} and outputting the corresponding Boolean associated to X. Abundance of X_{lo} encourages the conversion of Y_{hi} into Y_{lo}, and vice versa.

Copies of this motif under certain compositions build up the basic logic elements. For example, we can combine two such motifs in the following way:

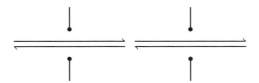

Note that we have omitted the species labels at the end of each line.

We can draw the corresponding reaction network:

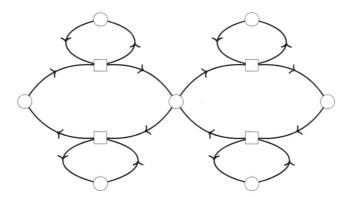

An implementation of AM with four bimolecular reactions (Mertzios *et al.* 2014) is given by

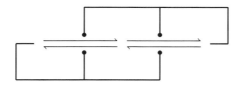

Such autocatalytic reactions can be realized as a composition in RxNet, gluing a portion of a reaction network to itself. From the previous diagram, we can pass to the reaction network for this AM motif by gluing together the appropriate species:

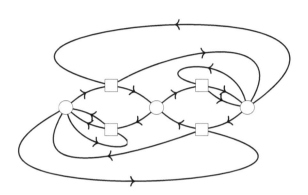

Cardelli, Kwiatkowska, and Whitby (2018) use a slightly modified version of this network in their construction of a Muller C-element:

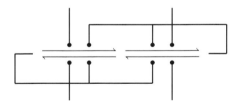

which corresponds to adding a loop to each transition in the previous reaction network. At low molecular counts, this implementation of a C-element is still not satisfactorily robust, as fluctuations can lead to undesired behavior. To fix this, the authors compose this with another layer of AM, arriving at their final design for the C-element:

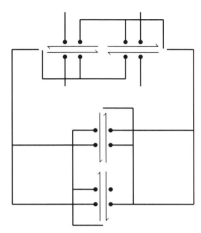

This circuit can again be viewed as a composite of two copies of the previous circuit, glued together in such a way that the outputs of one AM amplify the result of the other, providing a robust implementation of the C-element.

We refer the reader to Cardelli, Kwiatkowska, and Whitby (2018) for the specific CRN implementations of the various logic gates AND, OR, NAND, and so on, but each of these basic circuit elements can be viewed as composites built up via the basic operations of identifying a catalyst with a reactant/product and/or gluing the outputs of one reaction onto the inputs of another. Thus we see that there is a certain subcategory of RxNet whose morphisms correspond to CRNs that implement asynchronous logic elements built up from a few simple composition operations.

Conclusions

Such composition operations can change the structural properties of the constituent reaction networks, in particular altering the topology by creating new cycles and coupling existing ones. There are a number of theorems relating structural properties of reaction networks to properties of their associated rate equation, for instance, the existence, or lack thereof, of multiple equilibria (Banaji and Craciun 2010; Craciun, Tang, and Feinberg 2006; Shinar and Feinberg 2012). Recent work on the thermodynamics of open chemical reaction networks also highlights the coupling between the topology of the network and its ability to transduce free energy (Polettini and Esposito 2014).

In that work, the authors analyze the thermodynamics of open reaction networks in which boundary species serve as "chemostats," meaning that their concentrations are held fixed at certain prescribed values either via some external coupling or because they can be treated as effectively constant over the time intervals of interest. Chemostatting a species can induce *emergent* cycles, cycles not present in the closed network. Such cycles can have nontrivial *affinities*, particular linear combinations of the chemical potentials. They elucidate the connection between emergent cycles and broken conservation laws in such open systems, decomposing the steady-state entropy production in the spirit of Schnakenberg's work on the thermodynamics of networked master equation systems (Schnakenberg 1976).

Here we see the connection between three parallel lines of work:

1. chemical reaction networks as a platform for asynchronous computation;

2. a compositional framework for open chemical reaction networks; and

3. the thermodynamics of open chemical reaction networks.

CRNs provide the common thread. Properly woven, these three strands provide a framework for reasoning about the nonequilibrium thermodynamics of distributed computation as implemented by coupled biochemical networks found in nature. ✦

Acknowledgments

This material is based upon work supported by the National Science Foundation under Grant No. CMMI 1746077.

REFERENCES

Aris, R. 1965. "Prolegomena to the Rational Analysis of Systems of Chemical Reactions." *Archive for Rational Mechanics and Analysis* 19 (2): 81–99.

Baez, J. C., and B. S. Pollard. 2017. "A Compositional Framework for Reaction Networks." *Reviews in Mathematical Physics* 29 (09): 1750028.

Banaji, M., and G. Craciun. 2010. "Graph-theoretic Criteria for Injectivity and Unique Equilibria in General Chemical Reaction Systems." *Advances in Applied Mathematics* 44 (2): 168–184.

Cardelli, L. 2014. "Morphisms of Reaction Networks That Couple Structure to Function." *BMC Systems Biology* 8 (1): 84.

Cardelli, L., and A. Csikász-Nagy. 2012. "The Cell Cycle Switch Computes Approximate Majority." *Scientific Reports* 2:656.

Cardelli, L., M. Kwiatkowska, and M. Whitby. 2018. "Chemical Reaction Network Designs for Asynchronous Logic Circuits." *Natural Computing* 17 (1): 109–130.

Cardelli, L., M. Tribastone, M. Tschaikowski, and A. Vandin. 2017. "Comparing chemical reaction networks: A categorical and algorithmic perspective." *Theoretical Computer Science* (December).

Craciun, G., Y. Tang, and M. Feinberg. 2006. "Understanding Bistability in Complex Enzyme-Driven Reaction Networks." *Proceedings of the National Academy of Sciences of the United States of America* 103 (23): 8697–8702.

Doty, D., and S. Zhu. 2018. "Computational Complexity of Atomic Chemical Reaction Networks." In *SOFSEM 2018: Theory and Practice of Computer Science*, edited by A. M. Tjoa, L. Bellatreche, S. Biffl, J. van Leeuwen, and J. Wiedermann, 212–226. Cham, Switzerland: Springer International.

Fong, B. 2015. "Decorated Cospans." *Theory and Applications of Categories* 30 (33): 1096–1120.

Fong, B., and D. I. Spivak. 2018. "Hypergraph Categories." arXiv 1806.08304.

Kempes, C. P., D. Wolpert, Z. Cohen, and J. Pérez-Mercader. 2017. "The thermodynamic efficiency of computations made in cells across the range of life." *arXiv preprint arXiv:1706.05043*.

Mertzios, G. B., S. E. Nikoletseas, C. L. Raptopoulos, and P. G. Spirakis. 2014. "Determining Majority in Networks with Local Interactions and Very Small Local Memory." In *International Colloquium on Automata, Languages, and Programming*, 871–882. New York: Springer.

Peterson, J. L. 1977. "Petri Nets." *ACM Computing Surveys (CSUR)* 9 (3): 223–252.

Polettini, M., and M. Esposito. 2014. "Irreversible Thermodynamics of Open Chemical Networks. I. Emergent Cycles and Broken Conservation Laws." *Journal of Chemical Physics* 141 (2): 07B610_1.

Rao, R., and M. Esposito. 2016. "Nonequilibrium Thermodynamics of Chemical Reaction Networks: Wisdom from Stochastic Thermodynamics." *Physical Review X* 6 (4): 041064.

Schnakenberg, J. 1976. "Network Theory of Microscopic and Macroscopic Behavior of Master Equation Systems." *Reviews of Modern physics* 48 (4): 571.

Shinar, G., and M. Feinberg. 2012. "Concordant Chemical Reaction Networks." *Mathematical Biosciences* 240 (2): 92–113.

Soloveichik, D., M. Cook, E. Winfree, and J. Bruck. 2008. "Computation with Finite Stochastic Chemical Reaction Networks." *Natural Computing* 7 (4): 615–633.

Soloveichik, D., G. Seelig, and E. Winfree. 2010. "DNA as a Universal Substrate for Chemical Kinetics." *Proceedings of the National Academy of Sciences of the United States of America* 107 (12): 5393–5398.

Spars, J., and S. Furber. 2002. *Principles of Asynchronous Circuit Design*. New York: Springer.

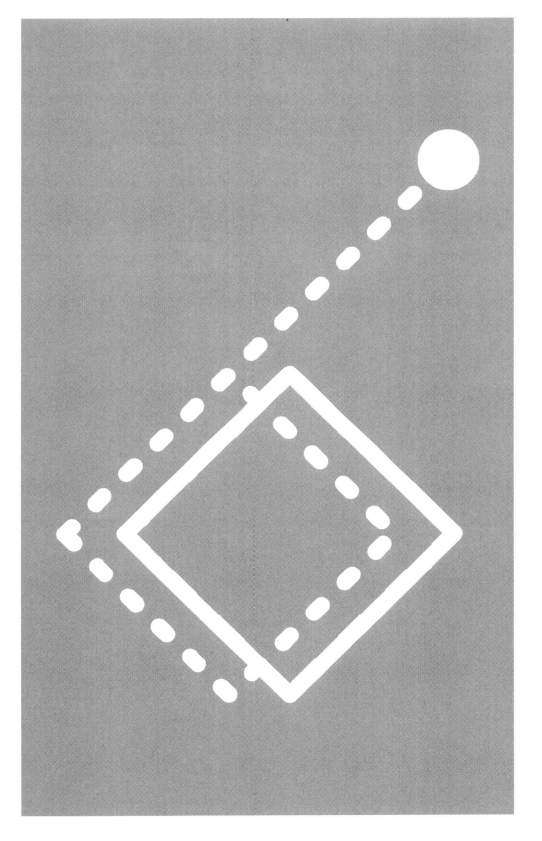

INFORMATION PROCESSING IN CHEMICAL SYSTEMS

Jakob L. Andersen, University of Southern Denmark
Christoph Flamm, University of Vienna
Daniel Merkle, University of Southern Denmark
Peter F. Stadler, University of Leipzig

Introduction

Chemical systems, that is, molecules and their reactions, are just time-dependent multiparticle quantum systems. As such, they are completely described by fundamental principles of physics, expressed in terms of quantum field theory (QFT) (Weinberg 2005). This level of description, however, is of little practical use to a chemist because the computational efforts to obtain answers to essentially all questions that a chemist might ask exceed, by far, the limits of present-day technology. It is possible, however, to arrive at more useful levels of description by means of a hierarchy of approximations and simplifications, making use of specific properties that distinguish chemical reactions for arbitrary quantum systems (see, e.g., fig. 3.1; Andersen *et al.* 2017a). These include the immutability of atomic nuclei and the idea that chemical reactions comprise only a redistribution of electrons. Furthermore, the Born–Oppenheimer approximation (Born and Oppenheimer 1927) postulates a complete separation of the wave function of nuclei and electrons due to the large difference in their masses, which leads to the concept of potential energy surfaces that determine the geometry of molecules and make

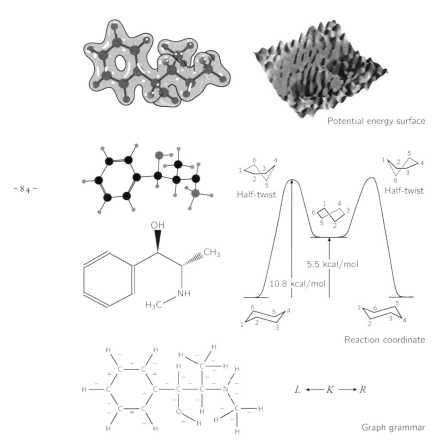

Figure 3.1. Coarse graining of molecules and chemical reactions from (top) quantum mechanics to (bottom) graph grammars. The transition from one level to a coarser is defined by principled approximations such as Born–Oppenheimer approximation or the replacement of spatial coordinates by neighborhood relations on graphs. Each step sacrifices accuracy and introduces empirical parameters but incurs a substantial reduction of computational effort. Figure modified from Andersen *et al.* (2017a).

it possible to view chemical reactions as classical paths on this surface (Mezey 1987; Heidrich, Kliesch, and Quapp 1991). The computation of potential energy surfaces by solving the Schrödinger equation is one of the key problems in quantum chemistry, for which a wide variety of approximation methods have been developed with different trade-offs between accuracy and computational effort. Semiempirical methods capitalize on the empirical fact that chemical bonds are usually formed by pairs of electrons to further simplify the electronic wave function. Molecular modeling and molecular dynamics (McCammon, Gelin, and Karplus 1977; Burkert and Allinger 1982) abandon quantum mechanics and model the potential energy surface as a sum of empirical contributions for pair bonds and electrostatic effects. This simplifies the computations sufficiently to treat macromolecules and supramolecular complexes that are intractable with quantum-chemical methods.

Even coarser approximations have been developed for particular classes of molecules. Aromatic ring systems, for instance, are well described in terms of purely graph-theoretical models known as Hückel theory (Hückel 1931; Hoffmann 1963). Nucleic acids can be coarse grained even further by aggregating their molecular building blocks (nucleotides) into single vertices. Watson–Crick base pairs then become edges in the graph representation known as *secondary structure* (Zuker and Stiegler 1981).

In this contribution, we adopt labeled graphs as the level of description of choice, that is, the level of chemical formulas and reaction schemes most familiar to chemists. We use this formalism to develop an algebraic description of chemical reaction networks that is consistent with traditional, flux-based methods, such as flux balance analysis or elementary mode analysis, and, at the same time, is sufficiently detailed to capture the constructive aspects of reorganizing atoms through arbitrary reaction systems.

There are at least two levels of description at which chemical reaction networks can be interpreted as formal systems of computation. At the coarser level, the focus is on transformation of multisets of molecules interpreted as abstract types. More precisely, a chemical reaction ρ transforms a multiset of molecules (the educts) into a different multiset (the products):

$$\sum_x s^-_{x\rho} x \to \sum_x s^+_{x\rho} x. \qquad (3.1)$$

The *stoichiometric coefficients* $s^-_{x\rho}$ and $s^+_{x\rho}$ count the number of educt and product molecules of type x in reaction ρ, respectively. Interpreted as a computation, the reactions form the instruction of a program that is executed concurrently. Input and output are specified by the number of particles of given types. There is a rich literature on this topic (see, e.g., Chen, Doty, and Soloveichik 2012; Cardelli, Kwiatkowska, and Laurenti 2018, and references therein).

Here we are concerned with a more fine-grained level of description, where molecules are treated not as featureless formal types but as objects with explicit internal structure— atoms and chemical bonds, which are captured mathematically as labeled graphs. The possible chemical reactions are subject to stringent rules imposed by fundamental laws of physics, such as the conservation of charge and mass, which in chemistry imply the conservation of atom types and impose tight constraints on changes in bond patterns. This algebraic structure, which effectively describes chemistry itself, can be viewed as a computational system in its own right.

Generative Chemistry: Reactions as Graph Transformation

We represent molecules as labeled graphs. In this picture, atom types (and charges) become vertex labels, and the different types of chemical bonds become edge labels. A chemical reaction, then, is a transformation of a (usually not connected) graph by rearranging and relabeling some of its edges:

$$\begin{array}{c}\text{[diagram]}\end{array} \quad (3.2)$$

A concrete reaction ρ is not only associated with its stoichiometry but also with a rule that captures the rearrangement of the bonds. The importance of rules lies in the fact that they readily generalize to large sets of structurally related molecules. In practice, this further restricts chemically meaningful reactions: most bond rearrangements are energetically and/or kinetically impossible, leaving a rather restricted set of reaction rules to fill the textbooks of chemistry. Chemical reactions are therefore modeled as graph transformations (Benkö, Flamm, and Stadler 2003).

Although several competing mathematical frameworks for graph transformation exist, there are good reasons to choose the double-pushout (DPO) formalism. In DPO, the rewriting of the educt pattern L into the resulting graph R is specified as a span $L \leftarrow K \rightarrow R$, where K denotes the subgraph of L that remains unchanged during the rewriting operation (whence it is also a subgraph of R). The rule $L \leftarrow K \rightarrow R$ can be applied to a graph G if and only if (1) the pattern (precondition) L can be embedded in G (by means of a suitable graph morphism m) and (2) there are objects D and H with so-called pushouts such that the diagram

$$\begin{array}{ccccc} L & \xleftarrow{l} & K & \xrightarrow{r} & R \\ {\scriptstyle m}\downarrow & & \downarrow & & \downarrow \\ G & \longleftarrow & D & \longrightarrow & H \end{array} \quad (3.3)$$

commutes (Ehrig *et al.* 2006). Diagrams such as the one in equation (3.3) assert that the compositions of morphisms along two paths with the same start and end points yield the same result. Here the application of $L \leftarrow K \rightarrow R$ to G implies that there must be a graph D obtained from G by removing the parts corresponding to L that are not present in K, into which the parts present in R but not in K (i.e., $R \setminus K$) can then be inserted to obtain the rewritten graph H. In the category **Graph**, the objects D and H are guaranteed to be unique if they exist. The existence of D is contingent on the so-called *gluing condition*, which deals with conflicts stemming from deletion of vertices. As chemical rules must preserve all atoms, this condition is trivially fulfilled. The right-hand graph H, finally, exists whenever there are no edge collisions upon insertion of $R \setminus K$. For instance, it is not allowed to insert an edge (u, v) into D if D already contains the edge (u, v). The graph H is thus the unique product obtained by rewriting the educt G with respect to the rule $L \leftarrow K \rightarrow R$ and the matching morphism m. We do not require the graphs to be connected. The connected components of G and H represent the educt and product molecules that are transformed in a chemical reaction. In figure 3.2, the reaction of equation (3.2) is shown as a DPO transformation using a rule for the Diels–Alder reaction.

For a graph transformation rule to be chemical, we require in addition that (1) all graph morphisms are injective (i.e., they describe subgraph relations), (2) the restriction of l and r to the vertices is bijective (ensuring the atoms are preserved), and (3) changes in edges (chemical bonds) and charges conserve the total number of electrons. The first two conditions suffice to ensure that a reaction is applicable whenever there is a matching morphism $m \colon L \rightarrow G$; they also ensure that all reactions are logically reversible (Andersen *et al.* 2013). Furthermore, they ensure that given $m \colon L \rightarrow G$, the *atom map* of every reaction is well defined; that is, given $m \colon L \rightarrow G$, there is a bijection

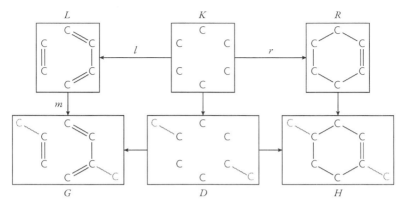

Figure 3.2. A double-pushout (DPO) diagram with the direct derivation corresponding to the reaction in equation (3.2), using a rule p for the Diels–Alder reaction.

$\varphi\colon V(G) \to V(H)$ between the vertices (atoms) of the educt graph G and the product graph H. The third condition captures much of the semantics of chemistry, which views chemical reactions as rearrangements of the electron pairs that form the chemical bonds.

From a chemist's point of view, the salient feature of a chemical reaction is the breaking and forming of bonds around a relatively localized set of atoms, known as the reaction center. From a physicist's point of view, this process can be broken down into elementary operations of how electrons are pushed around to achieve this redistribution of atoms into a different bonding contexts. Which of the elementary electron pushing operations is preferable over other ones depends on the local context, as indicated in equation (3.3). Considerations of energy, and thus of thermodynamics, enter the picture at this point. Formally, this is achieved by associating the total enthalpy of atomization (TAE), that is, the total energy contained in the chemical bonds, with each molecule. In practice, this is not at all trivial to compute. For some applications, it is possible to design reasonable approximations in terms of the labeled graphs only (Benkö, Flamm, and Stadler 2003). The energetics of a chemical reaction depends not only on the differences in the TAEs but also on environmental parameters, such as the interaction of the individual educts and products with the solvents and their concentrations. The constraints defined by the low-level energetic patterns inherent in chemical bonds propagate up to higher levels of description. In particular, they restrict which reaction patterns are allowed in specific molecular contexts and under given environmental conditions. We will return to these issues later.

Chemical reactions can be composed. Algebraically, this amounts to considering linear combinations of the form $\sum_\rho a_\rho \rho$, which (for nonnegative coefficients a_ρ) translates to the "net reactions"

$$\sum_x \left(\sum_\rho a_\rho s^-_{x\rho}\right) x \to \sum_x \left(\sum_\rho a_\rho s^+_{x\rho}\right) x. \qquad (3.4)$$

Compositions are also available at the level of reaction rules. Given two rules $L_1 \leftarrow K_1 \to R_1$ and $L_2 \leftarrow K_2 \to R_2$ with R_1 and L_2 overlapping in a common subgraph D, one can construct a composite rule $L \leftarrow K \to R$ by means of the commutative diagram

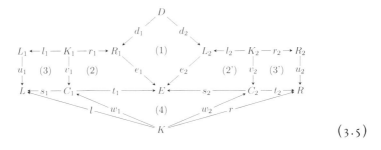

$$(3.5)$$

Given D and its embeddings of d_1 into R_1 and d_2 into R_2, the composite rule $L \leftarrow K \to R$ is uniquely defined when all numbered squares in equation (3.5) commute. In particular, this allows multiple copies of the same rule to be executed concurrently (in which case, one sets $D = \emptyset$). Composing $L \leftarrow K \rightarrow R$ with its inverse $R \leftarrow K \rightarrow L$ along $D = R$ yields an "identity reaction." Hence rule composition can explain any integer linear combinations of chemical reactions in equation (3.4).

Chemical reactions are transformations among multisets of molecules. However, the traditional formulation of graph transformation presented in the preceding paragraph defines one-to-one transformations. When considering the enumeration of all reactions originating in a given set of molecules, one must theoretically consider all selections of multisets from these molecules for the construction of the educt graph G. Not all multisets can in practice be successfully transformed by every rule. Higher-level algorithms based on rule composition have therefore been devised to enumerate reactions efficiently (Andersen *et al.* 2013). This operation can then be used to define a domain-specific programming language for computing reaction networks (Andersen *et al.* 2014). Additional constraints that are not captured by the transformation rules can be used in this setting to guide the expansion of networks. Note that the computation of reaction networks in this sense is not necessarily an actual simulation of any particular chemical process but is an enumeration of potential reactions.

The graph transformation system described here, along with rule composition and the domain-specific programming language, is implemented in the software package MØD (Andersen 2018; Andersen *et al.* 2016). An extension of the graph transformation system provides modeling support for stereo-information in both molecules and transformation rules (Andersen *et al.* 2017b).

Chemical Hyperflows

Directed hypergraphs, with molecules as vertices and reactions as hyperedges, are the natural topological representations of chemical reaction networks (Klamt, Haus, and Theis 2009). The directed hypergraph of a chemical network is completely specified by the stoichiometric coefficients $s_{x\rho}^+$ and $s_{x\rho}^-$. In the absence of catalysts, at least one of $s_{x\rho}^+$ or $s_{x\rho}^-$ vanishes. In this case, it is sufficient to consider the *stoichiometric matrix* $\mathbf{S}$ with entries $\mathbf{S}_{x\rho} = s_{x\rho}^+ - s_{x\rho}^-$. Many of the traditional approaches to the analysis of chemical reaction networks, such as flux balance or elementary mode analysis, focus on stationary fluxes J, that is, on solutions of $\mathbf{S} \cdot J = 0$ (see, e.g., Schuster, Fell, and Dandekar 2000; Larhlimi and Bockmayr 2009; Price *et al.* 2003; Orth, Thiele, and Palsson 2010). A wide array of quantities and substructures of interest can be specified in terms of integer hyperflows on directed hypergraphs (Andersen *et al.* 2017c). Integer-valued solutions are necessary to support subsequent mechanistic interpretations of chemical processes.

Pathway approaches, whether insisting on integer hyperflows or allowing real-valued fluxes, in their core use only the topology of the reaction network and the balance conditions implied by the stoichiometry. They are not necessarily realistic with respect to thermodynamic constraints. Assuming that an estimate of the Gibbs free energy G_x is available for each species x, the methods can be extended with constraints for the second law of thermodynamics (Henry, Broadbelt, and Hatzimanikatis 2007). The reaction ρ may appear in a pathway only if it is thermodynamically favorable, that is, if

$$\sum_x \mathbf{S}_{x\rho} G_x + RT \cdot \sum_x \mathbf{S}_{x\rho} \ln[x] < 0. \quad (3.6)$$

As is usual in the chemistry literature, $[x]$ denotes the concentration of molecular-type x, and G_x is the free energy of x, which combines its TAE with contributions specific to the reaction conditions. Thus a solution encodes not only a pathway but also a set of steady-state concentrations that make the pathway feasible. The estimation of G_x is not at all easy. For biochemical systems, usually machine learning and graph-theoretical methods (Noor *et al.* 2013), which rely on measured data for a training set of molecules, are used.

The temporal behavior of concentrations in a reaction network is linked to elementary chemical reactions by the rules of mass-action kinetics. Each chemical reaction ρ is endowed with a rate constant k_ρ that depends on the molecules involved and on environmental parameters such as temperature, pressure, or solvent. The change of the concentration $[x]$ through a single reaction ρ is proportional to the product of the educt concentrations. Because different reactions run concurrently, the temporal behavior of a network of elementary reactions follows the system of ordinary differential equations

$$\frac{\mathrm{d}}{\mathrm{d}t}[x] = \sum_\rho k_\rho s^+_{x\rho} \prod_y [y]^{s^-_{y\rho}}. \tag{3.7}$$

Algorithms for the simulation of these equations are generally based on either ordinary differential equations or discrete event simulation due to Gillespie (1976). In their usual formulation, both types of algorithms require the underlying reaction network for the system to be known beforehand. However, with rule-based calculi, such as the DPO formalism described in the previous section, it becomes possible to reformulate the Gillespie algorithm as a "network-free algorithm," where reactions are computed as needed by the simulation instead of by prior enumeration (Suderman *et al.* 2018). In the last decade, the theory for Markovian stochastic processes at the ensemble level has been worked out in detail (Jarzynski

2011; Seifert 2012; Van den Broeck and Esposito 2015; Van den Broeck, Sasa, and Seifert 2016). This theory of *stochastic thermodynamics* describes the behavior of a closed or open small-scale system, including the effects of fluctuations. Furthermore, the description of the systems is not restricted to near-equilibrium dynamics. All these approaches do not make explicit use of the structure of the molecules, however.

Atom Redistribution as Computation

The energetic constraints distinguishing chemically meaningful rules from arbitrary graph transformation have a crucial impact on the topology of chemical reaction networks. It is very difficult, and maybe impossible in practice, to infer the elementary energetic patterns at the level of molecules from observed reaction networks, because at this level of description the molecular graphs are relegated to mere vertex labels in a directed hypergraph. The energetics of a reaction, however, depends on the individual molecules that take part in a reaction and on the precise manner in which bonds are rearranged (as well as environmental parameters). The same abstract stoichiometric reaction scheme can therefore be implemented with vastly different thermodynamic properties. It is indeed an interesting research question how chemically feasible implementations can be found for a given stoichiometric network, and whether this is possible at all for any network. The question of how a reactive chemical system computes thus requires an intermediate level of modeling in which the molecule graphs, and thus the rules of chemical bonds and their energetics, are still visible.

In the context of flux balance analysis, an approach to support atom tracing in metabolic networks has been introduced that includes information on how atoms are transported through a reaction network. These *atom transition networks* (Weitzel,

Wiechert, and Nöh 2007) are defined as directed graphs: all atoms of all compounds define the set of vertices, and the edge set is defined based on all possible individual atom transitions within the metabolic network. Although this representation as a directed graph makes the network amenable to efficient algorithms, it also has significant disadvantages. In particular, it is not possible to capture the intrinsic *concurrency* of atom transitions in individual chemical reactions. More specifically, the transfer of an individual atom does not reflect the mechanistically correct process.

To address this issue, we construct the *atom transition hypergraph* by augmenting the directed hypergraph representing the chemical reaction networks with possible atom transitions for each of the reactions as well as with symmetry information. The additional information is naturally described by morphisms and hence is compatible with the graph transformation approach: (1) the symmetries within each molecule are described as automorphisms, and (2) the atom transitions in a reaction are captured by the bijections connecting the educt to the product graphs described earlier. Each of the morphisms in this model of atom tracing maintains the concurrency of atom transitions properly. Because all these morphisms are bijective on atoms, it is natural to interpret them as permutations and to employ group-theoretical approaches to analyze atom traces. To this end, a linearization is defined by uniquely labeling all atoms of all compounds with labels 1 to n. The two types of morphisms can then be viewed as actions of corresponding permutation groups. Figure 3.3 shows the atom transition hypergraph for the citric acid (TCA) cycle, a key reaction network in most living organisms.

A permutation group by definition requires an inverse element for each group action (i.e., for each morphism). This is naturally true for all the automorphisms within a molecule. The inverse of a morphism describing a chemical reaction is not

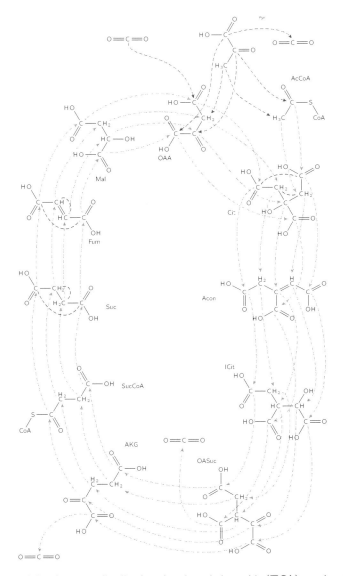

Figure 3.3. Atom redistribution in the citric acid (TCA) cycle. The reaction network disregards stereochemical constraints, which are implied by enzymes. Automorphisms are depicted as dotted edges within the only symmetric molecules: succinate (Suc), fumarate (Fum), and citrate (Cit). The morphisms describing atom transitions are shown as dashed edges. Each round of the cycle adds two carbon atoms in the form of acetyl-coenzyme A, and two carbon atoms are removed in the form of CO_2. The explicit atom maps reveal that the first CO_2 is formed from carbon that was added in a previous round, whereas the second CO_2 may be from a newly added carbon atom.

necessarily part of the reaction network, however. The group-theoretical approach thus has to assume that all chemical reactions are reversible. Although this is logically the case, it is very useful to model explicitly the fact that many chemical reactions are thermodynamically irreversible; that is, it is impossible to obtain measurable fluxes in the reverse direction for any plausible choice of concentrations and reaction conditions. It is possible, however, to use transformation semigroups, more specifically the monoids that are the semigroup analogues of permutation groups (Howie 1995), to answer questions about atom tracing. In the atom transition hypergraph of figure 3.3, consider a specific reaction. The dotted edges of a specific reaction represent a transformation (in contrast to a permutation) in the transformation semigroup.

An example of considerable practical impact is to follow the path of "labeled" atoms (i.e., rare isotopes placed at specific positions of an input compound) through a complex reaction network. Mathematically, this corresponds to computing the orbit of a vertex (atom) in the monoid that represents the atom transition hypergraph. As an example, consider the atom transition network of the TCA cycle in figure 3.3. Its net effect is to convert a two-carbon atom (C_2) unit into two C_1 units that exit the cycle as CO_2; the biological function of this process produces ATP. The conversion is achieved by fusing the C_2 unit to a C_4 unit that, in the global picture, acts as a catalyst. The details of the cycle are more complex, however, because the two C atoms that leave the TCA cycle are not the ones that entered in the same round. Likewise, the C_4 catalyst does not remain constant throughout the turn of the cycle. Instead, it is regenerated in each round in a way that changes the identity of its atoms. How the atoms mix depends on the reaction rules, the symmetries of the molecules, the relative reaction velocities, and the reversibility of the reactions in the network. For instance, if the sequence of reactions transforming Fum to OAA is reversible and fast compared to the fusion reaction AcCoA + OAA → Cit, then a ^{13}C label of the atom in the CH_3 group of Pyr would end up distributed uniformly over the carbon atoms of all C4 compounds in the TCA cycle.

Because the concurrency property is respected, the formalism also supports the tracing of k-tuples of atoms. Semigroup algorithms are available that make it possible to perform these computations efficiently. While very efficient algorithms have become available to compute the automorphism group of an unlabeled undirected graph, a generalization that makes it possible to include additional information, such as stereochemical features of a molecular representation, has only been introduced recently (Andersen and Merkle 2018). Some issues also complicate the picture, however. There may be more than one transformation rule transforming a multiset of educt molecules into the same multiset of products. In addition, it is possible that the same transformation rule can be applied with different matching morphisms, which may result in different atom traces.

Concluding Remarks

In this chapter, we have briefly reviewed an essentially algebraic view on chemical reactions and chemical reaction networks. It captures fundamental features of chemistry by focusing on the rearrangement of chemical bonds and the identity of atoms. It also emphasizes the nature of chemistry as a constructive system akin to formal calculi. Abstractions of chemistry have indeed been introduced as formal models of computation, for example, in Berry's Chemical Abstract Machine (Berry and Boudol 1992). At the same time, chemistry is highly constrained by conservation of mass and energy, providing a very direct link to thermodynamic considerations. Each chemical transformation is associated with an enthalpy that is entirely determined by the underlying physics of the chemical bonds in products and educts. While changes in environmental conditions, such as solvents, temperature, and pressure, and the choice of concentrations affect the pertinent free energies, these are also constrained in practice by the solubility of the material, the thermal stability of the molecules, and minimal concentration necessary, for example, for practical readout in a "chemical computer." Together, these constraints imply practical bounds on the dissipation of free energy in chemical systems that can be estimated at least in principle given the chemical reaction network.

Chemical reaction systems can implement computational models that may be of practical interest in many different ways. For example, the mass action kinetics in equation (3.7) shows that chemical reaction networks cover a large class of systems of ordinary differential equations. The exploration of chemical spaces with the aim of realizing molecules with particular structural properties corresponds to a class of satisfiability problems. The reorganization of atoms along reaction cycles can be seen as a nonstandard model of computation in its own right that combines concurrency with the algebraic operations of monoids. Practical thermodynamic limitations likely arise in different ways in each of these computational interpretations of chemical networks because the computational paradigms may require different readouts (e.g., concentration variables vs. the presence of particular molecules vs. particular isotope patterns) and different environments. The constructive, atom- and bond-centered model of chemical reaction networks outlined here at least provides a coherent framework to address such questions in a systematic manner that scales to large system sizes. ✦

Acknowledgments

The work was supported in part by the National Science Foundation (INSPIRE 1648973). Furthermore, it is partially supported by the Danish Council for Independent Research, Natural Sciences, grant DFF-7014-00041. J. L. A. additionally acknowledges the University of Vienna, to which he was affiliated at the time of writing.

REFERENCES

Andersen, J. L. 2018. *MedØlDatschgerl (MØD)*. http://mod.imada.sdu.dk.

Andersen, J. L., C. Flamm, D. Merkle, and P. F. Stadler. 2013. "Inferring Chemical Reaction Patterns Using Graph Grammar Rule Composition." *Journal of Systems Chemistry* 4:4.

———. 2016. "A Software Package for Chemically Inspired Graph Transformation." In *Graph Transformation, ICGT 2016*, edited by Rachid Echahed and Mark Minas, 9761:73–88. Lecture Notes in Computational Science. Berlin: Springer.

———. 2017a. "An Intermediate Level of Abstraction for Computational Systems Chemistry." *Philosophical Transactions of the Royal Society of London, Series A* 375:20160354.

———. 2017b. "Chemical Graph Transformation with Stereo-information." In *10th International Conference on Graph Transformation (ICGT 2017)*, edited by Juan de Lara and Detlef Plump, 10373:54–69. Lecture Notes in Computational Science. Heidelberg, Germany: Springer.

———. 2017c. "Chemical Transformation Motifs—Modelling Pathways as Integer Hyperflows." *IEEE/ACM Transactions on Computational Biology*.

Andersen, J. L., C. Flamm, D. Merkle, and P. Florian Stadler. 2014. "Generic Strategies for Chemical Space Exploration." *International Journal of Computational Biology and Drug Design* 7:225–258.

Chapter 3: Information Processing in Chemical Systems

Andersen, J. L., and D. Merkle. 2018. "A Generic Framework for Engineering Graph Canonization Algorithms." In *Proceedings of the Twentieth Workshop on Algorithm Engineering and Experiments, ALENEX 2018*, 139–153.

Benkö, G., C. Flamm, and P. F. Stadler. 2003. "A Graph-Based Toy Model of Chemistry." *Journal of Chemical Information and Modeling* 43 (4): 1085–1093.

Berry, G., and G. Boudol. 1992. "The Chemical Abstract Machine." *Theoretical Computer Science* 96:217–248.

Born, M., and J. R. Oppenheimer. 1927. "Zur Quantentheorie der Molekülen." *Annales Physik* 389:457–484.

Burkert, U., and N. L. Allinger. 1982. *Molecular Mechanics*. Vol. 177. Washington, DC: American Chemical Society.

Cardelli, L., M. Kwiatkowska, and L. Laurenti. 2018. "Programming Discrete Distributions with Chemical Reaction Networks." *Natural Computing* 17:131–145.

Chen, H., D. Doty, and D. Soloveichik. 2012. "Deterministic Function Computation with Chemical Reaction Networks." *Natural Computing* 7433:25–42.

Ehrig, H., K. Ehrig, U. Prange, and G. Taenthzer. 2006. *Fundamentals of Algebraic Graph Transformation*. Berlin: Springer.

Gillespie, D. T. 1976. "A General Method for Numerically Simulating the Stochastic Time Evolution of Coupled Chemical Reactions." *Journal of Computational Physics* 22:403–434.

Heidrich, D., W. Kliesch, and W. Quapp. 1991. *Properties of Chemically Interesting Potential Energy Surfaces*. Vol. 56. Lecture Notes in Chemistry. Berlin: Springer.

Henry, C. S., L. J. Broadbelt, and V. Hatzimanikatis. 2007. "Thermodynamics-Based Metabolic Flux Analysis." *Biophysical Journal* 92:1792–1805.

Hoffmann, R. 1963. "An Extended Hückel Theory. I. Hydrocarbons." *Journal of Chemical Physics* 39:1397–1412.

Howie, J. M. 1995. *Fundamentals of Semigroup Theory*. Oxford: Clarendon.

Hückel, E. 1931. "Quantentheoretische Beiträge zum Benzolproblem. I. Die Elektronenkonfiguration des Benzols und verwandter Verbindungen." *Zeitschrift für Physik* 70:204–286.

Jarzynski, C. 2011. "Equalities and Inequalities: Irreversibility and the Second Law of Thermodynamics at the Nanoscale." *Annual Review of Condensed Matter Physics* 2:329–351.

Klamt, S., U. Haus, and F. Theis. 2009. "Hypergraphs and Cellular Networks." *PLoS Computational Biology* 5 (5): e1000385.

Larhlimi, A., and A. Bockmayr. 2009. "A New Constraint-Based Description of the Steady-State Flux Cone of Metabolic Networks." *Discrete Applied Mathematics* 157:2257–2266.

McCammon, J. A., B. R. Gelin, and M. Karplus. 1977. "Dynamics of Folded Proteins." *Nature* 267:585–590.

Mezey, P. G. 1987. *Potential Energy Hypersurfaces.* Amsterdam: Elsevier.

Noor, E., H. S. Haraldsdóttir, R. Milo, and R. M. T. Fleming. 2013. "Consistent Estimation of Gibbs Energy Using Component Contributions." *PLoS Computational Biology* 9:1–11.

Orth, J. D., I. Thiele, and B. Ø. Palsson. 2010. "What Is Flux Balance Analysis?" *Nature Biotechnology* 28:245–248.

Price, N. D., J. L. Reed, J. A. Papin, S. J. Wiback, and B. Ø. Palsson. 2003. "Network-Based Analysis of Metabolic Regulation in the Human Red Blood Cell." *Journal of Theoretical Biology* 225:185–194.

Schuster, S., D. A. Fell, and T. Dandekar. 2000. "A General Definition of Metabolic Pathways Useful for Systematic Organization and Analysis of Complex Metabolic Networks." *Nature Biotechnology* 18:326–332.

Seifert, U. 2012. "Stochastic Thermodynamics, Fluctuation Theorems and Molecular Machines." *Reports on Progress in Physics* 75:126001.

Suderman, R., E. D. Mitra, Y. T. Lin, K. E. Erickson, S. Feng, and W. S. Hlavacek. 2018. "Generalizing Gillespie's Direct Method to Enable Network-Free Simulations." *Bulletin of Mathematical Biology.*

Van den Broeck, C., and M. Esposito. 2015. "Ensemble and Trajectory Thermodynamics: A Brief Introduction." *Physica A* 418:6–16.

Van den Broeck, C., S. Sasa, and U. Seifert. 2016. "Focus on Stochastic Thermodynamics." *New Journal in Physics* 18:020401.

Weinberg, S. 2005. *The Quantum Theory of Fields.* Cambridge: Cambridge University Press.

Weitzel, M., W. Wiechert, and K. Nöh. 2007. "The Topology of Metabolic Isotope Labeling Networks." *BMC Bioinformatics* 8:315.

Zuker, M., and P. Stiegler. 1981. "Optimal Computer Folding of Larger RNA Sequences Using Thermodynamics and Auxiliary Information." *Nucleic Acids Research* 9:133–148.

NATIVE CHEMICAL AUTOMATA AND THE THERMODYNAMIC INTERPRETATION OF THEIR EXPERIMENTAL ACCEPT/REJECT RESPONSES

Marta Dueñas-Díez, Repsol Technology Center
Juan Pérez-Mercader, Harvard University

Introduction

Computation—defined as the pathway for information to be input, to be processed mechanically, and to be output in a useful way (Evans 2011)—takes place not only in the myriad of electronic devices we use daily but also in living systems. Life carries out computations mostly by using chemical support: inputs are chemical substances, the mechanical processing occurs via chemical reaction mechanisms, and the result is chemical as well. Machines carrying out computations are typically referred to as automata (Hopcroft, Motwani, and Ullman 2006); hence, to a large extent, living systems can be viewed as chemical automata (Bray 2009). Classic automata are arranged hierarchically from simplest to most powerful (Hopcroft, Motwani, and Ullman 2006): finite automata, then pushdown automata, and, at the top of the hierarchy, Turing machines (Turing 1936).

Although the subject of this contribution already has an interesting history, we give here a brief, personal, and short summary of some interesting developments in the field of chemical computation. Interest in chemical computing dates back to the early 1970s, when Conrad (1972) studied information processing in molecular systems and how it differs from electronic digital computing. A theoretical chemical diode was first suggested by Okamoto, Sakai, and Hayashi (1987), an idea that Hjelmfelt, Weinberger, and Ross (1991) further developed to suggest that neural networks and chemical automata could be constructed connecting such chemical diodes. In the 1990s, Magnasco (1997) studied the Turing completeness of chemical kinetics. The first experimental realization of chemical AND and OR logic gates using reaction diffusion was achieved in 1995 by Tóth and Showalter (1995), followed by XOR gates (Adamatzky and Lacy Costello 2002) and counters (Górecki, Yoshikawa, and Igareshi 2003), and still is an active area of research due to the difficulties associated to linking many gates to carry out more advanced computations. Computations carried out in a more native way, without requiring diffusion, have been suggested using complex biomolecules such as DNA (Adleman 1994; Benenson 2009) or chromatin (Prohaska, Stadler, and Krakauer 2010; Bryant 2012). In summary, most artificial approaches to chemical computing, inspired by living systems, focus on reaction–diffusion systems mostly representing logic gates or use complex biomolecules to solve very specific problems.

Our approach (Pérez-Mercader, Dueñas-Díez, and Case 2017) differs from the aforementioned work in that we use the power of chemistry, and the molecular recognition associated with the occurrence of chemical reactions, in a one-pot reactor, that is, a single well-mixed container where multiple rounds of reactions can take place, without using external geometrical aids or complex biomolecules and relying fully on the power of molecular recognition and the robustness associated with Avogadro's number to carry out computations. We have recently demonstrated experimentally that this approach, without using biochemistry, can recognize a language that only automata at the Turing machine level of the hierarchy can recognize (Dueñas-Díez and Pérez-Mercader 2019).

In this contribution, we apply the well-known natural connection between chemistry and thermodynamics (Donder 1927; Kondepudi and Prigogine 2014) to study and interpret the chemical reject/accept signatures of chemical automata in thermodynamic terms. We do this for three examples, one at each main level of the three-level hierarchy in computing automata theory (Hopcroft, Motwani, and Ullman 2006). Of course, this connection is only a first step toward quantifying the thermodynamic cost of chemical computation and, more importantly, toward its optimization (Bennett 1982; Landauer 1961). Indeed, the same thermodynamic metrics we apply as the reject/accept signatures after a word is processed can be applied during the course of the computation, not just at its end. If we apply our metrics continuously during the computation of a complete sequence, we can assess the thermodynamic cost of computation as each symbol is processed and therefore determine how the thermodynamic cost evolves as the input sequence length grows or even compare the cost of different types of rejects. We suggest using the three languages chosen below, or other similar well-known languages and their associated automata, as minimal complete examples to run quantitative studies of the thermodynamic cost of computation.

One-Pot Native Chemical Computation

Our work focuses on demonstrating experimentally how computations of different complexity (in the sense of classical automata theory) can be carried out by chemical means exclusively, in a homogeneous reactor, and without requiring complex biomolecules. In our approach, the input to be computed is represented by a sequence of symbols from a chemical alphabet in which each letter corresponds to a certain constant amount, or aliquot, of a carefully chosen reacting chemical species. The input is sequentially added letter by letter, that is, aliquot by aliquot, to a one-pot reactor at constant time intervals (Pérez-Mercader, Dueñas-Díez, and Case 2017). The processing of each letter consists in selectively activating specific pathways in the chemical mechanism and, correspondingly, altering the resulting chemical state/landscape in a systematic way. Finally, the output of the computation is in the form of a distinct chemical response; that is, for a given automata/language combination, the chemical behavior associated with a rejected sequence is different from the chemical behavior associated to that of an accepted sequence (Dueñas-Díez and Pérez-Mercader 2019). Naturally, we expect that such distinct chemical responses correspond to some distinct thermodynamic signatures as well.

To show that chemistry can carry out computations of different complexity levels, we carry out the following steps. First, we choose a specific language of interest that a tailored chemical automaton should recognize, that is, a specific problem to be solved. From classic automata theory, we then identify the class of automata needed to recognize it and the computational requirements as defined by the corresponding automata tuple. We translate this into specific requirements for the chemical reactions and their reactants, products, and intermediates, leading us to select the alphabet description appropriately. Then, the specific quantitative recipes for initial conditions and alphabet aliquots are selected so that the chemical reaction monitoring system allows detection of the machine's response with sufficient/reasonable precision.

Chemical Finite Automaton Recognizing the Regular Language L_1 of All Words Containing at Least One "a" and One "b"

Regular languages are the simplest languages in automata hierarchy (Hopcroft, Motwani, and Ullman 2006). They do not require counting: their words all contain or all exclude certain patterns of the alphabet symbols or affixes (Hopcroft, Motwani, and Ullman 2006; Cohen 1991). Regular languages are recognized by a finite automaton (FA), an abstract device that at each given time is in one of a finite number of states, and the device transitions states depending on the input using a finite set of rules. At the end of a computation, that is, after a word is processed by the FA, the device terminates in either an accept state or a reject state, depending on whether the input word belongs to the regular language recognized by the FA.

Following the intuitive notion that simple chemistries can recognize regular languages, we reverse-formulate the question as follows: what can a single bimolecular reaction of the type $A + B \rightarrow C + D$ compute? If we represent the letters, **a** and **b**, in a language L_1 by the aliquots of A and B corresponding to this reaction, we see that such a bimolecular reaction recognizes the regular language L_1 of all words that contain at least one **a** and one **b**. For an illustrative and visual implementation, we can choose a precipitation reaction in an aqueous medium such as

$$KIO_3 + AgNO_3 \rightarrow AgIO_3(s) + KNO_3. \qquad (4.1)$$

If, during computation, a white precipitate of silver iodate is observed, the input string has been recognized and accepted; if the solution is clear from precipitate, the string has been rejected because there was no reaction and the input string was therefore not recognized. The only requirement in this example to choose the recipes of alphabet symbols **a** (potassium iodate) and **b** (silver nitrate) quantitatively is that the product of their concentrations, once one aliquot of each is added to the reactor, exceeds the solubility product constant of silver iodate at the operating reactor temperature, thus guaranteeing the appearance of a precipitate. Figure 4.1 shows the chemical representation of symbols *a* and *b*, the bimolecular precipitation reaction, the corresponding theoretical FA transition graph to recognize L_1, and the results of testing two sequences experimentally. Sequence *aab* gives a white precipitate, corresponding to the final state q_f in the abstract FA. Sequence *aaa* shows no precipitate, corresponding to state q_1, and thus the input is rejected by the chemical automaton.

Because this reaction is exothermic, the accept and reject states can also be detected by monitoring temperature (if the temperature remains constant, the sequence is rejected, but if the temperature increases, then the sequence is accepted). Hence we see that the heat of reaction is the thermodynamic equivalent to the chemical precipitate response.

Chemical 1-Stack PDA Recognizing the Context-Free Language L_2 of Balanced Parentheses

Next, we go one important step up in the hierarchy and consider the case of context-free languages. We show how one-pot native chemistry recognizes a language in which both counting and sequence order are relevant. Context-free languages (CFL) are those whose words involve matching of substrings, affixes, or symbols and therefore require counting to one arbitrarily high integer (Hopcroft, Motwani, and Ullman 2006). We choose the Dyck language (Weisstein 2009), the language of balanced parentheses, as L_2: a sequence of parentheses is balanced if, during its processing, the number of closed parentheses never exceeds the number of open parentheses, and at the end of the computation, the number of open parentheses matches exactly the number

Chapter 4: Native Chemical Automata

L_1 = {Language of all words that have at least one a and at least one b}.
L_1 is a regular language, and therefore is recognized by the FA

Chemically implemented by an acid/base reaction.

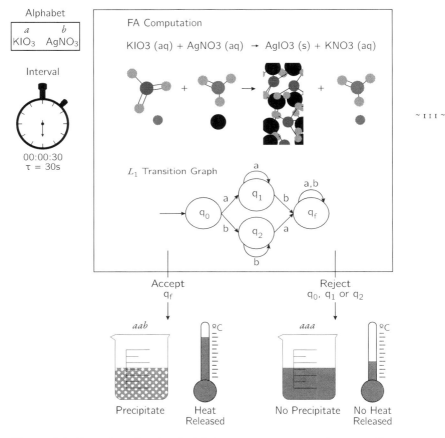

Figure 4.1. Operation of a chemical finite-state automaton: language L_1, described by the regular expression $(a+b)^* \, a(a+b)^*b(a+b)^* + bb^*aa^*$, is recognized by a FA and is realized chemically by a precipitation reaction. In this example, once the full sequence has been processed, if the solution contains a visible precipitate of $AgIO_3$, or, equivalently, heat has been released during computation, the input string has been accepted as a word in L_1. In contrast, if the solution does not contain any visible precipitate, or, equivalently, no heat was released during computation, then the input string was rejected. Input sequences *aab* and *aaa* were tested experimentally and in the former a precipitate was observed, while no precipitate was observed in the latter.

of closed parentheses (Hopcroft, Motwani, and Ullman 2006; Cohen 1991).

In theoretical computer science, a CFL is recognized by a one-stack pushdown automaton (PDA). This automaton differs from a FA by being endowed with an additional element, the stack, in which to store a string of arbitrary length. Furthermore, it can be read and modified only at its top, in a last-in-first-out fashion (Hopcroft, Motwani, and Ullman 2006; Cohen 1991), just as in a cafeteria "stack of trays." The transitions in a PDA depend not only on the current input symbol and state but also on the current symbol at the top of the stack. A transition may result not only in changing the state of the automaton but also in pushing (adding) an element to the top of the stack or popping (removing) an element from the stack. The "accept" criterion is often associated with the set of transitions leading to an empty stack at the end of the computation. For our chosen language L_2, the stack keeps track of the excess of open parentheses with respect to the closed parentheses and, indeed, it has to be empty at the end of the computation.

The requirement of a "stack" translates in native chemical computing into the condition of having a pathway in the reaction mechanism in which there is an intermediate species that is produced (pushed) in one subreaction and consumed (popped) in another subreaction. This in turn leads us to select as an example for the actual implementation of a chemical L_2-PDA the language of Dyck words by means of a pH reaction and with the following alphabet assignment: "(" is an aliquot of the base (NaOH), ")" is an aliquot of the weak diprotic acid ($CH_2(COOH)_2$), and "#"—the symbol that delimits the beginning- and end-of-sequence—is an aliquot of a pH indicator. The quantification of the recipes of the symbol aliquots is carried out so that one aliquot of "(" and one aliquot of ")" neutralize each other to the midpoint in the

pH curve (Petrucci *et al.* 2011) and the pH indicator is selected to change color around the midpoint (methyl red indicator in our implementation).

At the beginning of a computation, the L_2-PDA reactor contains deionized water and an aliquot of the pH indicator. The processing of the symbols sequentially fed to the reactor leads to changes in pH whose value we assign to the stack. The L_2-PDA **Accepts** the input string if during the computation the $pH \geq$ midpoint-pH *but* is at midpoint-pH (empty stack) at the end of computation, that is, after adding "#." Conversely, the L_2-PDA *rejects* an input string if the pH falls below the midpoint-pH at any stage during computation (excess of ")", and attempting to "pop" from an already empty stack), or if the pH is larger than the midpoint-pH value at the end of computation (excess of "(", or the stack is "not empty") (cf. fig. 4.2).

The response given by the L_2-PDA can again be interpreted in terms of a thermodynamic measure: the enthalpy yield of the computation $Y_{\Delta H}$ (%). This is defined as the ratio between the enthalpy produced or consumed during computation divided by the total formation enthalpy of the chemical input:

$$Y_{\Delta H}\ (\%) = \frac{\text{reaction heat during computation}}{\text{formation heat of input string}} \times 100 \quad (4.2)$$

$$Y_{\Delta H}\ (\%) = \frac{\sum_1^R \int_0^{t_{end}} v_i\, \Delta H^o_{r,i} dt}{\sum_1^n [j]_{\text{input}} \Delta H^o_{f,j}} \times 100 \quad (4.3)$$

Here R is the number of the reactions in the kinetic mechanism (at the level of coarse graining associated with the time between symbol processing), v_i is the velocity of reaction i (mol/($dm^3 \cdot s$)), $\Delta H^o_{r,i}$ represents the enthalpy of reaction i, J is the number of symbols in the language alphabet, $[j]_{\text{input}}$ is the total change in concentration of species j due to the input string, and $\Delta H^o_{f,j}$ is the formation enthalpy of chemical species j.

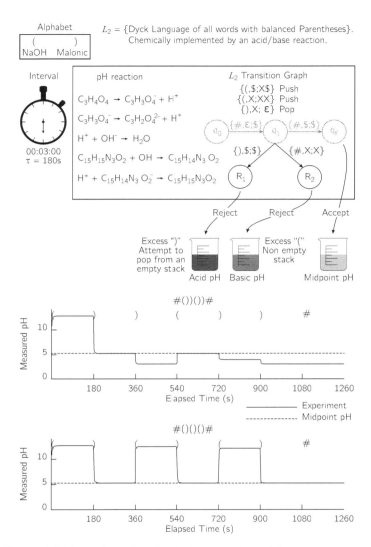

Figure 4.2. Operation of a chemical one-stack pushdown automata: L_2 is recognized by a one-stack PDA. Here the reaction pH acts as the stack. If $pH \geq$ midpoint pH (intermediate gray tone or lightest gray tone, respectively) during computation, and $pH =$ midpoint pH (lightest gray tone) at the end of computation, then the input string is accepted. Otherwise, if $pH <$ midpoint pH (darkest gray tone) any time during computation, the string is rejected (attempting to "pop" from an empty stack). Also, if the $pH >$ midpoint pH at the end of computation (intermediate gray tone), there is an excess of open parentheses and the string is rejected. Above are the experimental results for rejected ())()) and accepted ()()() words.

During computation, the dominant contribution to reaction heat occurs whenever a pair of parentheses is compensated via reaction R_3 (third reaction) in the mechanism (cf. fig. 2): $OH^- + H^+ \to H_2O$. Hence the enthalpy yield can be approximated as follows: $Y_{\Delta H}$ (%) =

$$\frac{\int_0^{tend} v\,\Delta H_r^o dt}{[\text{malonic}]_{\text{input}} \Delta H_{f,\text{malonic}} + [OH^-(aq)]_{\text{input}} \Delta H_{f,OH^-(aq)}} \times 100$$
$$\approx \frac{n_{\text{pairs}} c\, \Delta H_r^o}{[\text{malonic}]_{\text{input}} \Delta H_{f,\text{malonic}} + [OH^-(aq)]_{\text{input}} \Delta H_{f,OH^-(aq)}} \times 100 \qquad (4.4)$$

where n_{pairs} is the number of pairs of parentheses that have been balanced and c is the change in molarity of the solution due to the addition of each aliquot of malonic acid. For our *pH* reaction, the heat of the acid–base neutralization reaction is $\Delta H_r^o = -55.89$ kJ/mol, and the formation enthalpies can be found in standard thermodynamic databases (Haynes 2014).

By chemical engineering design of our 1-PDA, *Dyck* words maximize the enthalpy yield, whereas input strings that have excess of either open or closed parentheses will result in smaller enthalpy yields than strings with balanced parentheses. This again provides us with a thermodynamic metric to assess the result of the computation.

Chemical 2-Stack PDA/TM Recognizing Context-Sensitive Language $L_3 = \{a^n b^n c^n$, Where $n > 0\}$

Finally, we demonstrate that our native chemical computing approach can be used successfully to recognize a language that only a Turing machine (TM) can recognize. A TM is an automaton equipped with an infinite tape and a read-write head working together with a finite set of transition rules. Initially, the input is written on the tape, each letter of the string written on one cell. The head can move left or right, reading and erasing and writing symbols on the tape based on the finite set of rules. As a part of the state transition, the TM decides if the next cell to be scanned is to the right or the left of the current scanned cell. The infiniteness of the tape and the possibility of moving either to the left or to the right of the tape are the factors that make the TM capable of recognizing all computable languages (Hopcroft, Motwani, and Ullman 2006; Turing 1936; Cohen 1991).

For our experimental implementation, we choose a well-defined and decidable language (Dueñas-Díez and Pérez-Mercader 2019). The language $L_3 = \{a^n b^n c^n,$ where $n > 0\}$, is made up of words consisting of n-repeats of **a**, followed by n-repeats of **b**, followed by n-repeats of **c**. Note that L_3 is a context-sensitive language not recognizable by either a FA or a one-stack PDA (Cohen 1991), and though it is not the most complex language a theoretical TM can recognize, it is quite convenient for an experimental implementation, as it brings into play all the features of a TM. Context-sensitive languages are recognized by a subclass of TMs, linearly bounded automata (LBA), in which only the cells occupied by the input are used for computation (Linz 2012).

TMs are equivalent to two-stack PDAs (Minsky 1961) because two stacks can emulate the function of moving right and left on an infinite (or arbitrarily long) tape. Taken together with the requirement of two interrelated stacks leads us to translate this into the chemical requirement of interrelated redox reactions and to oscillatory chemistry (Pérez-Mercader, Dueñas-Díez, and Case 2017; Dueñas-Díez and Pérez-Mercader 2019). We have chosen arguably the best-known oscillatory chemistry, the Belousov–Zhabotinsky (Belousov 1959; Zhabotinsky 1964) reaction. As was the case before, alphabet symbols are carefully chosen to map into distinct pathways in the reaction mechanism and, consequently, to have distinct systematic effects on the measured oscillatory behavior (Dueñas-Díez and Pérez-Mercader 2019): *a* is transcribed as an aliquot of sodium bromate affecting dominantly the autocatalytic production of $HBrO_2$ and catalyst oxidation, b is transcribed as an aliquot of malonic acid dominantly affecting the bromination of the weak acid and the reduction of catalyst, c is transcribed as an aliquot of NaOH affecting the pH-dominated subset of reactions, and # is transcribed as an aliquot of catalyst affecting the redox-dominated subset of reactions. The quantitative recipes were selected and engineered to maintain the oscillatory regime for as long a word as possible, while simultaneously providing measurable changes in the oscillations. To implement this in a reactor, we used a combination of simulation and experimental studies.

The results of this experimental implementation have been reported in detail elsewhere (Dueñas-Díez and Pérez-Mercader 2019). A key finding is that each state in the abstract TM transition graph has its own distinct chemical counterpart, for example, for our chosen language L_3, the reject due to the input containing ba has a different chemical signature than a reject due to an excess of a. There is a systematic clustering of chemical behaviors when mapping two basic phenomenological descriptors of the final oscillations (frequency and an oscillation amplitude-related difference measure). Experimentally, we find that words in the language are placed in a locus in this map, while rejected sequences (same as words, of course) lie either above or below (Dueñas-Díez and Pérez-Mercader 2019).

To find a more intuitive criterion for acceptance/rejection, we introduce a metric based on the integral of the final oscillations (which we call the area) $A^{(\text{Word})}$ associated with the word undergoing processing in the computation:

$$A^{(\text{Word})} = V_{\max} \times \tau' - \int_{t_\#+30}^{t_\#+\tau} V_{\text{osc}}(t)\, dt, \quad (4.5)$$

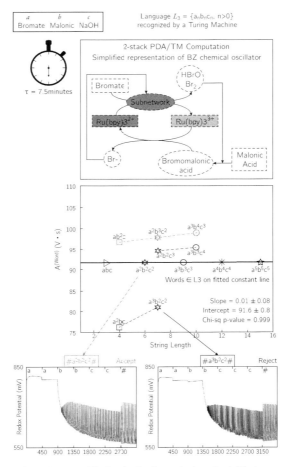

Figure 4.3. A Belousov–Zhabotinsky-based chemical Turing machine for L_3 at work: L_3 is recognized by a two-stack PDA/TM, as shown by the constant area represented by the thick black line joining words in L_3 in the upper right-hand corner graph. Words not in L_3 lie elsewhere in the plot. During computation, if certain alphabet patterns in the redox potential V are detected, the strings are rejected. The rejection is specific for each type of reject. Here five words (in black font) were accepted and seven strings (in gray font) were rejected (two due to excess as, three due to excess bs, and two due to excess cs). The bottom panels compare the evolution of V for rejected $a^3b^2c^2$ (bottom right panel) and accepted $a^2b^2c^2$ (bottom left panel).

where $t_\#$ is the time in reaction coordinates at which the end-of-expression symbol is added, τ' is the time interval between symbols minus thirty seconds (the first thirty seconds are discarded in the integration to allow for fast transients to dissipate), $V_{\max}$ is the maximum redox potential (all catalyst in oxidized form), and V_{osc} is the measured redox potential, which can be well approximated by Nernst equation:

$$V_{\text{osc}} = V_0 + \frac{RT}{n_e F} \ln\left(\frac{[Ru(bpy)_3^{3+}]}{[Ru(bpy)_3^{2+}]}\right), \qquad (4.6)$$

with $[Ru(bpy)_3^{2+}]$ and $[Ru(bpy)_3^{3+}]$, respectively, denoting the reduced and oxidized form of the catalyst, which can in turn be written in terms of the extent of reaction for the elementary redox reactions in the oxidation and reduction subsets as appropriately coarse grained. The quantity n_e denotes the number of electrons involved in the reduction–oxidation process and is $= 1$ for this reaction. The redox potential is related to the Gibbs free energy ΔG (Kuhn and Försterling 2000) as

$$\Delta G_{\text{osc}} = -n_e F V_{\text{osc}}. \qquad (4.7)$$

We can thus rewrite the area defined above in terms of the Gibbs free energy corresponding to full oxidation $\Delta G'$ and the redox Gibbs free energy ΔG_{osc}:

$$A^{(\text{Word})} = -\frac{1}{n_e F}\left(\Delta G' \times \tau' - \int_{t_\#+30}^{t_\#+\tau} \Delta G_{\text{osc}}(t)\, dt\right). \qquad (4.8)$$

The recipes for the alphabet aliquots can now be optimized to achieve a constant (i.e., n-independent or word length-independent) $A^{(\text{Word})}$ for words in L_3, while rejected sequences lie either above or below this value (cf. fig. 4.3, top right). Hence, if the area $A^{(\text{Word})}$ is constant and independent of string length for the words in L_3, so is the integral of ΔG_{osc}. Finally, we point out that the dimensions of this area are the same as those of the action in physics and that its origin reminds one of the mass-action law in chemistry.

Conclusions

We have demonstrated experimentally that nonbiochemical chemistry in a homogeneous one-pot reactor, where the chemical inputs to be computed are fed sequentially at constant time intervals, has the capability to run successful computations at the three fundamental levels in the hierarchy of classical automata theory.

Our approach allows tailoring a chemical reactor to run a specific computation, that is, recognizing a specific language of interest, identifying an appropriate chemical transcription/translation of the alphabet and the chemistry of the automaton so that the reactor provides a distinctive thermodynamic/chemical response for those inputs that belong to said language. The design and operation of each chemical automaton follow similar principles, as the examples L_1, L_2, and L_3 illustrate. The elements of an automaton's tuple have chemical counterparts; for example, there are as many types of chemical "reject" states in the practical chemical automaton as "reject" states in the abstract automaton.

There are of course differences between abstract and actual automata. Any experimental chemical (or otherwise actual) realization of a TM necessarily has a noninfinite chemical tape; for that reason, it is most practical to implement the TM in the subclass of linear bounded automata. Note, however, that this is not too restrictive: by optimizing the operational strategy, including choice of reactor type and recipes, one can extend the tape to the needed length.

Computational versatility can also be enhanced by combining different chemical automata. The earlier discussed chemistries, including the Belousov–Zhabotinsky oscillatory chemistry, can be reconfigured to solve other languages of computational interest by appropriately selecting the alphabet symbols and their recipes such that all abstract-tuple elements have their chemical counterparts. The richness of time scales and nonlinearity in the Belousov–Zhabotinsky chemistry (or any other oscillatory chemistry) can be further exploited for computation. The reactions selected in this chapter are meant to provide illustrative examples for each class of automata. Other reactions in the appropriate classes can of course be used. Furthermore, other instances of specific computations can also be designed and carried out based on other specific chemistries, such as pH oscillators or biochemical oscillators. Finally, because abstract automata can be connected to create new automata (Hopcroft, Motwani, and Ullman 2006), we can imagine that chemical automata can likewise be interconnected to carry out more complex computations or generalizations of our automata (e.g., from a TM to a universal TM) if the underlying chemistries are compatible with each other, share common chemical species, and can be deployed in the same solvent media.

For each of the three implemented languages, we have identified a thermodynamic interpretation of the accept/reject states that is equivalent to the chemical response. Translating the criterion from purely chemical to its thermodynamic equivalent may simplify the interpretation of the acceptance/rejection of native chemical automata, as clearly seen for the case of the context-sensitive language L_3. In the examples discussed here, the thermodynamic interpretation involves thermodynamic potentials like enthalpy (languages L_1 and L_2) or Gibbs energy (language L_3). Such thermodynamic potentials were introduced in equilibrium thermodynamics to describe how closed systems approach equilibrium because, according to the extremum principles (Kondepudi and Prigogine 2014; Callen 1985), these potentials reach an optimal value at equilibrium. For example, in a closed system at constant pressure and temperature, the Gibbs free energy is at a minimum in equilibrium. Our chemical automata are open systems in nonequilibrium due to the semibatch feed of the chemical input, and hence these extremum principles do not apply. However, in the same way that an open system can be maintained at a (nonequilibrium) stationary state by the influx of matter and energy, we can direct the thermodynamic potentials to certain values and even to some (nonequilibrium) optimum values by feeding it with specific sequences in which matter and/or energy are inputs to the system. In this case, when the system reaches a (nonequilibrium) optimum, it is not as an (unavoidable) result from an extremum principle but driven or directed by our sequential chemical inputs of the language of interest and the specific recipes used. The chemical input directs the dominant reaction pathways, which in turn direct the thermodynamic pathways as well. This form of bootstrapping brings with it chemical control of the nonlinear, out-of-equilibrium chemistry itself.

Connecting the topology of complex reaction networks, their dynamics, and their thermodynamics is a recent and growing area of research (Rao and Esposito 2016). In our native chemical automata, the connection between chemistry and thermodynamics can contribute to better study and understanding of the energetic cost of computation, and probably how to control this cost in the quest to approach Landauer's limit (Landauer 1961), at least for the important case of chemical automata. Furthermore, our approach does use liquid phase (dense) chemistry and kinetics and is not restricted by any approximations relying on the dilute gas approximation.

Chapter 4: Native Chemical Automata

Nothing in the aforementioned precludes the extension of these results to biochemistry and biology. In particular, one can begin to think about the application of the preceding information-processing thermodynamics in the coming wave of new biochemistry-based oscillators (Novák and Tyson 2008), DNA-based oscillators (Srinivas *et al.* 2017), and computing ecologies of natural and synthetic bacteria—and finally, also to the study of the metabolic efficiency and the cost of chemical information processing and computation in extant living systems.

REFERENCES

Adamatzky, A., and B. de Lacy Costello. 2002. "Experimental Logical Gates in a Reaction-Diffusion Medium: The XOR Gate and Beyond." *Physical Reviews E* 66 (4): 046112.

Adleman, L. M. 1994. "Molecular Computation of Solutions to Combinatorial Problems." *Science* 266 (5187): 1021–1024.

Belousov, B. P. 1959. "A Periodic Reaction and Its Mechanism." *Compilation of Abstracts on Radiation Medicine* 147 (145): 1.

Benenson, Y. 2009. "Biocomputers: From Test Tubes to Live Cells." *Molecular Biosystems* 5:675–685.

Bennett, C. H. 1982. "The Thermodynamics of Computation—A Review." *International Journal of Theoretical Physics* 21 (12): 905–940.

Bray, D. 2009. *Wetware: A Computer in Every Living Cell.* New Haven, CT: Yale University Press.

Bryant, B. 2012. "Chromatin Computation." *PLoS One* 7 (5): e35703.

Callen, H. B. 1985. *Thermodynamics and an Introduction to Thermostatistics.* Hoboken, NJ: John Wiley.

Cohen, D. I. A. 1991. *Introduction to Computer Theory*. 2nd ed. Hoboken, NJ: John Wiley.

Conrad, M. 1972. "Information Processing in Molecular Systems." *Biosystems* 5 (1): 1–14.

Donder, T. de. 1927. *Affinité*. Paris: Gauthier-Villars.

Dueñas-Díez, M., and J. Pérez-Mercader. 2019. "How Chemistry Computes: Language Recognition by Non-Biochemical Chemical Automata." *ChemRxiv*.

Evans, D. 2011. "Introduction to Computing: Explorations in Language, Logic, and Machines." http://computingbook.org.

Górecki, J., K. Yoshikawa, and Y. Igareshi. 2003. "On Chemical Reactors That Can Count." *Journal of Physical Chemistry A* 107 (10): 1664–1669.

Haynes, W. M., ed. 2014. *CRC Handbook of Chemistry and Physics*. 95th ed. Boca Raton, FL: CRC Press.

Hjelmfelt, A., E. D. Weinberger, and John Ross. 1991. "Chemical Implementation of Neural Networks and Turing Machines." *Proceedings of the National Academy of Sciences of the United States of America* 88 (24): 10983–10987.

Hopcroft, J. E., R. Motwani, and J. D. Ullman. 2006. *Introduction to Automata Theory, Languages, and Computation*. 3rd ed. Boston, MA: Addison-Wesley Longman.

Kondepudi, D., and I. Prigogine. 2014. *Modern Thermodynamics: From Heat Engines to Dissipative Structures*. Hoboken, NJ: John Wiley.

Kuhn, H., and H. D. Försterling. 2000. *Principles of Physical Chemistry*. Hoboken, NJ: John Wiley.

Landauer, R. 1961. "Irreversibility and Heat Generation in the Computing Process." *IBM Journal of Research and Development* 5 (3): 183–191.

Linz, P. 2012. *An Introduction to Formal Languages and Automata*. 5th ed. Burlington, MA: Jones / Bartlett Learning.

Magnasco, M. O. 1997. "Chemical Kinetics Is Turing Universal." *Physical Review Letters* 78 (6): 1190.

Minsky, M. L. 1961. "Recursive Unsolvability of Post's Problem of Tag and Other Topics in Theory of Turing Machines." *Annals of Mathematics* 74 (3): 437–455.

Novák, B., and J. J. Tyson. 2008. "Design Principles of Biochemical Oscillators." *Nature Reviews Molecular Cell Biology* 9:981–991.

Okamoto, M., T. Sakai, and K. Hayashi. 1987. "Switching Mechanism of a Cyclic Enzyme System: Role as a 'Chemical Diode'." *Biosystems* 21 (1): 1–11.

Pérez-Mercader, J., M. Dueñas-Díez, and Daniel Case. 2017. *Chemically-Operated Turing Machine.* US Patent 9582771B2.

Petrucci, R. H., F. G. Herring, J. D. Madura, and C. Bissonette. 2011. *General Chemistry: Principles and Modern Applications.* 10th ed. Pearson Prentice Hall.

Prohaska, S. J., P. F. Stadler, and D. C. Krakauer. 2010. "Innovation in Gene Regulation: The Case of Chromatic Computation." *Journal of Theoretical Biology* 265 (1): 27–44.

Rao, R., and M. Esposito. 2016. "Nonequilibrium Thermodynamics of Chemical Reaction Networks: Wisdom from Stochastic Thermodynamics." *Physical Reviews X* 6:041064.

Srinivas, N., J. Parkin, G. Seelig, E. Winfree, and D. Soloveichik. 2017. "Enzyme-Free Nucleic Acid Dynamical Systems." *Science* 358 (6369): eaal2052.

Tóth, Á., and K. Showalter. 1995. "Logic Gates in Excitable Media." *Journal of Chemical Physics* 103 (6): 2058–2066.

Turing, A. M. 1936. "On Computable Numbers, with an Application to the Entscheidungsproblem." *Proceedings of the London Mathematical Society* 2:230–265.

Weisstein, E. W. 2009. *CRC Encyclopedia of Mathematics.* 3rd ed. Boca Raton, FL: CRC Press.

Zhabotinsky, A. M. 1964. "Periodic Oxidation of Malonic Acid in Solution (Investigation of the Kinetics of the Reaction of Belousov)." *Biofizika* 9:306–311.

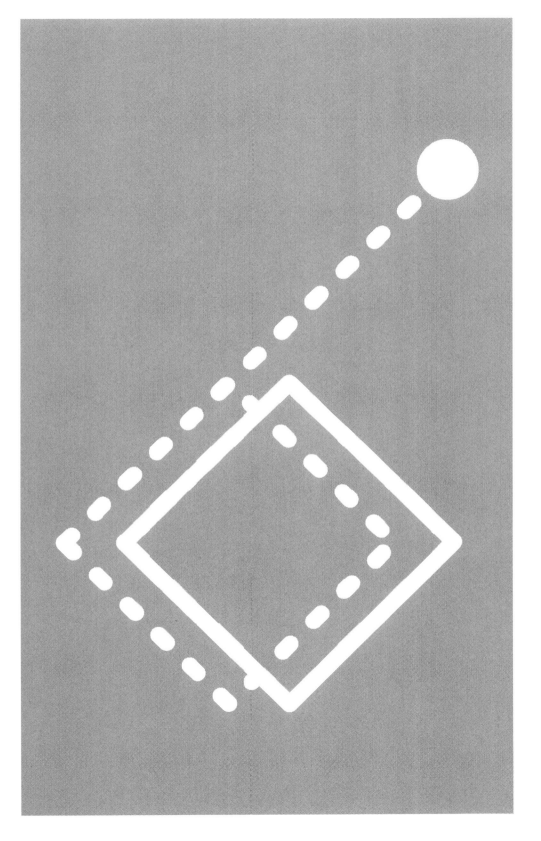

INTERGENERATIONAL CELLULAR SIGNAL TRANSFER AND ERASURE

GW C. McElfresh, University of Kansas
J. Christian J. Ray, University of Kansas

Biological substrates for computation have been considered since before the advent of modern deterministic computers (McCulloch and Pitts 1943; von Neumann 1956; Bennett 1982). Technological advances in measuring cellular responses to molecular signals have again raised the question of how stochastic networks compute.

Signaling pathways enable living cells to process responses to stimuli from the extracellular environment. The uncertainty of signal transmission in a single cell has prompted various research efforts to quantify how much a cell knows about its environment. Advances in nonequilibrium thermodynamics have arrived alongside analyses of biological signaling. Often, models of signaling that consider only the time scale of molecular fluctuations have been considered (see, among others, Cheong *et al.* 2011; Barato, Hartich, and Seifert 2014; Govern and ten Wolde 2014; Bo, Del Giudice, and Celani 2015; Hartich, Barato, and Seifert 2016), especially in relation to the bacterial chemotactic response (Lan *et al.* 2012).

We suggest that an important time scale for biological signaling should be on the order of gene expression (in the case of bacteria, potentially multiple generations). Growing cells invest energy to grow and divide, thereby diluting the results of previous computations. Because the remnants of previous responses are reduced but not necessarily completely erased, gradual dilution imparts a memory effect: a daughter cell is predisposed to respond in a qualitatively similar manner to its mother cell. Quantifying thermodynamic costs of molecular receptor signaling on short time scales reveals much about the extreme limits of the biological cost of computation, but such energy use is ultimately minor compared to the massive costs of gene expression that can arise as a result of such a signal. Here we seek to explore the effects of those costs on cellular information processing.

We analyze cellular memory in a broad class of bacterial information transfer systems: two-component system (TCS) modules. TCSs respond to information about modulation of the physicochemical environment in and around the cell. Our analysis places nongenetic intergenerational information transfer in a computational context and raises the question of the appropriate scales for analyzing the thermodynamics of information in living systems.

Signaling Dynamics on the Time Scale of Generations

Two relevant time scales of cellular signaling responses are molecular kinetic fluctuations and gene expression programs. In a bacterial cell, the time scale of protein turnover (and thus shifts in gene expression) is set by the generation time for the majority of protein types. This is because most proteins are quite stable; the relevant quantity for protein kinetic activity is concentration, and growth of the cell is the fastest process that reduces the concentration. Considering a signal that activates a transcription factor, the loss of the signal depends on the elimination of the responding proteins. Thus, for the mean-field birth–death process with constant production α/τ and constant generation time τ, we have dynamics of protein concentration $x(t)$ as

$$x(t) = e^{-t/\tau}\left(x(0) + \alpha(e^{t/\tau} - 1)\right),$$

and protein half-life after loss of signal is $\tau \ln 2$, or about 70% of a cell's lifetime, due to growth-mediated dilution. Positive feedback on the activation signal can promote the transcriptionally activated state of the cell, further exaggerating the effect. Many studies have explored the implications of such phenotypic memory (e.g., Nevozhay *et al.* 2012; Kaufmann *et al.* 2007; Frick *et al.* 2015; Burrill *et al.* 2012; Inniss and Silver 2013; Lambert and Kussell 2014; Ray 2016).

To make the conditions underlying cost and benefit more concrete, we introduce a common signaling pathway in bacteria: the TCS. Our goal here is to create a biologically realistic model that allows numerical determination of thermodynamic and informational quantities.

Models of Bacterial Two-Component Signaling

TCSs are a common sensing mechanism in bacteria that have a notable level of conservation across phyla (for a review of TCS evolution, see Capra and Laub 2012). Though many variations on the core motif exist, the canonical TCS has a dimeric sensor histidine kinase (SHK) and a cognate response regulator (RR). The sensor responds to stimuli by increasing phosphorylated RR (Zschiedrich, Keidel, and Szurmant 2016). Once phosphorylated, RR dimerization is stabilized, allowing it to become a transcription factor for genes that are typically relevant to the original stimulus. In many TCSs, one of the operons regulated by the RR is the TCS operon itself, providing feedback and potentially affecting the regulatory activity of the TCS (Batchelor and Goulian 2003; Shin *et al.* 2006; Shinar *et al.* 2007; Groban *et al.* 2009; Ray and Igoshin 2010). TCS operons have strong gene expression polarity, meaning the expression level of the gene closest to the transcription start site is higher than expression of the subsequent gene(s). Because of this effect, [RR] exceeds [SHK] by orders of magnitude to maintain a sensitive, yet reproducible, response to stimuli (Batchelor and Goulian 2003; Aiso and Ohki 2003). There are multiple distinct TCSs in most characterized bacterial species, each responding to distinct stimuli and inducing distinct responses (Skerker *et al.* 2008; Laub and Goulian 2007; Rowland and Deeds 2014). However, TCSs are integrated into global responses. For example, phosphate limitation depends on a complex between multiple sensors, including a TCS sensor called PhoR (Gardner *et al.* 2014). We developed coarse-grained models for the TCS core motif that were parameterized to represent a large class of them approximately, but with special reference to the PhoBR system in *Escherichia coli*, which has been extensively studied (Hoffer *et al.* 2001; Gao and Stock 2013; Gao and Stock 2015; Gao and Stock 2017, and references therein).

COARSE-GRAINED KINETIC MODEL

The sensor of a TCS is a dimer composed of two inactive monomers. It matures into a dimeric form that is usually in the cell membrane and senses changes in environment. The mature sensor has two reaction pathways: one that favors creating the active regulator and one that favors the inactivation of the regulator. The result is a dynamic balance between the competing processes of activation and deactivation. Which one dominates at a given time depends on how much stress signal is present. Figure 5.1A depicts this process. Conformational states in figure 5.1A represent ensembles of protein structure conformations that are functionally equivalent in terms of the reaction kinetics, which is why we refer to this as a coarse-grained kinetic model. All of the depicted reaction rates follow mass action kinetics at this scale.

Chapter 5: Intergenerational Cellular Signal Transfer and Erasure

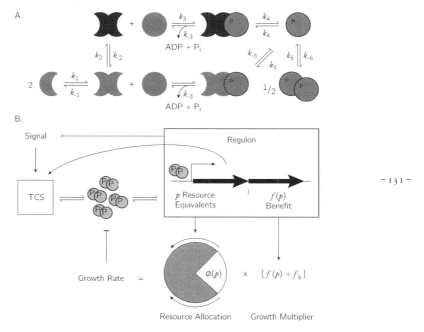

Figure 5.1. Coarse-grained multiscale model of a two-component system (TCS). (a) ATP associates with the sensor histidine kinase (SHK), ✖, along with fast interchange between ADP and ATP. External signal stimulates the SHK conformational switch (k_2). Physical interaction between SHK and response regulator (RR) allows phosphotransfer to RR, stablizing the dimeric RRP$_2$, ●●, an active transcription factor. SHK phosphatase activity is an ATP dissipative step. (b) Nested feedback loops involved in signals from TCSs. Signal stimulates production of RRP$_2$, ●●, which modulates a regulon (upregulation of several genes). Often, transcription of the TCS operon itself is induced: feedback that may affect the signal level. The regulon typically counteracts the signal, another feedback loop. Expression of the regulon entails a metabolic investment, reducing the fraction of resources devoted to growth. Growth dilutes the molecules, affecting bimolecular reaction propensities. The TCS maintains responsiveness by constantly dissipating ATP energy, but the major cost of the TCS during the signal is in the regulon.

We have inferred that the SHK component of *E. coli* PhoBR switches between kinase-active and kinase-inactive conformational ensembles because phosphatase activity is unaffected in mutants lacking kinase activity (Carmany, Hollingsworth, and McCleary 2003). Because ATP or ADP is bound in close proximity to the phosphorylated site on the SHK, the kinase and phosphotransfer reactions are reversible. The step in a TCS that truly dissipates energy is phosphatase activity: effectively irreversible dephosphorylation of a phosphorylated RR monomer.

The cytoplasm contains ATP at approximately one hundredfold excess over ADP (Qian 2007). Our model assumes that ATP quickly replaces ADP in the binding pocket of SHK molecules. SHK reversibly binds its cognate RR. SHK is then capable of reversibly transferring the phosphoryl group to the RR.

In the limit of large numbers of molecules, the steady-state fraction of active SHK is $\frac{k_2}{k_2-k_{-2}}$. In this model, we can say that the rate of kinase phosphorylation is $\frac{k_3 k_2}{k_2+k_{-2}}$. We can find the potential difference (Qian 2007), $\Delta\mu = k_b T \ln \frac{\mathcal{J}^+}{\mathcal{J}^-}$, where $\mathcal{J}^+$ represents the flux toward transcriptionally active RRP$_2$ (●), $\mathcal{J}^-$ represents the reverse flux toward the inactive state, and $k_b T$ is the Boltzmann constant times the temperature. In the equilibrium state, the two fluxes balance, and we have detailed balance. Deviations of $\Delta\mu$ from zero quantify how far out of equilibrium the system is being driven by mass and energy input from the rest of the cell.

Our TCS model has the following mean-field fluxes:

$$\mathcal{J}^+ = k_1[\text{SHK}m]^2 \times \frac{k_2 k_3}{k_{-2} + k_2}[\text{SHK}] \times (k_{-5} + k_4)[\text{SHK.RRP}] \times k_6[\text{RRP}]^2$$

$$\mathcal{J}^- = k_{-1}[\text{SHK}] \times k_{-3}[\text{SHK.RRP}] \times (k_5 + k_{-4})[\text{RRP}] \times k_{-6}[\text{RRP}_2],$$

where [SHKm] represents SHK monomers, [SHK] is SHK dimers, [SHK.RRP] is the SHK + RR complex, [RRP] is phosphorylated RR monomer, and [RRP$_2$] is transcriptionally active, phosphorylated RR dimer. We have identified specific parameter values for each rate constant that reflect the PhoBR system (table 5.1).

Table 5.1. Calibrated parameters

Parameter	Estimated value	Notes/reference
k_1	10 $(\mu M\ s)^{-1}$	Fast SHK dimerization
k_{-1}	0.00001 s^{-1}	Rare SHK dedimerization
k_2	Conditional, s^{-1}	$k_2 \in [0.001, 10]$
k_{-2}	0.1 s^{-1}	Assumed fast
k_3	0.0004 $(\mu M\ s)^{-1}$	Model calibration
k_{-3}	0.0087 s^{-1}	Gao and Stock (2013)
k_4 and k_{-5}	1 s^{-1}	Model calibration
k_{-4} and k_5	0.036 $(\mu M\ s)^{-1}$	Gao and Stock (2013)
k_6	1 $(\mu M\ s)^{-1}$	Model calibration
k_{-6}	4 s^{-1}	Mack, Gao, and Stock (2009)
k_{txnb}	0.00001 s^{-1}	Model calibration
k_{txn}	0.00025 s^{-1}	Model calibration
k_{txni}	0.15 s^{-1}	TCS transcription initiation rate when RRP_2 is bound
K_{mtxn}	2.5 μM	P_{TCS} half-sat; Gao and Stock (2015)
k_{pb}	1.66 $(\mu M\ s)^{-1}$	P_{TCS} binding rate; Elf, Li, and Xie (2007)
k_{pu}	3.86 s^{-1}	P_{TCS} unbinding rate; inferred from K_{mtxn} and k_{pb}
k_{degb}	0.027 s^{-1}	Aiso and Ohki (2003)
k_{degr}	0.0044 s^{-1}	Aiso and Ohki (2003)
k_{tsn}	0.05 s^{-1}	Model calibration
χ	0.37	Marzan and Shimizu (2011)
a_{fit}	$\approx 1.123 \times 10^{-4}$	"
b_{fit}	≈ 1.77	"
c_{fit}	≈ 3.75	"

Note. P_{TCS} refers to the promoter of the two-component system operon.

In practice, living cells constantly produce ATP; the TCS has a constant source of energy in ATP and a sink in ADP + P_i. The gene regulatory activity of the TCS, including its autoregulation, also contributes to the total energy in the system. The steady state of a functional TCS is intrinsically out of equilibrium: $\Delta\mu > 0$.

CONNECTIONS TO CELLULAR PHYSIOLOGY

Activation of a TCS upregulates a regulon, the set of genes that are the target of the regulator. The cell pays a metabolic cost for the response, but also benefits from the ameliorative activities of the regulon. For example, in the case of phosphate starvation, the PhoBR TCS induces expression of alkaline phosphatase (*phoA*), recovering phosphorus from phosphate ester. However, the complete regulon of PhoBR consists of approximately forty upregulated genes; the metabolic cost of expressing it is significant. Lynch and Marinov (2015) give a sense of the scale of a regulon. They estimated the absolute cost per gene to be $10^3 - 10^8$ hydrolyzed phosphate bonds in bacteria. This is likely to be the majority of the metabolic cost of TCS activation.

We consider the fraction of the growth budget dedicated to the TCS to be $1 - \phi(\rho) = 1 - \chi\frac{\rho}{\rho_{\max}}$, where ρ represents the size of the total regulon, $\rho_{\max}$ is the maximal hypothetical induction, and χ is the maximal fraction of the growth budget that the regulon can take. We have defined the "growth budget" somewhat amorphously so that we can use $1 - \phi(\rho)$ as a multiplier to limit growth rate. Then we have a growth multiplier that determines the growth benefit from expressing the TCS regulon, $f(\rho) + f_b$, where f_b is the basal growth rate without the benefit of the TCS signal, while $f(\rho)$ is the ameliorative contribution of the regulon. We take a linear

Chapter 5: Intergenerational Cellular Signal Transfer and Erasure

benefit, $f(\rho) = \alpha \frac{\rho}{\rho_{max}}$, where $\alpha + f_b$ represents the maximal growth rate attainable in a given condition without accounting for TCS regulon cost. The net growth rate accounting for both cost and benefit is then

$$\gamma = (1 - \phi(\rho))(f(\rho) + f_b) = \left(1 - \chi \frac{\rho}{\rho_{max}}\right)\left(\alpha \frac{\rho}{\rho_{max}} + f_b\right). \quad (5.1)$$

The trade-off effect naturally arises because this form is quadratic in ρ, with a predicted optimal regulon size at the point where $\frac{\partial \gamma}{\partial \rho}\big|_{\chi,\alpha,f_b,\rho_{max}} = 0$, which gives $\rho = \frac{\rho_{max}(\alpha + f_b \chi)}{2\alpha\chi}$.

The situation is not that simple, however, because both α and f_b depend on the same conditions that determine the activation state of the TCS: kinetic parameter k_2. The relationship could take a variety of forms. We estimated the relationship empirically using biomass in a chemostat experiment in an *E. coli* strain that has had the *phoB* gene (response regulator) deleted (Marzan and Shimizu 2011). This strain does not produce the TCS regulon. Its steady-state biomass in a chemostat at various levels of phosphate starvation therefore gives f_b for the case of the PhoBR system. The biomass data happen to fit an inverse logistic function with $r^2 > 0.999$. Assuming that TCS activation rate, k_2, is proportional to the degree of phosphate starvation in PhoBR, we have

$$f_b = \frac{a_{fit}}{b_{fit} + e^{-c_{fit} + k_2}}.$$

In the preceding equation, $\alpha + f_b$ is the maximum possible recovery from the signal-induced growth rate: with γ_u as the upper limit of the growth rate and ϵ (≤ 1) as the efficiency of the regulon to recover the growth rate,

$$\alpha + f_b = \epsilon \left(\gamma_u - \frac{a_{fit}}{b_{fit} + e^{k_2 - c_{fit}}}\right) + \frac{a_{fit}}{b_{fit} + e^{k_2 - c_{fit}}}.$$

For the PhoBR system, we have a growth model with free parameters ϵ, γ_u, χ, and ρ_{max}. The same study that gave data for the logistic fit of f_b (Marzan and Shimizu 2011) also measured relative expression of selected PhoBR regulon genes in wild-type cultures. From this, we estimated $\chi \approx 0.37$. We assumed that the genes upregulated by the TCS were mostly capable of reducing phosphate stress ($\epsilon = 0.95$) and that the growth medium without phosphate starvation is relatively favorable ($\gamma_u = 0.0004$/s $= 1.44$/h). The hypothetical maximum induction of the regulon ($\rho_{max} = 150$ μM) was set by calibration with the average regulon transcription and translation rates, k_{txnR} and k_{tsnR} (table 5.1).

Using the growth model, we create two multiscale models of a TCS embedded in cellular physiology—one representing the average of many cells and a stochastic simulation that tracks the dynamics of signaling in single cells. We first describe the mean-field model TCS dynamics. We then use this to develop a stochastic model. We calibrate both models with the mean-field model, explore average responses with it, and then use the stochastic model to simulate the dynamics of signal transfer as the population recovers from signal loss.

MEAN-FIELD MODEL

We represent two types of processes: reversible chemical reactions and irreversible reactions that represent dissipative processes, such as transcription, translation, cellular growth, and, in the TCS, the irreversible step in hydrolysis of ATP—phosphatase activity of SHK. We allow transcription and translation to be governed by mass-action kinetics. The complete model is a set of differential equations with twelve variables: bicistronic messenger RNA (mRNA), monocistronic RR mRNA, downstream regulon mRNA, downstream regulon protein, and the species represented in figure 5.1A.

The equations are omitted for brevity, but all interactions are assumed to be mass action, except for gene regulation processes, which take Michaelis–Menten form, with $V_{\max}$ given by k_{txn} (TCS operon) or k_{txnR} (regulon operons) and K_m given by K_{mtxn} (table 5.1). (We assume that most promoters of the TCS regulon are calibrated to typical concentrations of RRP$_2$, and therefore we assign the same K_m for the TCS and all regulon promoters.) mRNA is unstable and actively degraded by cells; degradation of mRNA is taken to be a mass-action process. On the basis of the work of Aiso and Ohki (2003), our model has an unstable bicistronic TCS mRNA species capable of initiating translation of both RR and SHK as well as a more stable monocistronic mRNA species that only initiates translation of RR. Dilution of molecules depends on the previously described growth model: loss of protein has a rate $\gamma[\text{Protein}] = (1 - \phi(\rho))(f(\rho) + f_b) \times [\text{Protein}]$, and loss of mRNA has a rate $(k_{\text{degRNA}} + \gamma) \times [\text{mRNA}]$ for degradation rate constants that depend on the specific mRNA species.

STOCHASTIC MODEL

The stochastic model is based on the mean-field model with the following additions. Reactions occur in individual cell agents that have a volume growing according to the growth model described earlier, based on Bandyopadhyay, Wang, and Ray (2018). Increments of stochastic simulation occur at approximately constant volume intervals, then the volume is updated based on the resulting growth rate. Increments in reaction volume affect any bimolecular interactions (Gillespie 1976). We chose a quasi-constant volume interval of 1 s, which is less than the expected time to add a single phospholipid in a cell that is growing relatively quickly.

For the stochastic growth model, we assume that the mean-field growth model holds, with the exception of regulon fluctuations. The downstream regulon of a TCS potentially undergoes significant fluctuations that are entrained to RRP_2 fluctuations. However, there is still an independent stochastic component: between the expression of multiple genes, upward fluctuations in expression in some genes may be counterbalanced by downward fluctuations in expression in other genes. We therefore represent gene expression from $n = 40$ independent loci, all assumed to have identical binding and gene expression kinetics, producing mRNA into a common pool that results in a common regulon.

In the stochastic model, we represent explicit promoters for the regulated genes, with binding/unbinding and irreversible transcription initiation events. We set the binding constants and transcription initiation constants to be equal to the Michaelis–Menten form of the mean-field model (table 5.1).

Each cell agent grows at a rate set by the growth model (γ), and, when the initial volume has been doubled, it divides, partitioning all non-DNA species into two daughter cells with a binomial distribution. Jun *et al.* (2018, and references therein) suggest that the "adder" principle is an excellent phenomenological representation of cell volumes during the *E. coli* cell cycle: a constant cell volume is added before division. In our model, each cell agent has a volume of 1 femtoliter and doubles to 2 femtoliters before division. Promoters/DNA are all deterministically inherited into both daughter cells. The cellular simulation is implemented in Python, with the stochastic simulations run in StochKit and using GillesPy (Abel *et al.* 2016) to interface the Python cell script with the stochastic simulations.

Chapter 5: Intergenerational Cellular Signal Transfer and Erasure

Results

AVERAGE CELLULAR GROWTH AND SIGNAL DYNAMICS

Simulations using the mean-field model reveal the effects of induction and shutoff of a TCS in *E. coli* (fig. 5.2). The model suggests that intermediate levels of induction have a slightly lower growth rate than the fully induced system when the stress becomes more severe (fig. 5.2C). The reason for the effect is clear looking at the model variants lacking transcriptional feedback with constant low and high TCS gene expression (shaded lines in figure 5.2). The system with transcriptional feedback switches from being nearly equivalent to the low-TCS-expression feedbackless case to being nearly equivalent to the high-TCS-expression feedbackless case. It is the transcriptional feedback that allows the system to adapt to higher signal levels. Constant high TCS expression causes grossly more ATP hydrolysis (which is the same as the phosphatase flux, $k_{-3} \times$ ([SHKa.RRP] + [SHK.RRP])) than the case with low expression or transcriptional feedback (fig. 5.2D). This demonstrates a trade-off between cost and benefit: in the autoregulated TCS, it is possible to sacrifice large investments in stress responses at the cost of a slightly lower growth rate, unless the stress becomes severe.

Our model predicts that the TCS has a potential difference $\Delta\mu \approx 15\,k_bT$, varying slightly depending on signal level (fig. 5.2B). The same is not true for the ATP dissipation rate of the TCS, which increases dramatically at the largest induction levels (fig. 5.2D).

We find that shutting off the signal (initial conditions at the $k_2 = 10$ steady state, instantaneously switched to $k_2 = 10^{-3}$) reveals three relevant time scales (figs. 5.2E–5.2F). On the generational time scale, $\gg 1000$ s, the regulon is diluted, and

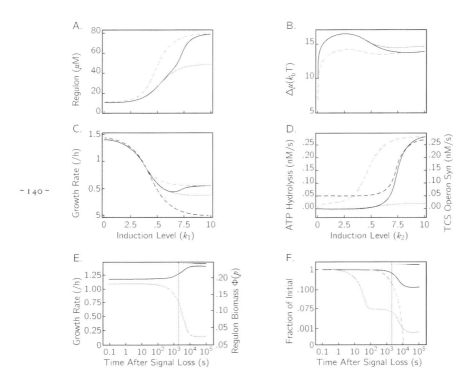

Figure 5.2. Predicted steady-state and dynamical physiological outcomes of activating a TCS, with parameters calibrated to represent the *E. coli* PhoBR system having a regulon containing approximately forty genes. Line styles represent the intact system (black solid) and transcriptional feedbackless system with basal (shaded solid) or maximal (shaded dashed) expression. (a) The level of induction is related to the size of the signal—in this case, phosphate limitation. (b) Potential difference $\Delta\mu$ in units of $k_b T$. (c) TCS induction recovers a fraction of the growth rate lost to the stress condition. (d) Rates of ATP hydrolysis by the TCS ($k_{-3}([\text{SHKa.RRP}] + [\text{SHK.RRP}])$) and TCS operon synthesis ($k_{txnb} + \frac{k_{txn}[\text{RRP}_2]}{K_{mtxn}+[\text{RRP}_2]}$). (e) Dynamics of growth rate and biomass on recovery from fully induced to uninduced conditions. The solid black line is growth rate; the shaded line is biomass; and the vertical line is the first generation of growth. Dots represent subsequent generations. (f) Dynamic loss of TCS activity. The solid black line is the regulon; the shaded solid line is SHKa; and the shaded dashed line is RRP_2.

normal growth resumes. Loss of TCS transcriptional activity (RRP_2) occurs at a faster rate. At very short time scales, the sensor shuts off to an intermediate quasi–steady state before being driven even lower by the effects of growth dilution (fig. 5.2F).

Chapter 5: Intergenerational Cellular Signal Transfer and Erasure

INTERGENERATIONAL SIGNAL TRANSFER IN A TWO-COMPONENT SYSTEM

The stochastic cell growth framework captures the rate of signal loss and the interaction between cell division and dynamics of the signal (fig. 5.3). We used the same switch from high to low signal as earlier. Figure 5.3A confirms the mean-field results that signal shutoff is faster than loss of the regulon. The half-life of RRP_2 is less than half of a generation, the regulon half-life is more than one generation, and a purely growth-diluted molecule half-life would be ≈70% of a generation. Both species follow nearly deterministic trajectories. The same is not true for TCS total protein expression, where protein dilution is highly lineage dependent (fig. 5.3B). The difference in time scales between the signal shutoff and the residual response illustrates an intergenerational memory effect.

Discussion

It is increasingly feasible to model time scales of cellular information processing that are relevant to fitness and evolution without them being oversimplified toy models. The disadvantage of this approach is the loss of generality: the necessary quantity of empirical information requires that they simulate a specific system. This is a small problem in the face of ever-increasing high-resolution physiological data. The ability to capture interactions accurately between the short time scales of molecular fluctuations and the global physiological shifts in a cell is an unmistakable advantage. Here we have demonstrated how such models can address questions of energetics and cellular information processing, setting up a framework for future, more thorough studies of information flow.

Our model of TCS suggests that the vast majority of metabolic (ATP) cost lies in the production of the regulon, which has a higher ATP investment compared to the signaling

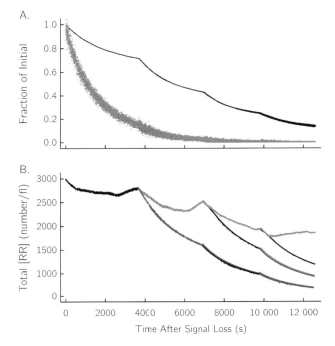

Figure 5.3. Stochastic dynamics of intergenerational signal loss. The simulation was started in a high activated steady state ($k_2 = 10$) and allowed to relax to an inactive state ($k_2 = 10^{-3}$). (a) Levels of the TCS regulon (black line) and transcriptionally active regulator RRP_2 (shaded line) follow different time scales. Cell division times are evident. These results superimpose the levels in all of the cell agents in the simulation. The results closely follow the expected deterministic mean. (b) Levels of the TCS proteins display striking heterogeneity that arises at cell birth. Different individual cells are represented by different shades of gray or black. Some cells are nearly to basal levels of the protein, while others still have substantial residual protein several generations later.

system itself. Monte Carlo sampling of the TCS kinetic parameters shows that our empirical parameter set lies in the middle of possible responses (not shown). Though not precisely quantitative of any particular system, the numerical results are reliable.

In the intact system, the constant source of ATP along with material influx of TCS proteins maintains the TCS out of thermodynamic equilibrium in all conditions (fig. 5.2B). At the same time, the system is driven by global physiological variables coupled to stochasticity effects, which diversifies the level of memory in a lineage-dependent manner: some cells and all of their daughters undergo rapid loss of TCS proteins, while other cells maintain a longer-lived high expression level that may be metastable (fig. 5.3). 🌿

Acknowledgments

This work was supported by the National Institutes of Health award numbers P20GM103418 and P20GM103638. The content is solely the responsibility of the authors and does not necessarily represent the official views of the National Institutes of Health.

REFERENCES

Abel, J. H., B. Drawert, A. Hellander, and L. R. Petzold. 2016. "GillesPy: A Python Package for Stochastic Model Building and Simulation." *IEEE Life Sciences Letters* 2 (3): 35–38.

Aiso, T., and R. Ohki. 2003. "Instability of Sensory Histidine Kinase mRNAs in *Escherichia coli*." *Genes to Cells* 8 (2): 179–187.

Bandyopadhyay, A., H. Wang, and J. C. J. Ray. 2018. "Lineage Space and the Propensity of Bacterial Cells to Undergo Growth Transitions." *PLOS Computational Biology* 14 (8): e1006380.

Barato, A. C., D. Hartich, and U. Seifert. 2014. "Efficiency of Cellular Information Processing." *New Journal of Physics* 16:103024.

Batchelor, E., and M. Goulian. 2003. "Robustness and the Cycle of Phosphorylation and Dephosphorylation in a Two-Component Regulatory System." *Proceedings of the National Academy of Sciences of the United States of America* 100 (2): 691–696.

Bennett, C. H. 1982. "The Thermodynamics of Computation—A Review." *International Journal of Theoretical Physics* 21 (12): 905–940.

Bo, S., M. Del Giudice, and A. Celani. 2015. "Thermodynamic Limits to Information Harvesting by Sensory Systems." *Journal of Statistical Mechanics* P01014.

Burrill, D. R., M. C. Inniss, P. M. Boyle, and P. A. Silver. 2012. "Synthetic Memory Circuits for Tracking Human Cell Fate." *Genes and Development* 26 (13): 1486–1497.

Capra, E. J., and M. T. Laub. 2012. "Evolution of Two-Component Signal Transduction Systems." *Annual Review of Microbiology* 66:325–347.

Carmany, D. O., K. Hollingsworth, and W. R. McCleary. 2003. "Genetic and Biochemical Studies of Phosphatase Activity of PhoR." *Journal of Bacteriology* 185 (3): 1112–1115.

Cheong, R., A. Rhee, C. J. Wang, I. Nemenman, and A. Levchenko. 2011. "Information Transduction Capacity of Noisy Biochemical Signaling Networks." *Science* 334 (6054): 354–358.

Elf, J., G.-W. Li, and X. S. Xie. 2007. "Probing Transcription Factor Dynamics at the Single-Molecule Level in a Living Cell." *Science* 316 (5828): 1191–1194.

Frick, P. L., B. B. Paudel, D. R. Tyson, and V. Quaranta. 2015. "Quantifying Heterogeneity and Dynamics of Clonal Fitness in Response to Perturbation." *Journal of Cellular Physiology* 230 (7): 1403–1412.

Gao, R., and A. M Stock. 2015. "Temporal Hierarchy of Gene Expression Mediated by Transcription Factor Binding Affinity and Activation Dynamics." *mBio* 6 (3): e00686–15.

Gao, R., and A. M. Stock. 2013. "Probing Kinase and Phosphatase Activities of Two-Component Systems In Vivo with Concentration-Dependent Phosphorylation Profiling." *Proceedings of the National Academy of Sciences of the United States of America* 110 (2): 672–677.

———. 2017. "Quantitative Kinetic Analyses of Shutting Off a Two-Component System." *MBio* 8 (3): e00412–17.

Gardner, S. G., K. D. Johns, R. Tanner, and W. R. McCleary. 2014. "The PhoU Protein from *Escherichia coli* Interacts with PhoR, PstB, and Metals to Form a Phosphate-Signaling Complex at the Membrane." *Journal of Bacteriology* 196 (9): 1741–1752.

Gillespie, D. T. 1976. "A General Method for Numerically Simulating the Stochastic Time Evolution of Coupled Chemical Reactions." *Journal of Computational Physics* 22 (4): 403–434.

Govern, C. C., and P. R. ten Wolde. 2014. "Optimal Resource Allocation in Cellular Sensing Systems." *Proceedings of the National Academy of Sciences of the United States of America* 111 (49): 17486–17491.

Groban, E. S., E. J. Clarke, H. M. Salis, S. M. Miller, and C. A. Voigt. 2009. "Kinetic Buffering of Cross Talk between Bacterial Two-Component Sensors." *Journal of Molecular Biology* 390 (3): 380–393.

Hartich, D., A. C. Barato, and U. Seifert. 2016. "Sensory Capacity: An Information Theoretical Measure of the Performance of a Sensor." *Physical Review E* 93:022116.

Hoffer, S. M., H. V. Westerhoff, K. J. Hellingwerf, P. W. Postma, and J. Tommassen. 2001. "Autoamplification of a Two-Component Regulatory System Results in "Learning" Behavior." *Journal of Bacteriology* 183 (16): 4914–4917.

Inniss, M. C., and P. A. Silver. 2013. "Building Synthetic Memory." *Current Biology* 23 (17): R812–R816.

Jun, S., F. Si, R. Pugatch, and M. Scott. 2018. "Fundamental Principles in Bacterial Physiology-History, Recent Progress, and the Future with Focus on Cell Size Control: A Review." *Reports on Progress in Physics* 81:056601.

Kaufmann, B. B., Q. Yang, J. T. Mettetal, and A. van Oudenaarden. 2007. "Heritable Stochastic Switching Revealed by Single-Cell Genealogy." *PLOS Biology* 5 (9): e239.

Lambert, G., and E. Kussell. 2014. "Memory and Fitness Optimization of Bacteria under Fluctuating Environments." *PLOS Genetics* 10 (9): e1004556.

Lan, G., P. Sartori, S. Neumann, V. Sourjik, and Y. Tu. 2012. "The Energy–Speed–Accuracy Trade-off in Sensory Adaptation." *Nature Physics* 8:422–428.

Laub, M. T., and M. Goulian. 2007. "Specificity in Two-Component Signal Transduction Pathways." *Annual Review of Genetics* 41:121–145.

Lynch, M., and G. K. Marinov. 2015. "The Bioenergetic Costs of a Gene." *Proceedings of the National Academy of Sciences of the United States of America* 112 (51): 15690–15695.

Mack, T. R., R. Gao, and A. M. Stock. 2009. "Probing the Roles of the Two Different Dimers Mediated by the Receiver Domain of the Response Regulator PhoB." *Journal of Molecular Biology* 389 (2): 349–364.

Marzan, L. W., and K. Shimizu. 2011. "Metabolic Regulation of *Escherichia coli* and Its *phoB* and *phoR* Genes Knockout Mutants under Phosphate and Nitrogen Limitations as well as at Acidic Condition." *Microbial Cell Factories* 10 (1): 39.

McCulloch, W. S., and W. Pitts. 1943. "A Logical Calculus of the Ideas Immanent in Nervous Activity." *Bulletin of Mathematical Biophysics* 5 (4): 115–133.

Nevozhay, D., R. M. Adams, E. Van Itallie, M. R. Bennett, and G. Balázsi. 2012. "Mapping the Environmental Fitness Landscape of a Synthetic Gene Circuit." *PLOS Computational Biology* 8 (4): e1002480.

Qian, H. 2007. "Phosphorylation Energy Hypothesis: Open Chemical Systems and Their Biological Functions." *Annual Review of Physical Chemistry* 58:113–142.

Ray, J. C. J. 2016. "Survival of Phenotypic Information during Cellular Growth Transitions." *ACS Synthetic Biology* 5 (8): 810–816.

Ray, J. C. J., and O. A. Igoshin. 2010. "Adaptable Functionality of Transcriptional Feedback in Bacterial Two-Component Systems." *PLOS Computational Biology* 6 (2): e1000676.

Rowland, M. A., and E. J. Deeds. 2014. "Crosstalk and the Evolution of Specificity in Two-Component Signaling." *Proceedings of the National Academy of Sciences of the United States of America* 111 (15): 5550–5555.

Shin, D., E.-J. Lee, H. Huang, and E. A. Groisman. 2006. "A Positive Feedback Loop Promotes Transcription Surge That Jump-Starts Salmonella Virulence Circuit." *Science* 314 (5805): 1607–1609.

Shinar, G., R. Milo, M. R. Martínez, and U. Alon. 2007. "Input–Output Robustness in Simple Bacterial Signaling Systems." *Proceedings of the National Academy of Sciences of the United States of America* 104 (50): 19931–19935.

Skerker, J. M., B. S. Perchuk, A. Siryaporn, E. A. Lubin, O. Ashenberg, M. Goulian, and M. T. Laub. 2008. "Rewiring the Specificity of Two-Component Signal Transduction Systems." *Cell* 133 (6): 1043–1054.

von Neumann, J. 1956. "Probabilistic Logics and the Synthesis of Reliable Organisms from Unreliable Components." *Automata Studies* 34:43–98.

Zschiedrich, C. P., V. Keidel, and H. Szurmant. 2016. "Molecular Mechanisms of Two-Component Signal Transduction." *Journal of Molecular Biology* 428 (19): 3752–3775.

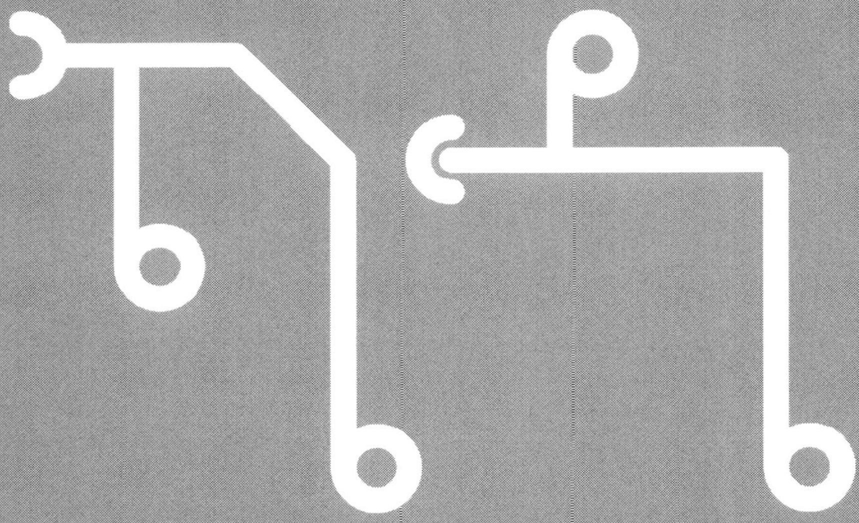

PROTOCELL CYCLES AS THERMODYNAMIC CYCLES

Bernat Corominas-Murtra, Institute of Science and Technology Austria
Harold Fellermann, Newcastle University
Ricard Solé, Pompeu Fabra University

Introduction

The rise of cellular systems pervades the evolution of complex life beyond simple catalytic networks of molecules (Szathmáry and Smith 1995; Schuster 1999; Lane 2015). The physical conditions under which such systems appear are largely unknown. Especially intriguing is that the cycles of growth and self-replication, along with the increase of complexity, seem to be in contradiction to the laws of thermodynamics as we know them. Because real systems display an enormous degree of complexity, researchers have proposed the study of much simpler, even artificial, chemical systems that mimic the properties of living beings and can be much more easily understood. These systems are known as *protocellular systems*. In addition, protocellular entities are the starting point of a biological landscape where a spatially well-defined structure provides a scaffold for efficient, confined chemical reactions to occur. Spatial compartmentalization creates a natural separation between an "inside" world and the external environment. The protocellular agent thus involves the emergence of a dynamical set of boundary conditions allowing the cooperative replication of a whole macromolecular assembly. Along with a reliable compartment (Deamer 2005), a metabolic component needs to be included. Such a metabolic part plays the role of building new materials from available energy and promoting compartment instabilities leading to self-replication. The compartment is usually assumed to be formed in a water-dominated environment and built from the spontaneous assembly of surfactant molecules (Mouritsen 2005). These molecules have a polar nature, with a hydrophilic and a hydrophobic terminal that are attracted or repelled by water molecules, respectively. In contrast with equilibrium aggregates defining closed vesicles as a natural energy minimization process, protocell replication requires an out-of-equilibrium context to allow for a process of growth followed by destabilization and, eventually, splitting into two new daughter cells (Fellermann *et al.* 2015; Zwicker *et al.* 2017).

Despite the modeling efforts and experimental trials, a reliable self-replication cycle as described earlier is still missing (Solé 2016; Serra and Villani 2017). Most examples require necessary extrinsic factors to trigger instability. An obvious obstacle here is the potentially vast parameter space involved (including, for example, reaction rates, temperature, and other physical variables or molecular properties, such as surfactant geometry) and our ignorance about the domains in that space where cell division is physically allowed. One potential path to reach such parameter domains has been recently obtained by means of directed evolution of lipid vesicles (Points *et al.* 2018) in a new form of what Lee Cronin dubbed *inorganic evolution* (Gutierrez *et al.* 2014).

A major obstacle in tackling the creation of these self-replicating agents is the absence of a thermodynamic theory of their cell cycles (Fellermann *et al.* 2015). Important advances have been made over the years regarding the thermodynamics of living processes (Morowitz 1968, 1993;

Deamer 1997), and recently, interesting, fresh approaches relating thermodynamics, information, and the essential biochemical reactions have been proposed, opening the door to a deeper understanding of the essential thermodynamics of biological processes (Smith 2008a, 2008b, 2008c; England 2013). In this chapter, we characterize a driven, out-of-equilibrium chemical system able to satisfy two crucial conditions for living systems: to capture material resources and turn them into building blocks (grow and divide) by the use of externally provided free energy (a metabolic machinery) and to keep its components together and distinguish itself from the environment (Kauffman 2003; Ganti 2003; Solé, Rasmussen, and M. A. Bedau 2007; Rasmussen *et al.* 2008; Ruiz-Mirazo, Briones, and Escosura 2014). We show that the successive cycles of growth and division have the formal properties of a thermodynamic cycle, where all the parameters can be computed from its defining energy landscape.

Characterizing the thermodynamic life cycle sheds light on one of the most crucial questions of theoretical biology: How could primitive systems undergo successive cycles of growth and division in a way that seems to contradict the laws of thermodynamics? We show through a realistic system that the self-replication process is bounded by entropic constraints defined by free energy gradients. The local conditions enabled by the entropic constraints for duplication suggest that the framework under which the spontaneous cycles of self-replication appear could be much more subtle than previously thought, and that an accurate thermodynamic exploration could find unexpected conditions under which they are possible.

Protocellular Building Blocks in Equilibrium

We here consider protocellular systems that are composed of surfactant molecules, also called *lipids*; precursors for such lipids; and water (Bachmann, Luisi, and Lang 1992). In particular, we will work with fatty acids as surfactants and their anhydrites as precursors, although the concepts of our analysis are not bound to this choice.

Like all surfactants, fatty acid molecules are amphiphilic in that they combine a hydrophilic "head" group with a hydrophobic "tail." Fatty acid anhydrites are molecules that can be derived from fatty acids via condensation: two fatty acid molecules join in their head groups by expelling one water molecule. The condensation reaction drastically reduces the hydrophilic property of the head groups, which form the center of the anhydrite molecule rather than its head. As a consequence, anhydrites are typically entirely hydrophobic and show very little solubility in water.

Ternary mixtures of fatty acids, anhydrites, and water are known to self-assemble into a plethora of structures (Evans and Wennerstrøm 1994). Of particular interest for protocellular systems are so-called *emulsions*, where the hydrophobic compounds (anhydrites) form small droplets that are shielded from the aqueous environment by means of a thin layer of amphiphilic (fatty acid) molecules on their surface. Size and stability of the emulsion compartments are dictated partly by the water–oil–surfactant ratios and partly by the molecular geometry of the surfactant molecules, whose conical shape can induce a preferred curvature into the surfactant film, thereby determining a preferred compartment diameter (Evans and Wennerstrøm 1994). We will refer to these emulsion compartments as *protocellular aggregates* or simply *aggregates*.

Our emulsion is conceived as being in a kind of reaction tank connected to a heat reservoir at inverse temperature $\beta = \frac{1}{k_\text{B}T}$ (see fig. 6.1). A *state* σ_n of our system will be described by a 5−tuple:

$$\sigma_n \equiv \sigma_n(L_\text{a}, P_\text{a}, L_\text{b}, P_\text{b}, n),$$

where L_a and P_a are the amount of lipids and precursors forming aggregates, respectively, L_b and P_b are the amount of lipids and precursors in bulk, and n is the number of aggregates. In general, and if no confusion can arise, we will refer to a state as σ_n instead of as $\sigma_n(L_\text{a}, P_\text{a}, L_\text{b}, P_\text{b}, n)$ for notational simplicity. We keep the label subscript "n" accounting for the number of aggregates only for notational convenience. When we introduce time dependence, we write $\sigma_n(t) \equiv \sigma_n(L_\text{a}(t), L_\text{b}(t), P_\text{a}(t), P_\text{b}(t), n(t))$. As we shall see in the next section, each state will have an equilibrium free energy associated $G(\sigma_n(t))$. The probability that the system is in the particular state $\sigma_n(t)$, $p(\sigma_n(t))$ will be given by the equilibrium distribution

$$p(\sigma_n(t)) \propto e^{-\beta G(\sigma_n(t))}. \tag{6.1}$$

Considering that $L_\text{tot} = L_\text{a} + L_\text{b}$ and $P_\text{tot} = P_\text{a} + P_\text{b}$, the *macrostate* $\tilde{\sigma}_n$ will be defined as the 3-tuple

$$\tilde{\sigma}_n \equiv \tilde{\sigma}_n(L_\text{tot}, P_\text{tot}, n).$$

This macrostate can be realized through any state containing $L_\text{tot}, P_\text{tot}$ and n protocellular aggregates.

FREE ENERGIES

The equilibrium thermodynamic landscape of our system will be given by the Gibbs free energy of the state σ_n, $G(\sigma_n)$. The Gibbs free energy will be always defined over states of the system. The complex nature of these types of emulsions results in a free energy functional with several blocks, which we construct step by step. First, we focus on the free energy contribution of a single protocellular aggregate, containing L lipids and P precursors, G_a:

$$G_a(L, P) = \Delta\mu_L L + \Delta\mu_P P + G_{geo},$$

where $\Delta\mu_L$ and $\Delta\mu_P$ are the changes in chemical potential when moving precursors and lipids from bulk into the aggregate and G_{geo} is a geometric term expressing shape and surface contributions of the aggregate. This geometric term accounts for the membrane properties of the system and is computed according to the existence of a minimum energy configuration or *perfect* protocellular aggregate, which can be directly computed as the optimal packing from the knowledge to the sizes and geometries of the precursor molecules. The geometrical term thus reads

$$G_{geo} = \gamma A + \frac{\alpha}{A} + \kappa \oint_A (H - H_0)^2 dA,$$

where γ is the surface tension, α is the compressibility coefficient, and κ is the elastic bending modulus of the lipid membrane. The integral is the second-order expansion of the contribution of the Helfrich Hamiltonian to the overall free energy, with H being the curvature of the membrane of the current aggregate—as a function of some coordinates parameterizing the membrane surface—and H_0 being the curvature of the perfect aggregate. The integral is computed over the whole area of the membrane, A.

Once we have properly characterized the free energies of a single aggregate, we proceed to construct the free energy of the whole state σ_n. In this computation, one assumes that all the n protocellular aggregates of the emulsion display approximately the same size. Accounting properly for the degeneracy of states, and thus for translational and configurational entropies, the free energy of a *system* in the state $\sigma_n = \sigma_n(L_a, P_a, L_b, P_b, n)$ becomes

$$G(\sigma_n) = \mu_L^\circ L_{\text{tot}} + \mu_P^\circ P_{\text{tot}} + n G_a\left(\frac{L_a}{n}, \frac{P_a}{n}\right) - TS(\sigma_n),$$

with the standard chemical potentials μ_L° and μ_P° of lipids and precursors, respectively, and

$$S(\sigma_n) = n k_B \log \frac{V}{V_P/e} + k_B \log\left[\binom{L_{\text{tot}}}{L_a}\binom{P_{\text{tot}}}{P_a}\right]$$

being the translational and configurational entropy of the system at state σ_n, where V describes the system volume per droplet and V_P—the molecular volume of the precursor—has been chosen as typical volume unit whose sole purpose is to render the argument to the logarithm dimensionless.

Stochastic movement of molecules from aggregate to bulk and back, as well as fusion/fission of aggregates in the equilibrium regime, follow detailed balance conditions according to the corresponding kinetic ratios. Specifically, the transition rates between states are given by

$$\sigma_n(L_a, P_a, L_b, P_b, n) \xrightarrow{k_L^+ L_b} \sigma_n(L_a + 1, P_a, L_b - 1, P_b, n),$$
$$\sigma_n(L_a, P_a, L_b, P_b, n) \xrightarrow{k_L^- L_a} \sigma_n(L_a - 1, P_a, L_b + 1, P_b, n),$$
$$\sigma_n(L_a, P_a, L_b, P_b, n) \xrightarrow{k_P^+ P_b} \sigma_n(L_a, P_a + 1, L_b, P_b - 1, n),$$
$$\sigma_n(L_a, P_a, L_b, P_b, n) \xrightarrow{k_P^+ P_a} \sigma_n(L_a, P_a - 1, L_b, P_b + 1, n),$$

and, according to the detailed balance condition, kinetic constants are related through a Boltzmann factor:

$$k_L^- = k_L^+ e^{\beta[G(\sigma_n(L_a-1, P_a, L_b+1, P_b, n)) - G(\sigma_n(L_a, P_a, L_b, P_b, n))]}$$
$$k_P^- = k_P^+ e^{\beta[G(\sigma_n(L_a, P_a-1, L_b, P_b+1, n)) - G(\sigma_n(L_a, P_a, L_b, P_b, n))]}.$$

As the basis of duplication, the process of fission/fusion is of special interest to us. It satisfies the transition rates

$$\sigma_n(L_a, P_a, L_b, P_b, n) \xrightarrow{k_n^+ n} \sigma_{n+1}(L_a, P_a, L_b, P_b, n+1)$$
$$\sigma_n(L_a, P_a, L_b, P_b, n) \xrightarrow{k_n^- n} \sigma_{n-1}(L_a, P_a, L_b, P_b, n-1),$$

where the kinetic constants relate as

$$k_n^- = k_n^+ e^{\beta[G(\sigma_n(L_a,P_a,L_b,P_b,n)) - G(\sigma_{n-1}(L_a,P_a,L_b,P_b,n-1))]}. \quad (6.2)$$

Protocellular Division Cycles

So far, we have described a system in equilibrium. We thus need to introduce a gradient force that can presumably be tuned to drive the system toward a nonequilibrium, stationary state in which processes of protocell aggregate duplication spontaneously occur. This is achieved by introducing a metabolic process into the system. Such a minimal metabolism is realized by a chemical conversion of precursors into lipids, that is, into surface molecules. For fatty acid anhydrites, this corresponds to the spontaneous hydrolysis of the molecule into its two constituting fatty acid molecules. The net result of this reaction is a slow and continuous change of the boundary conditions of the whole system, determined by L_{tot} and P_{tot}, which translates into a change of surface to volume within the aggregates, until it eventually triggers a division event. Zwicker et al. (2017) recently demonstrated the feasibility of such a division process by computer simulations of a detailed continuous surface model.

The irreversible metabolic reaction that transforms precursors into lipids is then given as

$$\text{anhydrite} + H_2O \xrightarrow{k_m} 2\,\text{surfactant},$$

where the anhydride precursor can be regarded as a food molecule whose consumption results in the creation of surfactant building blocks. The metabolic reaction continuously converts the hydrophobic volume of the aggregates into new lipids. As the nutrient is hydrophobic but depends on the availability of water, we expect the reaction to take place at the aggregate surface. In doing so, the membranes of the protocellular aggregates are expected to undergo a process of increasing frustration that may eventually lead to a duplication event, if that is thermodynamically favorable. The metabolic reaction that converts supplied precursors into additional surfactants is assumed to be essentially irreversible and drives the system out of equilibrium by slowly changing the boundary condition of the system, and thus the macrostate, as

$$L_{\text{tot}}, P_{\text{tot}} \xrightarrow{k_{\text{m}} P} L_{\text{tot}}+2, P_{\text{tot}}-1,$$

where we use the earlier defined convention $L_{\text{tot}} \equiv L_a + L_b$ and $P_{\text{tot}} \equiv P_a + P_b$, thereby leaving open the potential amount of material in bulk or forming aggregates. The metabolism is assumed to be much slower than the other equilibration processes. Thus $k_P^+, k_L^+, k_n^+ \gg k_m$.

To avoid the consumption of all precursor molecules, we must create a constant inflow of them, as well as a mechanism that eventually eliminates the wasted material, to keep the concentrations approximately constant throughout the process. In figure 6.1, we detail this experimental setup. We provide a constant inflow of anhydrite precursors to be used as building blocks of new protocellullar aggregates. Let us assume that, at $t = 0$, our system is a reactor in a state $\sigma_n(0) = \sigma_n(L_a(0), P_a(0), L_b(0), P_b(0), n(0))$, assumed to be the most probable state, according to equation (6.1). Our system is coupled to a huge molecule reservoir and to a heat bath at inverse temperature β. After a time unit much larger than the equilibration process described by the transition rates, we feed

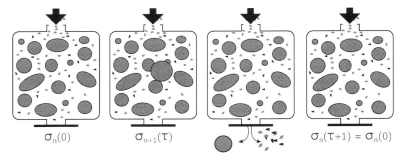

Figure 6.1. Experimental setup: at $t = 0$, a reactor connected to a particle and heat reservoir contains an emulsion with protocell aggregates composed of internal precursor molecules surrounded by lipids. The reactor is constantly fed with precursors that incorporate into existing aggregates. The spontaneous metabolic reaction converts precursors into lipids, thus changing the aggregates' surface-to-volume ratio until, at some time τ, they become unstable and divide. An outflow at the bottom removes newly created aggregates and waste, leaving the total aggregate density constant.

this reactor from the top with ΔP_{feed} precursors and water, such that it reaches a state $\sigma'_n(0)$, in which $L'(0), P'(0) = L(0), P(0) + \Delta P_{\text{feed}}$. Thanks to the metabolic turnover, the reaction will convert a fraction ΔP of the precursors into $\Delta L = 2\Delta P$ new lipid surfactants. This will drive the system into a new state, $\sigma_n(1) = (L_a(1), P_a(1), L_b(1), P_b(1), n(1))$, in which $L(1), P(1) = L'(0) + \Delta L, P'(0) - \Delta P$ lipids and precursors. This feeding strategy is repeated sequentially every time unit. If, after τ_{cycle} time units, a new aggregate is formed, $n(\tau_{\text{cycle}}) = n(0) + 1$—that is, there is a *division* event—we compensate the inflow of precursors by an outflow that removes from the system material proportional to one aggregate plus the proportional volume of aqueous solution. If feeding and metabolism are tuned correctly, extraction of newly created droplets will compensate the inflow of precursors, and the initial state $\sigma_n(0)$ will be recovered (see fig. 6.1). At the microscopic level, the creation of new lipids will introduce a perturbation to the surface of existing aggregates, whose size will grow until, in some aggregate, the frustration due to the geometric term will drive it into an unstable state and, presumably, break it into two smaller, more stable aggregates. Aggregates in this cycle are thus able to create new aggregates by division, and we say that aggregates exposed to these conditions *self-replicate* (see fig. 6.1).

Considering that equilibration processes run on fast time scales, the time evolution of the driven system described earlier can be considered as a succession of equilibrium states whose boundary conditions, given by L_{tot} and P_{tot}, change as long as we feed the system with precursors, which, in turn, are converted by the metabolism to produce new surfactants. We therefore consider that, at time $t = 0$, the state $\sigma_n(0) = \sigma_n(L_a(0), P_a(0), L_b(0), P_b(0), n(0))$ is *perfectly* in equilibrium. By *perfectly* in equilibrium we mean that the aggregates display, on average, a *perfect* surface coverage, resulting in a minimum energy contribution per aggregate, because in the perfect covering scenario there is no geometric contribution to the free energy. As long as the feeding process acts over the system, we will observe a trajectory of the system through different states

$$\sigma_n(0), \sigma_n(1), \ldots, \sigma_n(t), \ldots,$$

whose probability to appear follows equation (6.1), according to the $L_{\text{tot}}, P_{\text{tot}}$ present at each time step. The free energies increase in time, due to the increasing frustration in the surface of the aggregates, namely,

$$G(\sigma_n(0)) < G(\sigma_n(1)) < \cdots < G(\sigma_n(t)) < \ldots$$

This increasing trend is aborted when a duplication event occurs and the system is reset to the initial state. The conditions under which such a duplication event occurs are explored in the next section.

CONDITIONS FOR DUPLICATION: ENTROPIC CONSIDERATIONS

The constant creation of surface through the metabolic reaction makes the duplication event expected. To study the problem, we focus on the following situation: consider that, if the equilibrium configuration at time t contains n aggregates, at $t+1$, our system is *prepared* in a macrostate containing n cellular aggregates and a given amount of $L_{\text{tot}}, P_{\text{tot}}$, which have changed from t to $t+1$ due to the presence of the constant feeding and the metabolism. *Under which circumstances can one expect a spontaneous duplication event?* Based on the probability of transition between states inferred from equation (6.2), we derive an entropic inequality for the spontaneous protocell duplication in which the free energy gradients play a critical role. Such entropic inequality is of the same nature as the one derived in England (2013) based on the result on reversibility of trajectories in the phase space given by Crooks (1999).

To tackle the problem, it will be useful to introduce some notation. Assume that there are $\sigma_n^1, \ldots, \sigma_n^{M_n}$ states compatible with the macrostate $\tilde{\sigma}_n$, that is, states that can contain n aggregates with nonzero probability with a fixed number of $L_{\text{tot}}, P_{\text{tot}}$, and $\sigma_{n+1}^1, \ldots, \sigma_{n+1}^{M_{n+1}}$ states compatible with the macrostate $\tilde{\sigma}_{n+1}$ with the same $L_{\text{tot}}, P_{\text{tot}}$. We will refer to the states compatible with $\tilde{\sigma}_n$ with the letter k and the ones compatible with $\tilde{\sigma}_{n+1}$ with the letter j. Time dependence is introduced as follows: by writing, for example, $\tilde{\sigma}_n(t)$, we mean that *the system is in the macrostate $\tilde{\sigma}_n$ at time t*. The probability $p(k \to j)$ will represent the probability that the system jumps from macrostate $\tilde{\sigma}_n(t)$ to macrostate $\tilde{\sigma}_{n+1}(t+1)$ through the particular path starting at σ_n^k and ending at σ_{n+1}^j, both containing the same[1] $L_{\text{tot}}, P_{\text{tot}}$. The

---

[1] Notice that, in a little abuse of notation, we write the time $t+1$ to refer to the time just before the metabolism changes the boundary conditions of the

probability that macrostate $\tilde{\sigma}_n(t)$ is realized through a particular state σ_n^k will be written as $p(k|\tilde{\sigma}_n(t))$. The probability of jumping from macrostate $\tilde{\sigma}_n(t)$ to macrostate $\tilde{\sigma}_{n+1}(t+1))$ will therefore be given by

$$p(\tilde{\sigma}_n(t) \to \tilde{\sigma}_{n+1}(t+1)) = \sum_{j \leq M_{n+1}} \sum_{k \leq M_n} p(k|\tilde{\sigma}_n(t)) p(k \to j).$$

We can go further, because the detailed balance condition for fusion/fission events provided in equation (6.2) can be used to approach the quotient $p(j \to k)/p(k \to j)$:

$$\frac{p(j \to k)}{p(k \to j)} \approx \frac{k_n^-}{k_n^+} = e^{\beta \Delta_{jk} G} \; ; \quad \Delta_{jk} G \equiv G(\sigma_{n+1}^j) - G(\sigma_n^k).$$

With the preceding equation, one can write the reverse probability $p(\tilde{\sigma}_{n+1}(t) \to \tilde{\sigma}_n(t+1))$ as

$$p(\tilde{\sigma}_{n+1}(t) \to \tilde{\sigma}_n(t+1)) =$$
$$\sum_{k \leq M_n} \sum_{j \leq M_{n+1}} p(k|\tilde{\sigma}_n(t)) p(k \to j) \frac{e^{\beta \Delta_{jk} G}}{e^{\ln \frac{p(k|\tilde{\sigma}_n(t))}{p(j|\tilde{\sigma}_{n+1}(t))}}}, \quad (6.3)$$

which leads to

$$\frac{p(\tilde{\sigma}_{n+1}(t) \to \tilde{\sigma}_n(t+1))}{p(\tilde{\sigma}_n(t) \to \tilde{\sigma}_{n+1}(t+1))} =$$
$$\frac{\sum_{k \leq M_n} \sum_{j \leq M_{n+1}} p(k|\tilde{\sigma}_n(t)) p(k \to j) \frac{e^{\beta \Delta_{jk} G}}{e^{\ln \frac{p(k|\tilde{\sigma}_n(t))}{p(j|\tilde{\sigma}_{n+1}(t))}}}}{\sum_{k \leq M_n} \sum_{j \leq M_{n+1}} p(k|\tilde{\sigma}_n(t)) p(k \to j)}. \quad (6.4)$$

Rearranging terms, one finds that

$$\frac{p(\tilde{\sigma}_{n+1}(t) \to \tilde{\sigma}_n(t+1))}{p(\tilde{\sigma}_n(t) \to \tilde{\sigma}_{n+1}(t+1))} = \left\langle \frac{e^{\beta \Delta_{jk} G}}{e^{\ln \frac{p(k|\tilde{\sigma}_n(t))}{p(j|\tilde{\sigma}_{n+1}(t))}}} \right\rangle_{\tilde{\sigma}_n(t) \to \tilde{\sigma}_{n+1}(t+1)},$$

where the brackets $\langle \ldots \rangle_{\tilde{\sigma}_n(t) \to \tilde{\sigma}_{n+1}(t+1)}$ denote average over all the microscopic paths $\sigma_n^k \to \sigma_{n+1}^j$ from macrostate $\tilde{\sigma}_n(t)$ to

---

system. Strictly speaking, if we consider that the change in the boundary conditions happens at $t+1$, then both macrostates cannot share the same $L_{\text{tot}}, P_{\text{tot}}$. With this notation, we refer to the time $t+1-\epsilon$, with ϵ arbitrarily small.

macrostate $\tilde{\sigma}_{n+1}(t+1)$. Now, following a reasoning strictly analogous to the one found in England (2013), we arrive at the following inequality:[2]

$$\frac{p(\tilde{\sigma}_{n+1}(t) \to \tilde{\sigma}_n(t+1))}{p(\tilde{\sigma}_n(t) \to \tilde{\sigma}_{n+1}(t+1))} \geq e^{\beta \langle \Delta G \rangle_{\tilde{\sigma}_n(t) \to \tilde{\sigma}_{n+1}(t+1)} - \Delta H(t)}, \quad (6.5)$$

with $\Delta H(t) \equiv H(\tilde{\sigma}_{n+1}(t+1)) - H(\tilde{\sigma}_n(t))$, the increase of the Shannon entropy, defined as

$$H(\tilde{\sigma}_n(t)) = -\sum_{k \leq M_n} p(k|\tilde{\sigma}_n(t)) \ln p(k|\tilde{\sigma}_n(t)).$$

Equation (6.5) explicitly relates the imbalance between free energies to the statistical entropies as a key ingredient to understand duplication processes in cell-like systems. Now we observe that, by construction, at $t=0$, $p(\tilde{\sigma}_{n+1}(0) \to \tilde{\sigma}_n(0)) > p(\tilde{\sigma}_n(0) \to \tilde{\sigma}_{n+1}(0))$, which implies that

$$e^{\beta \langle \Delta G \rangle_{\tilde{\sigma}_n(0) \to \tilde{\sigma}_{n+1}(1)} - \Delta H(0)} > 1$$
$$\Rightarrow \beta \langle \Delta G \rangle_{\tilde{\sigma}_n(0) \to \tilde{\sigma}_{n+1}(1)} - \Delta H(0) > 0. \quad (6.6)$$

However, the actions of feeding and metabolic conversion will eventually create the conditions that reverse this inequality.

---

[2] Let A and B be two different macrostates, and let x and y be states compatible with A and B, respectively. Let $\frac{p(B \to A)}{p(A \to B)} = \left\langle \frac{e^{\beta \Delta G}}{e^{\ln \frac{p(x)}{p(y)}}} \right\rangle_{A \to B}$.

Accordingly, $1 = \left\langle e^{\beta \Delta G - \ln \frac{p(B \to A)}{p(A \to B)} - \ln \frac{p(x)}{p(y)}} \right\rangle_{A \to B}$. The Taylor expansion of the exponential ensures that $e^x \geq 1 + x$, so, if we know that $\langle e^x \rangle = 1$, then $1 + \langle x \rangle \leq 1$, so $\langle x \rangle \leq 0$:

$$\beta \langle \Delta G \rangle_{A \to B} - \ln \frac{p(B \to A)}{p(A \to B)} - \left\langle \ln \frac{p(x)}{p(y)} \right\rangle_{A \to B} \leq 0.$$

Finally, the equality $\left\langle \ln \frac{p(x)}{p(y)} \right\rangle_{A \to B} = H(B) - H(A)$ follows directly from basic probability reasoning. Therefore, rearranging that and the preceding equation, one is led to equation (6.5).

Whenever this inequality is reversed, a duplication cycle will be completed. We can therefore bound τ_{cycle} as

$$\tau_{\text{cycle}} \geq \min_{t} \left\{ \beta \langle \Delta G \rangle_{\tilde{\sigma}_n(t) \to \tilde{\sigma}_{n+1}(t+1)} - \Delta H(t) \leq 0 \right\}.$$

We observe that the preceding inequalities do not require equation (6.1) to hold. Therefore they are valid even in cases in which slow relaxation timescales prevent the system from reaching the equilibrium before changing the boundary conditions. In the case that equilibration dynamics is fast enough, equation (6.1) holds and equation (6.3) becomes an equality, defining an upper bound on the speed of the duplication process.

We reserve a final comment on the efficiency of the cycle. Assuming that at time τ_{cycle} the system is perfectly reset to $\sigma_n(0)$, that is, $\sigma_n(\tau_{\text{cycle}} + 1) = \sigma_n(0)$, one expects that the newly created aggregate has a free energy G_a as

$$G_a(0) = G_a\left(\frac{L_a(0)}{n}, \frac{P_a(0)}{n}\right),$$

that is, the expected energy per aggregate at $t = 0$. Since the product of the cycle is such a protocell aggregate, one can compute the amount of energy that we had to put into the system to produce it. In our process, the chemical potential of the precursors provided to the system is equivalent to the total energy provided to the system. Thus, if ΔP_{feed} is the precursors provided to the system per unit time,

$$E_{\text{cycle}} = \tau_{\text{cycle}} \cdot (\mu_P^\circ \Delta P_{\text{feed}}).$$

One can therefore define the efficiency of the cycle as η_{cycle}:

$$\eta_{\text{cycle}} = \frac{G_a(0)}{E_{\text{cycle}}}.$$

If $\eta_{\text{cycle}} \leq 1$, the amount of energy that must be invested to trigger a protocell cycle is bounded from below as

$$\tau_{\text{cycle}} \cdot (\mu_P^\circ \Delta P_{\text{feed}}) \geq G_a\left(\frac{L_a(0)}{n}, \frac{P_a(0)}{n}\right).$$

Discussion

In this study, we demonstrate that two of the core properties of living systems can be achieved by ternary emulsion systems.

Specifically, artificial *living* systems have the ability (1) to capture material resources and turn them into building blocks (grow and divide) by the use of externally provided free energy (a metabolic machinery) and (2) to distinguish itself from its environment via a system boundary that keeps its components, including the metabolic reaction centers, together. A system able to differentiate from the environment is therefore both driven out of equilibrium and expected to undergo a thermodynamic cycle every time it replicates. Both (1) and (2), absent in standard physical processes, are crucial properties of life (Kauffman 2003; Ganti 2003; Solé, Rasmussen, and M. A. Bedau 2007; Rasmussen *et al.* 2008; Ruiz-Mirazo, Briones, and Escosura 2014). Crucially, our system is not able to produce inheritable information. Nevertheless, we have shown that this is not a key requirement for self-replication, because free energy gradients in a driven system can generate the conditions for a sustainable life cycle to occur. Particular attention must be paid to equation (6.5), which identifies a relationship between the free energy of the self-replicating system and the statistical entropy. This shows explicitly, with free energies computed from realistic chemical systems, that in replicating systems the process of duplication can be much more subtle than expected. By defining and studying an entirely artificial, and radically simplified, model system for living organisms, one can circumvent the tremendous complexity of real living entities and focus on particular features of life. In our case, we demonstrate that a minimal process mimicking two key properties of living entities has all the properties of a thermodynamic cycle. Further work should explore how the understanding of thermodynamic constraints in the purely physical process of replication imposes conditions on the emergence of inheritable forms information. ✌

ACKNOWLEDGMENTS

R. S. acknowledges support by the Botín Foundation, by Banco Santander through its Santander Universities Global Division, by a MINECO grant FIS2016-77447-R, and by the Santa Fe Institute. B. C.-M. thanks the Section for the Science of Complex Systems of the Medical University of Vienna for supporting the stay at the workshop "Thermodynamics of Computation in Chemical and Biological Systems" held by the Santa Fe Institute in August 2017. B. C.-M. also thanks Artemy Kolchinsky for his helpful comments. B. C.-M., H. F., and R. S. thank Steen Rasmussen, John Hjort Ipsen, and Per Lyngs Hansen for their helpful hints during the beginnings of this work.

REFERENCES

Bachmann, P. A., P. L. Luisi, and J. Lang. 1992. "Autocatalytic Self-Replicating Micelles as Models for Prebiotic Structures." *Nature* 357:57–59.

Crooks, G. E. 1999. "Entropy Production Fluctation Theorem and the Nonequilibrium Work Relation for Free Energy Differences." *Physical Reviews E* 60 (3): 2721–2726.

Deamer, D. 1997. "The First Living Systems: A Bioenergetic Perspective." *Microbiology and Molecular Biology Reviews* 61:239.

———. 2005. "A Giant Step towards Artificial Life?" *Trends in Biotechnology* 23:336–338.

England, J. 2013. "Statistical Physics of Self-Replication." *Journal of Chemical Physics* 139:121923.

Evans, D. Fennell, and H. Wennerstrøm. 1994. *The Colloidal Domain: Where Physics, Chemistry, Biology, and Technology Meet.* Berlin: VCH.

Fellermann, H., B. Corominas-Murtra, P. L. Hansen, J. H. Ipsen, R. V. Solé, and S. Rasmussen. 2015. "Non-Equilibrium Thermodynamics of Self-Replication Protocells." arXiv 1503.04683.

Ganti, T. 2003. *The Principles of Life.* Oxford: Oxford University Press.

Gutierrez, J. M. P., T. Hinkley, J. W. Taylor, K. Yanev, and L. Cronin. 2014. "Evolution of Oil Droplets in a Chemorobotic Platform." *Nature Communications* 5:5571.

Kauffman, S. 2003. "Molecular Autonomous Agents." *Philosophical Transactions of the Royal Society of London, Series A* 361:1089–1099.

Lane, N. 2015. *The Vital Question: Energy, Evolution, and the Origins of Complex Life.* New York: W. W. Norton.

Morowitz, H. J. 1968. *Energy Flow in Biology.* New York: Academic Press.

———. 1993. *Beginnings of Cellular Life: Metabolism Recapitulates Biogenesis.* New Haven, CT: Yale University Press.

Mouritsen, O. 2005. *Life—as a Matter of Fat: The Emerging Science of Lipidomics.* Berlin: Springer.

Points, L. J., J. W. Taylor, J. Grizou, K. Donkers, and L. Cronin. 2018. "Artificial Intelligence Exploration of Unstable Protocells Leads to Predictable Properties and Discovery of Collective Behavior." *Proceedings of the National Academy of Sciences of the United States of America* 115:885–890.

Rasmussen, S., M. A. Bedau, L. Chen, D. Deamer, D. Krakauer, N. Packard, and P. Stadler, eds. 2008. *Proto-cells: Bridging Nonliving and Living Matter.* Cambridge, MA: MIT Press.

Ruiz-Mirazo, K., C. Briones, and A. de la Escosura. 2014. "Prebiotic Systems Chemistry: New Perspectives for the Origins of Life." *Chemical Reviews* 114 (1): 285–366.

Schuster, P. 1999. "How Does Complexity Arise in Evolution?" *Complexity* 2:22–30.

Serra, R., and M. Villani. 2017. *Modelling Protocells: The Emergent Synchronization of Reproduction and Molecular Replication.* Berlin: Springer.

Smith, E. 2008a. "Thermodynamics of Natural Selection I: Energy Flow and the Limits on Organization." *Journal of Theoretical Biology* 252:185–197.

———. 2008b. "Thermodynamics of Natural Selection II: Chemical Carnot Cycles." *Journal of Theoretical Biology* 252:198–212.

Chapter 6: Protocell Cycles as Thermodynamic Cycles

———. 2008c. "Thermodynamics of Natural Selection III: Landauer's Principle in Computation and Chemistry." *Journal of Theoretical Biology* 252:213–220.

Solé, R. V. 2016. "Synthetic Transitions: Towards a New Synthesis." *Philosophical Transactions of the Royal Society of London, Series B* 371 (1701): 20150438.

Solé, R. V., S. Rasmussen, and eds. M. A. Bedau. 2007. "Towards the Artificial Cell." *Philosophical Transactions of the Royal Society of London, Series B* 362.

Szathmáry, E., and J. M. Smith. 1995. *The Major Transitions in Evolution.* London: W. H. Freeman Spektrum.

Zwicker, D., R. Seyboldt, C. A. Weber, A. A. Hyman, and F. Jülicher. 2017. "Growth and Division of Active Droplets Provides a Model for Protocells." *Nature Physics* 13:408–413.

HOW AND WHAT DOES A BIOLOGICAL SYSTEM COMPUTE?

Sonja J. Prohaska, University of Leipzig
Peter F. Stadler, University of Leipzig
Manfred Laubichler, Arizona State University

Introduction

Information processing, or computation in a general sense, is a defining feature of living systems at all spatial and temporal scales. In a fundamental sense, all the structure and functioning of living systems is governed by genetic information that needs to be processed and executed at molecular, cellular, and organismal scales. Biological systems individually process information to maintain an inner homeostatic equilibrium and to respond to environmental clues and stimuli with often highly complex and coordinated behaviors. On the most complex scale, human culture, the evolutionary product of one biological species, has developed a whole "information economy." All these examples raise the question: how and what does a biological system compute?

In its most general form, a computational system is nothing more than a dynamical system defined on a system of "computational states" X. It may be understood in terms of a *flow* $\varphi : T \times X \to X$ describing the system's evolution along a time axis T, satisfying (1) $\varphi(0, x) = x$ and (2) $\varphi(t - s, x) = \varphi(t, \varphi(s, x))$. Most, but not all, models of computation assume a discrete time axis, $T = \mathbb{Z}$, which can be specified by a transition rule $\rho : X \to X$ that is defined as a single step, that is, $\rho(x) = \varphi(1,x)$. Most mathematical models of computation (Fernández 2009) are (discrete) dynamical systems:

1. State models, such as finite state machines, automata, and Turing machines, and parallel models, such as parallel random-access machines

2. Functional models, such as lambda calculus, recursive functions, and cellular automata, and rewriting systems, including graph grammars or membrane computing

3. Logical models, such as logic programming

4. Concurrent models, such as process calculi, Petri nets, and interaction nets

It seems natural to say that a physical system performs a given computation if its dynamics can be mapped to the state transitions of the computation. While this formal setup seems well suited to study the relationships of computation and (nonequilibrium) thermodynamics (Wolpert 2016), it appears to be too general to be of much interest in biology. After all, a cell, an organism, or even the evolution of the biosphere as a whole is a dynamical system. The issue, therefore, is not whether one can view (the dynamics of) a biological system as computation, as this is trivially true in the abstract sense outlined here.

Information processing or computation in the context of biological systems usually refers to particular processes that (1) conform, at least approximately, to an informal notion of computation that implicitly presupposes a particular, specific model of computation and that (2) produce output that is directly relevant for a cell's functioning or an organism's survival.

Chapter 7: How and What Does a Biological System Compute?

In the following section, we briefly outline, at a rather abstract level, some of the "computational" subsystems to highlight some common features related to our contention that biological systems are systems of "embodied computation." We deliberately set aside brains (neuronal systems) as well as eco- and social systems and focus on cells and organisms.

Biological Systems That "Compute"

REPLICATION, TRANSCRIPTION, AND TRANSLATION The subsystem that is probably most closely related to a "computer" is the ubiquitous *information metabolism*, comprising the replication of the DNA-based genome, the transcription of DNA into RNA, the maturation of RNA (by means of splicing), and the translation of the mature messenger RNA into a protein sequence. Replication and transcription can be considered as copying. Splicing deletes certain intervals (introns) from the original message. Translation, finally, implements a redundant map: the well-known genetic code.

Although the logic of this system is very simple, its practical implementation requires a complex machinery comprising several large macromolecular complexes that perform individual parts of replication, transcription, splicing, and translation with very high fidelity. As a mathematical model, very simple finite state automata (or, equivalently, regular expressions) suffice to capture the transition rules. Note that this model refers to the simple act of copying and translation and does not yet involve the more complex processes of regulation that govern this machinery in real-world cases—more on that later.

The machinery implementing replication, transcription, and translation can process arbitrary inputs, that is, arbitrary sequence information. As a consequence, it is possible to rewrite the genetic information either in the course of natural evolution or in the context of genetic engineering (Ma, Perli, and Lu 2016). This is possible because the rules underlying transcription and translation are almost universal; the information encoded in the genome is interpreted in the same way in all species. Small variations of the genetic code exist only in very special circumstances, in particular, for the tiny genomes of animal mitochondria.

REGULATION

Regulation (of gene expression) is commonly perceived as one of the key features of living systems and one of the obvious computational activities of a living organism. A computational account of regulation can found in Krakauer *et al.* (2016). Jacob and Monod (1961) already noted that DNA in itself does not provide the instructions necessary and sufficient for the synthesis of individual proteins. In addition, mechanisms are at work that promote or repress, that is, regulate transcription, RNA processing, and translation. Following Davidson and Erwin (2006) and Davidson (2006), gene regulatory networks (GRNs) can be described as wiring diagrams that represent logical functions, the execution of which captures the time evolution of protein concentrations. GRNs thus have been proposed to describe both the program of development and the rules by which cells react and adapt their internal states in response to environmental stimuli.

The implementation of GRNs in a cell is exceedingly complicated and relies on a plethora of different molecular mechanisms. We just very briefly mention a few. Transcription factors (TFs) are proteins that recognize particular sequence elements on the DNA. Each gene on the DNA is equipped with its own set of promoter and enhancer elements that determine which combination(s) of input TFs cause(s) the expression of the gene. External cues are processed by means of signaling cascades that consist of proteins capable of setting or removing specific phosphorylation marks on narrowly specified targets. Some of their targets, depending on their phosphorylation state, act as transcription factors. Still other molecules, often small or large RNAs that are not translated into proteins but act by complementary base pairing to their mRNA targets, determine the splicing or may shut off translation even if an mRNA has been transcribed.

GRNs are determined by the genes and regulatory elements, such as transcription factor binding sites, encoded in the genome. These molecular details determine the specificities and strengths of interactions and thus dynamical behavior of the system as a whole. There are at least two useful, largely complementary descriptions of GRNs. They may be viewed in terms of molecular concentration variables, described by the differential-equation models prevalent in systems biology (DiStefano 2013). This level of description provides a natural link to classical control theory. In this picture, the computational effort of the GRN is to ensure that expression profiles are close to one of a limited number of ideal concentration vectors that depend on the cell's environment and history. In a more coarsely grained approximation, expression values are just ON and OFF, leading to modeling the GRN as a Boolean network. Its attractors, for instance, have been interpreted as particular cell types (Kauffman 1969, 1993; Istrail, De-Leon, and Davidson 2007; Peter, Faure, and Davidson 2012; Peter and Davidson 2016, 2017). In a more modern account, cell types are characterized by core circuit GRNs that dominate their development (Arendt *et al.* 2016).

METABOLISM

The fluxes of chemical compounds—from the uptake of food; the synthesis of DNA, RNA, and proteins; and the excretion of waste products—of course also form a dynamical system. It is much harder to see, however, how a cell's metabolism is meaningfully interpreted as computation. It does have some obvious "computational" aspects, though. In metabolic networks the fluxes are not only regulated through the expression of the enzymes necessary to catalyze the individual chemical reactions, but also by means of direct feedback mediated by the metabolites themselves. Most prominently, many enzymes are inhibited by small molecules that are produced a step further down on a particular metabolic pathway. This implements control circuits that act at the time scales of chemical reactions and thus much faster than GRNs. Metabolic production also provides cues to the GRNs, because small molecules may also inhibit or activate proteins or RNAs that are part of the GRN.

It may be tempting to think of metabolism as "lower down," that is, as part of the substrate on which the GRN is built (in the same way that the CPU is "lower down" than the operating system, which in turn is "lower down" than a programming language). The intricate interdependencies of evolved biological systems, however, do not allow for such a neat hierarchy. Rather, what we find are the products of interwoven coevolutionary processes, whereby GRNs provide regulatory input into metabolism and the products of these metabolic reactions are an essential part of the execution of GRNs. Only if one ignores the actual mechanism can one think of a GRN—as it unfolds in four dimensions—as an appropriate coarse graining of specific biological processes, such as developmental differentiation (see later).

There is, however, another sense in which metabolism can be viewed as computational. Chemical reactions themselves can be thought of as rewriting the molecular structure of the educts into the molecular structure of the products. This aspect is discussed in some detail in Chapter 4 of this book.

EPIGENETICS

Several molecular mechanisms are responsible for *epigenetic* regulation that involves the establishment, change, and maintenance of a genome-wide gene expression state and its faithful transfer to the daughter cells during replication. They are tightly interfaced with the GRNs and can even be thought of as part of the GRNs, although they differ from other modes of gene regulation in that they have a persistent memory. In contrast to the genetic layer, though, this memory can not only be read during the lifetime of a cell or an organism but also extensively rewritten.

DNA methylation is the epigenetic mechanism closest to the genome itself (Smith and Meissner 2013). In vertebrates, DNA methyltransferases (DNMTs) attach a small chemical compound, a methyl-group, to a cytosine base (C) of DNA. This does not alter the way that C is read by the replication and transcription machinery but influences gene expression. DNA methylation usually silences genomic elements that would have an undesirable effect when getting expressed. Over the lifetime of an organism, the DNA methylation level changes dramatically. It is very low at early embryonic stages of development, when cells are totipotent, that is, able to differentiate into any type of cell and high in terminally differentiated cells. With cell fate decisions and lineage commitments along the way, more genes expressed earlier are shut off in a lineage-dependent manner. Although DNA methylation may not be responsible for the initiation of gene silencing, its major role is proposed to be permanent silencing of genes. As a long-lived silencing mark, DNA methylation serves as a memory and is involved in the maintenance of cell type identity. Most of this accumulated information is erased in gametogenesis and early embryogenesis, presumably to establish a gamete-specific methylation pattern and a state of totipotency for the development of the next generation, respectively.

A second layer of epigenetic regulation consists of the chemical modification of histones. Segments of genomic DNA are wrapped around histone octamers in about two turns, forming the basic structural units of chromatin, the nucleosomes. The latter are arranged in an almost regular, linear fashion on the genome. Chemical modifications of histones are therefore localized at specific genomic positions and thus specific genes or functional DNA elements. Similar to DNA methylation, histone modifications can regulate gene expression. The best-studied modifications are acetylation (attachment of an acetyl-group), phosphorylation (attachment of a phosphate-group), methylation (attachment of a methyl-group), and ubiquitylation (attachment of ubiquitin, a small protein). Enzymes that add or remove modifications are referred to as "writers" and "erasers," respectively. Recognition of specific modifications is carried out by "reader" domains that are often found as part of modification enzymes or members of larger protein complexes. For example, the polycomb PRC2 complex contains a protein (EED) whose protein domains (WD40-repeats) specifically bind trimethylated lysine-27 of histone H3 (denoted by H3K27me3) as well as a catalytically active protein (EZH2) that attaches methyl-groups to unmodified lysine-27 of histone H3. With the target lysine being on a neighboring nucleosome, binding of PRC2 to newly set H3K27me3 marks allows propagation of these marks onto neighboring nucleosomes. Similarly, components of the Mll complex (trithorax) can "read" H3K4me3 marks and "write" the same mark to neighboring nucleosomes. Both marks have very different effects on gene expression. Whereas H3K27me3 is known for its repressive effects, H3K4me3 is associated with gene activation. Some of these histone marks directly affect the physicochemical properties of the chromatin. Acetylation and phosphorylation, for instance, are negatively charged, just like the DNA backbone, and usually loosen the association of DNA and nucleosome to make the DNA more accessible. In other cases, such as the H3K4me3 and H3K27me3, it seems difficult to explain the opposing regulatory effects by the tiny physicochemical differences of methylating a lysine in different positions of the histone tail. Instead, proteins with specific reader domains seem to access the information stored in these marks and trigger distinct regulatory pathways (Prohaska, Stadler, and Krakauer 2010).

The marking states of histones thus have an important effect on gene expression and on the GRN at large. Conversely, the transcription and translation of the proteins involved in histone modifications are themselves regulated at the level of the GRN. The histone layer, however, provides a potentially persistent memory that may hold information independent of changes in concentrations of players in the GRN.

The large number of proteins and protein complexes containing combinations of reader, writer, and eraser domains strongly suggests that histone modifications can act as an elaborate rewriting system, with histone-modifying proteins or complexes playing the role of rewriting rules. In principle, this "chromatin computer" is computationally universal (Bryant 2012). Arnold, Stadler, and Prohaska (2013) have shown, furthermore, that it can perform computational tasks, such as the reconstruction of patterns of histone marks that are partially lost during cell division. To what extent living cells make use of this potentially extensive computational capacity remains a field of active research.

DEVELOPMENT

Development, the transformation of a fertilized egg into an adult organism, is another defining feature of biological systems and one that has often been associated with "computation" (Wolpert 1994; Rosenberg 1997; Laubichler and Wagner 2000, 2001). Development is a highly complex set of interconnected processes acting at the molecular, cellular, organismal, and environmental levels that has thus far defied simple theoretical and formal descriptions (Pavlicev and Wagner 2012; Hogeweg 2012; Fontana and Schuster 1998; Ancel and Fontana 2000; Stadler *et al.* 2001; Stadler and Stadler 2006). At a highly coarse-grained scale, we find attempts to describe development as a mapping from genotype to phenotype space (Omholt 2013; Salazar-Ciudad and Marín-Riera 2013; Benfey and Mitchell-Olds 2008; Ho 2017; Hansen 2011; Gutiérrez and Maere 2014; Pigliucci 2008, 2010; Ribeiro *et al.* 2016). Such a mapping would represent a form of computation as we defined it earlier. However, although this is a tempting approach that in some simple cases appears even to be possible, the empirical and formal challenges are formidable. These begin with the nontrivial problem of defining the genotype. Functionally, the genotype needed for such a mapping is not just the full genome sequence (which can easily be sequenced) but the totality of all hereditary elements that causally affect development, including all the systems of epigenetic control described previously, but also multiple layers of environmental causes, which, as constructed niches, represent complex coevolutionary phenomena. Furthermore, the various regulatory systems—GRNs, posttranscriptional regulation, egigenetic systems—lead to large degrees of redundancy and high complex mapping relationships, including numerous many-to-one and many-to-many relationships (Wagner and Altenberg 1996; Wagner

Chapter 7: How and What Does a Biological System Compute?

and Zhang 2011; Pavlicev and Hansen 2011; Ibáñez-Marcelo and Alarcón 2014). All these empirical phenomena make the attempt to establish genotype-to-phenotype maps extremely difficult, if not practically impossible for anything but the most simple and extreme cases.

And there is more. The whole concept of the genotype-to-phenotype map as a formal description of development is based on the assumption of a clear separation between generations as well as between genotype and phenotype (Weismann 1885; Buss 1987). Actually, it is based on Weissmann's old idea of the separation of germline and soma. In this view, the germ cells contain the hereditary information that is unaffected by developmental processes and modifications and thus forms the basis of intergenerational continuity governed by the rules of transmission genetics—the copying and recombination of genetic material (Griesemer 2000). However, this elegant abstraction is not grounded in reality. Rather, the hereditary material—the genotype—is a multilayered product of genomic, cellular, and environmental elements that is actively shaped and constructed by the actions of the previous generation(s).

If the formal conception of the genotype-to-phenotype map as a straightforward way of "computing the embryo" does not work, how can we describe development as a computational process? The answer here is that development is not a single process but a consequence of multiple interacting subsystems, each performing its own set of "computations" that both process and provide input into other systems. Furthermore, each of these systems is the product of intersecting evolutionary histories that have shaped the specific properties of these systems, generally optimizing their behavior and function, but always within multiple constraints (historical, structural, and physical).

Biological Computation Is Embodied

The brief survey in the previous section highlights that biological systems, even simple, single-cell organisms, comprise many different molecular mechanisms as well as different time scales that behave in a way that we perceive as computational. This heterogeneity also translates to a diversity of paradigms of computations that are naturally used to describe the different subsystems. We argue here that this is a salient property of biological computation that needs to be understood when addressing issues relating to the efficiency of biological information processing.

In each example, there is a separation of a molecular hardware that implements specific dynamical processes by means of interactions that are more or less completely determined by the properties of the particular molecules, supramolecular structures, or other entities involved. All these molecules are the result of evolutionary processes shaping them to their tasks and, presumably, optimizing them for efficiency toward the common goal of survival.

A useful framework for disentangling the influence and effect of the material basis of a computational system and the actual computation "running" on it comes from the field of embodied cognition or embodied intelligence. A key question asked there is, given a body and an environment, what brain complexity is needed to generate a desired behavior? A main insight of this line of research is that a suitable body can serve to dramatically simplify the necessary control tasks, because the physics of the body and the environment imposes drastic constraints on the degrees of freedom that need to be controlled. Within a given computational framework—for example, artificial neural networks—there is a clearly defined division of the necessary information into a contribution

Chapter 7: How and What Does a Biological System Compute?

constrained by the body and a residual that needs to be dealt with explicitly by a neutral controller (Keyan and Ay 2013; Montúfar, Ghazi-Zahedi, and Ay 2015), which can be quantified in computer experiments. Similar ideas of embodied computation have been investigated in some generality for collective, highly parallel information processing (Hamann and Wörn 2007).

The same idea pertains to information processing in all biological systems. The physicochemical instantiation of the system drastically restricts the set of attainable states and distorts the probabilities of transitions relative to an unbiased, combinatorial reference model. This view effectively implies a separation of the hardware and the process of computation it enacts. In translation, for example, the machinery of ribosomes, tRNAs, and aminoacyl-transferases provides a hardware that hard-codes the genetic code. The only control or computation required at this point is the decision whether to translate a particular message. In each of the examples of cellular processes discussed herein, the embodiment/hardware is reused many times; that is, the computational processes are much faster than the turnover of the physical entities involved. We argue, therefore, that most of the computation in biology is performed in the very limited control tasks required to operate and coordinate the various pieces of the molecular machinery.

Consider replication: DNA polymerases provide an environment in which physics dictates that arbitrary template sequences are nearly perfectly copied. The map from original to copy is almost entirely embodied in the machine. Therefore the *computation task* is not to copy the $2L$ bits of information of a genome with L letters but only the dynamical control of the copy process in response to environmental clues and the internal state of the cell, that is, the inputs provided to the replication machinery by its environment. In eukaryotes, the situation is a bit more complex than a simple ON/OFF switch (there are multiple origins of replication that are selected in a well-defined temporal order, and an error-correction machinery checks the result by homing in on mismatches), but still, the computational effort, we would argue, is much smaller than the $2L$ bit of genetic information that is copied in this manner. We argue that, probably, the overwhelming part of the "obvious biological computation," such as copying of bit strings, is embodied, accounting for the perceived efficiency of biological information processing.

Or take development: Here the multiple layers and temporal scales are embodied in a number of molecular and structural/organizational features that both restrict and enable computational control. While GRNs, including epigenetics, compute the decision to express a particular gene, which then gets translated into a protein, based on signaling input, the effect this protein has on the developmental process is mediated by layers of embodied structures comprising structural and molecular constraints.

Therefore cells and organisms are so difficult to understand in terms of our usual computational paradigms because the machinery at one level or time scale is the result of computation at another. That is, each of the molecular machines involved in replication, transcription, translation, epigenetic modification, or development is in turn the result of a computation, which again is embodied in particular molecular machines that have evolved to perform very specific, and very restricted, tasks, thus limiting the information processing that is actually required to a minimum.

Clearly biological systems are not the only systems that are composed of interconnected and interdependent subsystems that are the product of long coevolutionary processes.

Chapter 7: How and What Does a Biological System Compute?

Computers and computer networks, such as the internet, have a similar structure. In analyzing these technical computing systems, we often focus on interfaces, which then give us a layering into various elements, such as transistors, CPUs, operating systems, software, and protocols. This reliance on clearly defined interfaces justifies treating these as quasi-independent domains. Biologists have attempted something similar. However, as soon as we focus on the actual mechanisms and want to understand their peculiar features, it becomes harder to maintain a clear separation of domains. Maybe this reflects a still limited degree of formal theory in biology—in which case, finding the relevant interfaces that allow for a partitioning and theoretical understanding along the lines of computer science and physics would be a desired future state of theory in biology. Alternatively—and this might be an even more intriguing perspective—the historical nature and coevolutionary interdependencies characteristic of biological and social systems could also shed some new light on technical and computational systems, especially in their real-world applications, such as cybersecurity and related domains. In any case, continued interactions between biologists, physicists, and computer scientists are necessary.

Conclusion

Our brief survey of how biological systems compute raises many questions that, in our view, represent the contours of a productive future research program into biological computation. The first and most obvious question is, how are these systems different from standard computers? Intrinsically linked to this question is the challenge of efficiency (Wolpert 2016; Kempes *et al.* 2017). Is it true that biological systems are extremely efficient in processing information? And, if so, how is the structure of biological computation related to the thermodynamics of computation? Another challenge is to understand the specific nature of biological computation. It is our thesis that biological computation is best understood as a form of embodied computation that is itself the product of evolutionary processes. How can we best formalize and understand these aspects of biological computation? And can a better understanding of how life has successfully computed for more than 3.5 billion years help us build better computer systems today?

Clearly there is a lot to do and much to be gained from careful deliberations at the intersection of physics, computation, and biology.

For readers who are interested in a more detailed and more technical introduction to the computational aspects of biological systems, we recommend, as a starting point, the work of Wolpert (1994), Laubichler and Wagner (2001), Istrail, De-Leon, and Davidson (2007), and Peter, Faure, and Davidson (2012).

Acknowledgments

The work was supported in part by the National Science Foundation INSPIRE 1648973 (S. J. P. and P. F. S.) and RCN 1656284 (M. L.) and by the Smart Family Foundation (M. L.).

REFERENCES

Ancel, L., and W. Fontana. 2000. "Plasticity, Evolvability and Modularity in RNA." *Journal of Experimental Zoology, Part B* 288:242–283.

Chapter 7: How and What Does a Biological System Compute?

Arendt, D., J. M. Musser, Clare V. H. Baker, A. Bergman, C. Cepko, D. H. Erwin, M. Pavlicev, *et al*. 2016. "The Origin and Evolution of Cell Types." *Nature Reviews Genetics* 17:744–757.

Arnold, C., P. F. Stadler, and S. J. Prohaska. 2013. "Chromatin Computation: Epigenetic Inheritance as a Pattern Reconstruction Problem." *J. Theor. Biol.* 336:61–74.

Benfey, P. N., and T. Mitchell-Olds. 2008. "From Genotype to Phenotype: Systems Biology Meets Natural Variation." *Science* 320:495–497.

Bryant, B. 2012. "Chromatin Computation." *PLoS ONE* 7:e35703.

Buss, L. 1987. *The Evolution of Individuality*. Princeton, NJ: Princeton University Press.

Davidson, E. H. 2006. *The Regulatory Genome: Gene Regulatory Networks in Development and Evolution*. New York: Academic Press.

Davidson, E. H., and D. H. Erwin. 2006. "Gene Regulatory Networks and the Evolution of Animal Body Plans." *Science* 311:796–800.

DiStefano, J., III. 2013. *Dynamic Systems Biology: Modeling and Simulation*. New York: Academic Press.

Fernández, M. 2009. *Models of Computation: An Introduction to Computability Theory*. Berlin: Springer.

Fontana, W., and P. Schuster. 1998. "Shaping Space: The Possible and the Attainable in RNA Genotype–Phenotype Mapping." *Journal of Theoretical Biology* 194:491–515.

Griesemer, J. R. 2000. "Reproduction and the Reduction of Genetics." In *The Concept of the Gene in Development and Evolution: Historical and Epistemological Perspectives,* edited by P. Beurton, R. Falk, and H.-J. Rheinberger, 240–285. Cambridge: Cambridge University Press.

Gutiérrez, J., and S. Maere. 2014. "Modeling the Evolution of Molecular Systems from a Mechanistic Perspective." *Trends in Plant Science* 19:292–303.

Hamann, H., and H. Wörn. 2007. "Embodied Computation." *Parallel Processing Letters* 17:287–298.

Hansen, T. F. 2011. "Epigenetics: Adaptation or Contingency." In *Epigenetics: Linking Genotype and Phenotype in Development and Evolution,* edited by Benedikt Hallgrímsson and Brian K. Hall, 357–376. Berkeley: University of California Press.

Ho, W. C. 2017. "The Genotype–Phenotype Map: Origins, Properties, and Evolutionary Consequences." PhD diss., University of Michigan, Ann Arbor.

Hogeweg, P. 2012. "Toward a Theory of Multilevel Evolution: Long-Term Information Integration Shapes the Mutational Landscape and Enhances Evolvability." In *Evolutionary Systems Biology*, 195–224. New York: Springer.

Ibáñez-Marcelo, E., and T. Alarcón. 2014. "The Topology of Robustness and Evolvability in Evolutionary Systems with Genotype–Phenotype Map." *Journal of Theoretical Biology* 356:144–162.

Istrail, S., S. B. De-Leon, and E. H. Davidson. 2007. "The Regulatory Genome and the Computer." *Developmental Biology* 310:187–195.

Jacob, F., and J. Monod. 1961. "Genetic Regulatory Mechanisms in the Synthesis of Proteins." *Journal of Molecular Biology* 3:318–356.

Kauffman, S. A. 1969. "Metabolic Stability and Epigenesis in Randomly Constructed Genetic Nets." *Journal of Theoretical Biology* 22:437–467.

———. 1993. *The Origins of Order: Self-Organization and Selection in Evolution*. Oxford, UK: Oxford University Press.

Kempes, C. P., D. Wolpert, Z. Cohen, and J. Pérez-Mercader. 2017. "The Thermodynamic Efficiency of Computations Made in Cells Across the Range of Life." *Phil. Trans. R. Soc. A* 375:20160343.

Keyan, Z., and N. Ay. 2013. "Quantifying Morphological Computation." *Entropy* 15:1887–1915.

Krakauer, D. C., L. Müller, S. J. Prohaska, and P. F. Stadler. 2016. "Design Specifications for Cellular Regulation." *Th. Biosci.* 231–240.

Laubichler, M. D., and G. P. Wagner. 2000. "Organism and Character Decomposition: Steps towards an Integrative Theory of Biology." *Philosophy of Science* 67:S289–S300.

———. 2001. "How Molecular Is Molecular Developmental Biology? A Reply to Alex Rosenberg's Reductionism Redux: Computing the Embryo." *Biology and Philosophy* 16:53.

Ma, K. C., S. D. Perli, and T. K. Lu. 2016. "Foundations and Emerging Paradigms for Computing in Living Cells." *Journal of Molecular Biology* 428:893–915.

Montúfar, G., K. Ghazi-Zahedi, and N. Ay. 2015. "A Theory of Cheap Control in Embodied Systems." *PLoS Computational Biology* 11:e1004427.

Omholt, S. W. 2013. "From Sequence to Consequence and Back." *Progress Biophysics and Molecular Biology* 111:75–82.

Pavlicev, M., and T. F. Hansen. 2011. "Genotype–Phenotype Maps Maximizing Evolvability: Modularity Revisited." *Evolutionary Biology* 38:371–389.

Pavlicev, M., and G. P. Wagner. 2012. "A Model of Developmental Evolution: Selection, Pleiotropy and Compensation." *Trends in Ecology and Evolution* 27:316–322.

Peter, I. S., and E. H. Davidson. 2016. "Implications of Developmental Gene Regulatory Networks Inside and Outside Developmental Biology." 117:237–251.

———. 2017. "Assessing Regulatory Information in Developmental Gene Regulatory Networks." *Proceedings of the National Academy of Sciences of the United States of America* 114:5862–5869.

Peter, I. S., E. Faure, and E. H. Davidson. 2012. "Predictive Computation of Genomic Logic Processing Functions in Embryonic Development." *Proceedings of the National Academy of Sciences of the United States of America* 109:16434–16442.

Pigliucci, M. 2008. "Is Evolvability Evolvable?" *Nature Reviews Genetics* 9:75–82.

———. 2010. "Genotype–Phenotype Mapping and the End of the "Genes as Blueprint" Metaphor." *Philosophical Transactions of the Royal Society of London, Series B* 365:557–566.

Prohaska, S. J., P. F. Stadler, and D. C. Krakauer. 2010. "Innovation in Gene Regulation: The Case of Chromatin Computation." *Journal of Theoretical Biology* 265:27–44.

Ribeiro, A. H., J. M. Soler, E. C. Neto, and A. Fujita. 2016. "Causal Inference and Structure Learning of Genotype–Phenotype Networks Using Genetic Variation." In *Big Data Analytics in Genomics*, 89–143. New York: Springer.

Rosenberg, A. 1997. "Reductionism Redux: Computing the Embryo." *Biology and Philosophy* 12:445–470.

Salazar-Ciudad, I., and M. Marín-Riera. 2013. "Adaptive Dynamics under Development-Based Genotype–Phenotype Maps." *Nature* 497:361–364.

Smith, Z. D., and A. Meissner. 2013. "DNA Methylation: Roles in Mammalian Development." *Nature Reviews Genetics* 14:204–220.

Stadler, B. M. R., P. F. Stadler, G. Wagner, and W. Fontana. 2001. "The Topology of the Possible: Formal Spaces Underlying Patterns of Evolutionary Change." *Journal of Theoretical Biology* 213:241–274.

Stadler, P. F., and B. M. R. Stadler. 2006. "Genotype Phenotype Maps." *Biological Theory* 3:268–279.

Wagner, G. P., and L. Altenberg. 1996. "Perspective: Complex Adaptations and the Evolution of Evolvability." *Evolution* 50:967–976.

Wagner, G. P., and J. Zhang. 2011. "The Pleiotropic Structure of the Genotype–Phenotype Map: The Evolvability of Complex Organisms." *Nature Reviews Genetics* 12:204–213.

Weismann, A. 1885. *Die Continuität des Keimplasmas als Grundlage einer Theorie der Vererbung.* Jena, Germany: Fischer.

Wolpert, D. H. 2016. "The Free Energy Requirements of Biological Organisms: Implications for Evolution." *Entropy* 18 (4).

Wolpert, L. 1994. "Do We Understand Development?" *Science* 266:571–572.

TOWARD SPACE- AND ENERGY-EFFICIENT COMPUTATIONS

Anne Condon, University of British Columbia
Chris Thachuk, California Institute of Technology

Introduction

How might a simulation of computation that is both space and energy efficient be possible? If a Turing machine, on a problem instance of size n, requires $t(n)$ time and $s(n)$ space to complete, then a simulation of the computation is space efficient if it requires at most $poly(s(n))$ space, and energy efficient if it dissipates at most $\epsilon\, t(n)$ energy over the course of the computation, for sufficiently small $\epsilon \geq 0$. Lecerf (1963) and Bennett (1973) made significant progress on this question by introducing the notion of logically reversible computation—previously thought to be a prerequisite for energy-efficient computation[1]—and devising simulations of irreversible Turing machine computations by logically reversible Turing machines with a constant factor increase in time. Bennett (1989), Lange, McKenzie, and Tapp (2000), and others subsequently made progress on space-efficient simulations.

Recent work by Qian, Soloveichik, and Winfree (2011) and others made further significant progress by bridging the gap between logically reversible and physically realizable

---

[1] Logical reversibility is not a prerequisite for energy-efficient computation (Sagawa 2014; Grochow and Wolpert 2018). For a more nuanced view of this topic beyond logical reversibility, see Chapters 10 and 13 in this volume.

computations using DNA strand displacement systems. Their strand displacement simulations of Turing machines use arbitrarily little energy per step while incurring a quadratic slowdown in time. However, to complete, their simulations may require exponentially more molecules (physical space) than the space used by the Turing machine.

Building on these two earlier threads, we showed how computations that are logically reversible, with balanced, symmetric transitions, have energy-efficient implementations as DNA strand displacement systems and only require a quadratic increase in the number of molecules over the theoretical space required of the computation (Thachuk and Condon 2012).

Here we review these three lines of work with the goals of elucidating the current state of knowledge on the theory of space- and energy-efficient computations, and pointing to fruitful directions for future improvements.

Background

We start with a brief introduction to logical reversibility in the context of time-bounded computation as well as early work on space-bounded, logically reversible computations. We then describe a progression of ideas on how the abstract concept of logical reversibility might be implemented chemically, and how the energy efficiency of chemical implementations can be measured.

LOGICAL REVERSIBILITY AND TIME-BOUNDED COMPUTATION

Landauer (1961) asked: can we build computing devices that are energy efficient, that is, devices that dissipate arbitrarily little energy per step? His work suggested that energy must be dissipated when bits of information are

Chapter 8: Toward Space- and Energy-Efficient Computations

irreversibly erased. This led researchers to consider how logically reversible computation—where information is never irreversibly erased during the computation—may lead to energy-efficient computation. Note that to be reusable for another problem instance, a logically reversible computer must be "reset," a potentially energy-consuming process that includes erasure of the previous input and output.

Lecerf (1963) and Bennett (1973) formalized the notion of logically reversible, deterministic Turing machines. Transition functions of such machines not only have unique images that guarantee determinism, but also unique preimages, thereby ensuring that, when symbols on a tape are overwritten, those symbols can be reconstructed simply by reversing the transition. Thus any computation graph of such a machine is a line from a valid input configuration to a final configuration (i.e., no branching or merging). Turing machines that are logically reversible can simulate arbitrary (irreversible) Turing machine computations with a constant factor increase in the time needed (Bennett 1973); that is, for inputs of length n, if DTIME $(t(n))$ and ReversibleTIME $(t(n))$ are the classes of languages recognizable by deterministic and logically reversible Turing machines in $O(t(n))$ time, respectively, then DTIME $(t(n))$ = ReversibleTIME $(t(n))$.

LOGICALLY REVERSIBLE SPACE-BOUNDED COMPUTATION

Bennett also asked whether Turing machine computations can be simulated in a logically reversibly manner that is efficient with respect to the space of the machine being simulated. Using a recursive simulation of deterministic space-bounded Turing machines with recursion depth proportional to the space used, he showed that DSPACE $(s(n))$ is contained in ReversibleSPACE $(s^2(n))$ (Bennett 1989).

Whether the quadratic increase in space was necessary remained uncertain for another decade, when Lange, McKenzie, and Tapp (2000) showed that, indeed, logically reversible space equals deterministic space, that is,

$$\text{DSPACE}(s(n)) = \text{ReversibleSPACE}(s(n)).$$

Their construction performs an Eulerian tour of the configuration graph of a space-bounded computation.

CHEMICAL IMPLEMENTATIONS OF LOGICALLY REVERSIBLE COMPUTATION

Bennett (1973) envisioned Brownian computers that implement reversible computations chemically, that are "capable of dissipating an arbitrarily small amount of energy per step if operated sufficiently slowly." A single copy of a major reactant, such as a DNA molecule, encodes the current configuration. Reactions in the forward direction involve the major reactant plus additional minor reactants, which are present at "definite concentrations;" these reactions change the major reactant and generate additional (waste) products. The forward direction of each reaction corresponds to a computation step. We describe herein the relationship between the energy dissipated by a reaction and the ratio of the concentrations of minor reactants to products.

Bennett (1981) considered the possibility that reactions of a Brownian computer could arise purely as a result of "random thermal jiggling of its information-bearing parts," with concentrations of minor reactants and products at equilibrium. In this case, a logically reversible computation can proceed in the forward and reverse directions at the same rate while "high-potential-energy barriers" prevent the system from deviating from the computation path. However, he dismissed calling this possibility computation because, if "the major reactant initially corresponds to the initial state of a v-step computation, the system will begin a random walk through the chain of reactions, and after about v^2 steps will briefly visit the final state. This does not deserve to be called a computation" (Bennett 1973, 531).

Rather, to ensure high probability of completing the computation within reasonable time while keeping energy dissipation per step low, Bennett proposed that the concentrations of minor reactants driving the computation forward be maintained at a small percentage above their equilibrium concentrations. Reactions can still happen in both the forward and reverse directions in this scenario, but the forward bias that results from the higher concentrations of forward-driving reactants, compared with the reaction products, is sufficient to ensure that the final state is reached in a number of steps proportional to the computation length. The forward bias at the final step can be significantly higher to ensure that, once the final configuration is reached, the system will subsequently be in this configuration with high (say, at least 95%) probability.

Recently, researchers in the field of DNA computing and molecular programming have developed and implemented mechanisms for computing, with DNA strand displacement systems (DSDs), that are similar in principle to Bennett's proposal involving major and minor reactants (Soloveichik, Seelig, and Winfree 2010). We provide an example of DNA strand displacement later (see fig. 8.2). Qian, Soloveichik, and Winfree (2011) described how a logically reversible Turing machine could in principle be physically realized with DNA strand displacement operations, in a way that requires arbitrarily little energy per computation step. The worst-case upper bound on the time needed for this simulation is quadratic in the Turing machine time. An important assumption of this result is that the simulation occurs as an "open" system, where minor reactants are maintained at definite concentration by an external and energy-efficient metabolism process. Under this assumption, the physical space required to hold all major reactants of the simulation (i.e., the required volume) is also efficient with respect to Turing machine space. However, an external metabolism to maintain minor reactants for DSDs is as yet unknown. Operating this simulation in a "closed" system, where all minor reactants required for the simulation to complete are initially present in the same volume as the major reactants, can require exponentially more physical space than Turing machine space. The reasoning will be made clear in subsequent sections where we focus exclusively on simulations in closed systems to account fully for required (physical) space usage.

ACCOUNTING FOR ENERGY IN A CHEMICAL COMPUTATION

We follow the accounting of Bennett (1981) and Qian, Soloveichik, and Winfree (2011) for logically reversible computations where computation steps correspond to the forward direction of chemical reactions. We assume that the computation proceeds in a fixed volume that is maintained at constant temperature. Suppose that the concentrations of forward-driving reactants are X percent in excess of the equilibrium concentrations, relative to the concentrations of the reaction products. Then the energy dissipated (per step) is proportional to $\ln((100+X)/100)k_B T$, where k_B is Boltzmann's constant and T is temperature. The smaller X is, the less energy is dissipated, the higher is the ratio of reverse to forward reactions, and the smaller is the net forward bias of the reaction. If initial concentrations are sufficiently high relative to the length of the computation, the change in concentrations of reactants and products is negligible throughout, and so the reactions have approximately the same forward bias throughout. If the energy dissipated per step is ϵ for sufficiently small $\epsilon > 0$, the forward/reverse

Chapter 8: Toward Space- and Energy-Efficient Computations

ratio is $e^{\epsilon/(k_B T)}$, and the forward reaction bias is roughly proportional to ϵ.

In contrast, if the concentrations of reactants and products are in equilibrium, the system is driven only by Brownian motion, with forward and reverse reactions being equally likely. The energy dissipated per step is zero.

Although not discussed by Bennett (1981) or Qian, Soloveichik, and Winfree (2011), there is also an energetic cost to setting up the initial configuration of the computation—in our systems, this is the same cost associated with "resetting" the computation for a new input instance—and conducting a read-out of the current configuration at any given time. For instance, logically reversible simulations in closed systems typically rely on an initial out-of-equilibrium input state relaxing to equilibrium as the computation proceeds. This "distance from equilibrium" is characterized by the nonequilibrium generalized free energy (Esposito and Van den Broeck 2011; Parrondo, Horowitz, and Sagawa 2015). The point is that the system free energy increases over time until equilibrium is reached. For a logically reversible computation over N states, this difference in free energy is $k_B T \ln(N)$, which is the configurational entropy cost of beginning in a particular state of N states that are equally likely at equilibrium.

Challenges with Space-Efficient Chemical Implementations of Logically Reversible Computations

The simulation of Qian, Soloveichik, and Winfree (2011) in a closed system requires space (or volume) proportional to the number of steps of the computation, leaving open the question of how to do computations in both a space- and energy-efficient manner. We next provide an example to illustrate why a DSD-based chemical implementation of space-bounded, logically reversible computations may incur an exponential space blowup.

$$
\begin{aligned}
(1)\quad & 0_1 & \rightleftharpoons\ & 1_1 \\
(2)\quad & 0_2 + 1_1 & \rightleftharpoons\ & 1_2 + 0_1 \\
(3)\quad & 0_3 + 1_2 + 1_1 & \rightleftharpoons\ & 1_3 + 0_2 + 0_1
\end{aligned}
$$

$\{0_3,0_2,1_1\}$ $\quad$ $\{0_3,1_2,1_1\}$ $\quad$ $\{1_3,0_2,1_1\}$ $\quad$ $\{1_3,1_2,1_1\}$

1-for $\quad$ 2-for $\quad$ 1-for $\quad$ 3-for $\quad$ 1-for $\quad$ 2-for $\quad$ 1-for

1-rev $\quad$ 2-rev $\quad$ 1-rev $\quad$ 3-rev $\quad$ 1-rev $\quad$ 2-rev $\quad$ 1-rev

$\{0_3,0_2,0_1\}$ $\quad$ $\{0_3,1_2,0_1\}$ $\quad$ $\{1_3,0_2,0_1\}$ $\quad$ $\{1_3,1_2,0_1\}$

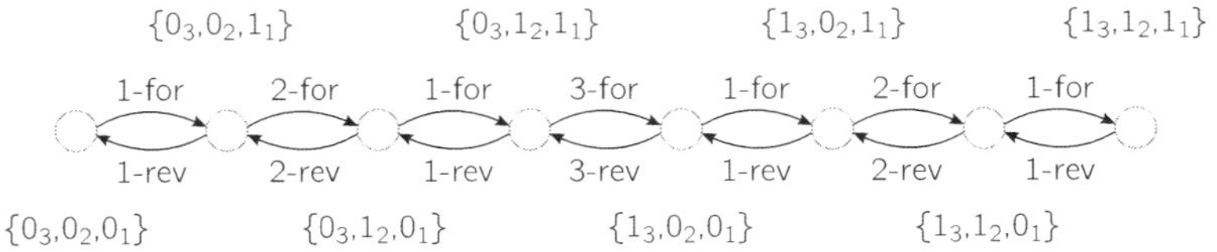

Figure 8.1. (a) Chemical reaction equations for a 3-bit standard binary counter. (b) The configuration graph of the computation performed by the 3-bit standard binary counter forms a chain and is logically reversible.

EXAMPLE: A BINARY COUNTER

For simplicity, we will use chemical reaction networks (CRNs) rather than Turing machines as our programming model. We choose CRNs because of the ease with which we can describe our binary counter, as CRNs are powerful enough to encompass general computation (Soloveichik *et al.* 2008) and can easily be "compiled" into DSDs (Soloveichik, Seelig, and Winfree 2010).

A standard n-bit binary counter begins at count 00..0, advances to 00..1, and so on, until reaching the count 11..1. A CRN that implements this counter for $n = 3$ is in figure 8.1A (Condon *et al.* 2012). The reactions involve six molecular species, with molecules 0_i and 1_i denoting that bit i has value 0 and 1, respectively, for $1 \leq i \leq 3$. We call 0_i and 1_i *signal* molecules; these correspond to Bennett's major reactants. The forward direction of the three (reversible) chemical reactions of figure 8.1A enable the counter to advance in the standard sequence if, initially, a single copy of molecules 0_3, 0_2, and 0_1 are present. The reactions can be generalized to n bits in a natural way, with reactions having up to n reactants and products. Because high-order reactions are atypical in chemical systems, it is worth noting that they can always be emulated by a sequence of binary reactions (Condon *et al.* 2012).

Figure 8.1B depicts all possible configurations of the counter as nodes, with edges between configurations that are reachable within one reaction step. Because the configuration graph forms a chain, it represents a logically reversible computation.

DSD IMPLEMENTATION OF A CRN REACTION

We do not know how to design a set of molecules that can directly implement all reactions of an arbitrary CRN without implementing additional (spurious) reactions. By "directly implement," we mean that there is a one-to-one correspondence between the designed molecular species and the species of the CRN.

However, seminal work by Soloveichik, Seelig, and Winfree (2010) demonstrated that an arbitrary CRN can be emulated by a DSD system, in which certain DNA strands represent the molecular species of the CRN (the major reactants) while additional *transformer* molecules (the minor reactants and products) also facilitate each reaction. A *formal* CRN reaction is often emulated by a sequence of *detailed* DSD reactions. See figure 8.2. Transformer molecules are often described as fuel molecules, because their concentrations can be set so as to bias a computation forward. Many DSD architectures now exist with this capability (cf. Cardelli 2013; Qian, Soloveichik, and Winfree 2011), and all make use of transformers.

We illustrate how the first reaction of our three-bit binary counter CRN can be implemented using DNA strand displacement (fig. 8.2). From top to bottom, the 0_1 signal strand interacts with a transformer first to become consumed—sequestered on a double-stranded complex—and ultimately the

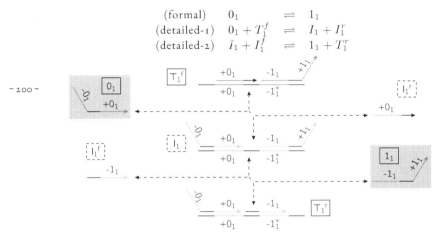

Figure 8.2. A strand displacement implementation of the reaction $0_1 \rightleftharpoons 1_1$ as proposed by Qian, Soloveichik, and Winfree (2011). (a) The single formal reaction is emulated as a sequence of two detailed reactions in the strand displacement implementation. (b) From top to bottom, the reactant signal strand 0_1 (shown in a shaded box on the left) and transformer molecule (labeled T_1^f, top middle) react, producing an unbound product transformer (labeled I_1^r, top right) and an intermediate complex (labeled I_1, middle). The intermediate complex, together with a second transformer reactant (labeled I_1^f, bottom left), reacts to produce the signal strand 1_1 (shaded box on the bottom right) and a final product transformer (labeled T_1^r, middle bottom). The product transformers can be applied in the opposite direction (from bottom to top) to consume signal 1_1 and produce signal 0_1 as well as the two original reactant transformers. Throughout, connected line segments represent DNA strands, with one-sided arrows denoting the 5' end. Black line segments represent short "toehold" sequences that enable strands to bind weakly to each other via Watson–Crick base pairing, thus enabling the reaction. Gray line segments represent "long domain" sequences that enable molecules to bind together more stably, while also displacing long domains of other strands. Long domains are labeled, and the Watson–Crick complement of a DNA sequence labeled x is labeled x^*.

1_1 signal strand is produced, released from a double-stranded complex. The strands contained within a shaded box are the signal strands. Two transformers (T_1^f and I_1^f) are reactants, and two transformers (I_1^r and T_1^r) are produced. The product transformers can perform reaction (1) in reverse (from bottom to top in figure 8.2). The important point is that the reactant transformers are not the same as the product transformers.

Note that this DSD scheme generally applies to any formal reaction: a sequence of detailed reactions each consumes a single reactant (of the formal reaction), and then a subsequent sequence of detailed reactions each produces a single product (of the formal reaction). The example in figure 8.2 has one reactant and one product and thus requires a sequence of two detailed reactions. Similarly, the reaction $0_2 + 1_1 \rightleftharpoons 1_2 + 0_1$ requires a sequence of four detailed reactions (Qian, Soloveichik, and Winfree 2011).

TAGGED CRNS

To capture this notion of transformer orientation at the level of a CRN, we can *tag* each side of a formal reaction to represent the set of transformers that is necessary to perform a reaction in the respective direction; intermediate reactions and intermediate species arising in a strand displacement emulation can be safely ignored as implementation details. In the case of reversible reactions, when considered as two separate reactions, the forward tag of one constitutes the reverse tag of the other. We call these *tagged chemical reaction equations*. This simple concept of tags allows us to account for the number of transformers and the minimum size of the reaction volume required to complete a computation if it were to be physically realized as a DSD system.

(1) $T_1^f + 0_1 \rightleftharpoons T_1^r + 1_1$
(2) $T_2^f + 0_2 + 1_1 \rightleftharpoons T_2^r + 1_2 + 0_1$
(3) $T_3^f + 0_3 + 1_2 + 1_1 \rightleftharpoons T_3^r + 1_3 + 0_2 + 0_1$

Figure 8.3. Tagged chemical reaction equations for a three-bit standard binary counter.

ACCOUNTING FOR SPACE IN A TAGGED CRN COMPUTATION

In a closed DSD system that is simulating a tagged CRN, we assume that the computation proceeds in a fixed volume that is at least proportional to the maximum number of molecules present at any step of the computation. Volume must be as large as the physical space requirement (Cook *et al.* 2009), and so when we account for space at the CRN level, we must include sufficient space to store not only the signal molecules but also the transformers that facilitate the reactions. Because tags denote sets of transformers, this accounting is straightforward. A stricter accounting would reflect the number of bases of molecules at the DSD level, but, for simplicity, we'll ignore this level of detail and simply count molecules at the tagged CRN level of abstraction, with one tag molecule representing a transformer set.

Returning to the tagged three-bit standard counter CRN, every configuration has exactly three signal molecules present, denoting the current bit sequence. However, the space requirement for tags is not nearly as succinct. Reaction (1) is used in the forward direction four times during the computation, corresponding to changes in the counter's lowest-order bit; reaction (2) is used twice; and reaction (3) is used once. In total, seven tags are required to advance through the eight states of the counter, and thus seven tags must be present in the initial configuration (and all subsequent configurations). In general, and despite the fact that all reactions are reversible, the progress of an n-bit counter relies on a sequence of $2^n - 1$ forward reactions and, consequently, $2^n - 1$ tags. In summary, molecular simulations of tagged CRNs in closed systems necessarily use space that grows at least proportionally with time, provided they do not have an external "metabolism" to recycle minor reactants, and always use reactions in the forward direction.

Space- and Energy-Efficient Implementation of Logically Reversible Systems

We continue with the example of a counter to illustrate how to avoid the exponential space blowup of the previous section. The idea is to use reactions in both the forward and reverse directions to drive the computation forward, to reuse transformer molecules. As we discuss later, our use of reactions in both directions resembles symmetric models of computation (Lewis and Papadimitriou 1982).

Reducing the space usage in this way relies on the computation to be unbiased; that is, concentrations of reactant and product transformers should be at equilibrium. However, Bennett asserts that an unbiased, logically reversible computation chain of this form, with output present only in the final state, does not constitute computation. We agree only in that the output of the computation must be observable with high probability. To address this concern, we show how any such chain implemented as a tagged CRN can be augmented to ensure that the output can be read with high probability without an asymptotic increase in time or space usage. We end this section with a summary of known results for space- and energy-efficient CRNs.

A GRAY CODE COUNTER

We review our space-efficient tagged CRN counter (Condon *et al.* 2012), which is based on the binary reflecting Gray code sequence (Savage 1997). The sequence is a *Gray code* as each successive value differs from the previous in exactly one bit position. It is called a binary *reflecting* Gray code (BRGC) because of its elegant recursive definition: the n-bit BRGC sequence is formed by reflecting the $(n-1)$-bit sequence across a line, then prefixing values above the line with 0 and those below the line with 1. A tagged CRN that implements a 3-bit BRGC counter is given in figure 8.4A. The recursive nature of the counter can be seen in the computation chain of figure 8.4B (and is reminiscent of the simulation of Bennett (1981)). For example, when the high-order bit is turned from a 0 to a 1 for the first time (reaction 3 in the forward direction), the sequence of reactions up until that configuration are next executed in reverse as the computation progresses. As a result, each specific reaction strictly alternates between its forward and reverse directions as the computation proceeds forward. Only a single transformer (the forward transformer) per reaction is initially needed to complete the whole computation. While a tagged CRN simulation of a standard n-bit counter requires $\Theta(2^n)$ transformer molecules, the BRGC variant requires only $\Theta(n)$. Note that both counters require only $\Theta(n)$ molecule types.

OBSERVING OUTPUT WITH HIGH PROBABILITY

In this running example of a counter, we define the *output* of the computation to be the counter's final value. As the computation performs an unbiased random walk along the symmetric and logically reversible computation configuration space, the steady state probability of observing the output is $p = 2^{-n}$ for an n-bit counter. This probability can be increased

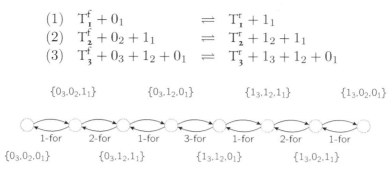

Figure 8.4. (a) Tagged chemical reaction equations for a 3-bit binary reflecting Gray code (BRGC) counter. (b) BRGC configuration graph. To reach the end configuration, the BRGC counter must perform a sequence of reactions that individually alternate in the forward and reverse directions, thus initially requiring only one transformer per reaction type.

in a number of ways. For instance, by adding one additional reaction type that produces a new signal and requires the final signal multiset of the original computation chain as catalysts, we can double the length of the new chain. The strategy is outlined in figure 8.5. In this case, the probability of observing the output increases to $p' > 0.5$, because the last state of the first half of the chain and all states of the second half of the chain contain the output. In this manner, for every new reaction added to the CRN, the probability of not observing an output signal is cut in half. Formally, the probability of observing the output becomes $p'' > 1 - 2^{-c}$ when $c \geq 0$ number of new reactions are added to extend the computation chain. Choosing $c = 5$ meets Bennett's benchmark to observe the output with at least 95% probability at steady state, all while maintaining the same asymptotic time and space usage.

ENERGY-EFFICIENT SIMULATIONS OF SPACE-BOUNDED COMPUTATION

Building on the principles of transformer reuse, we previously proposed a tagged CRN—implementable as a DSD—to solve

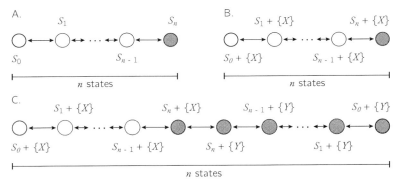

Figure 8.5. (a) A logically reversible computation chain of a tagged CRN that has a single configuration containing the output (shown shaded). S_i denotes the multiset of signals present in the ith configuration along the computation chain. (b) Adding a single copy of a new species, say, X, to the input configuration S_0, where X is not a reactant in the CRN, does not alter the progress of the computation. (c) The addition of a new reaction—$X + S_n \rightleftharpoons Y + S_n$, where S_n, the multiset of signals present in the original final configuration, catalyzes the new reaction—doubles the length of the computation and has the side effect that the output is now present in more than half of the configurations (shown shaded).

arbitrary instances of the PSPACE-complete quantified three-satisfiability problem (Thachuk and Condon 2012). For an instance consisting of m clauses over n variables, the tagged CRN requires $\Theta(m + n)$ space. With a polynomial increase in size, the same tagged CRN can be modified to solve any instance over n variables. In this way, our solution to simulating space-bounded computation is not unlike that of a uniform circuit family. Building on known computational complexity results, we show that it is possible to simulate any space-bounded computation with a space- and energy-efficient tagged CRN. While our computation is unbiased, and therefore does not dissipate energy per reaction step, we rely on the computation beginning out of equilibrium in the input state and then relaxing to equilibrium as the computation proceeds. Other simulations, such as those of Bennett (1973) and Qian, Soloveichik, and Winfree (2011), share the assumption of beginning in a particular input state. In our running example of an n-bit counter that steps through 2^n configurations, beginning in a particular input state requires setting n bits of information initially (i.e., fix all bits to the 0 value). In general, the energy expenditure of our simulation can be bounded as $\Theta(s(n))$ for an $s(n)$-space-bounded tagged CRN computation.

Theorem 1 (Thachuk 2013). *Any problem solvable in $s(n) \geq n$ space can be solved by a logically reversible tagged CRN using $O(s(n)^2 \log s(n))$ space and energy.*

Discussion

As noted earlier, Bennett raised two objections to chemical implementations of computations in which reactions (corresponding to forward computation steps) are unbiased. First, they are slow, because an unbiased random walk through v configurations requires v^2 steps, and second, the final configuration is visited rarely in such a walk. Here we have argued that, despite the first objection, unbiased chemical simulations are valuable to avoid blowup of space when simulating space-bounded computations. A key difference between our chemical simulations and those suggested by Bennett is that both the forward and reverse directions of reactions correspond to forward computation steps. We have also shown how the second objection can be overcome by ensuring that a constant fraction of configurations on the computation path contains the desired output.

Our Gray code counter simulation (and our more general simulations of space-bounded computations) are not only logically reversible but also symmetric, in that reactions can proceed in both the forward and reverse directions—in this sense, "symmetric" implies mechanistically reversible. As such, they are also related to symmetric models of computation, introduced by Lewis and Papadimitriou (1982), where, for a given symmetric Turing machine, every transition rule t in its set of transition rules $\mathcal{T}$ implies that the inverse transition rule $t^{-1} \in \mathcal{T}$. Chemical implementations of computations in which reactions are unbiased, while slow, seem to be essential for simulation of symmetric computations.

Our preceding analyses focused on closed systems with fixed volume. Might space-efficient DSD implementations of the standard counter, and space-bounded computations more generally, be possible if the reactant transformers flow into the volume as needed to ensure an excess over the product transformers, and waste, that is, product transformers, is removed from the system? In such a scenario, although the volume used by a computation could be linear in the space of the computation being simulated, the number of reactant (forward) transformers required overall and the amount of waste (number of product, or reverse transformers) produced are exponential in the space. For this reason, we argue that such a simulation still requires exponential space, unless a way is found (i.e., an energy-efficient external metabolism) to convert the waste back into forward transformers.

As the DSD systems we presented fall within the larger field of DNA computing, it is interesting to contrast our approach with the field's seminal work by Adleman (1994) on solving large combinatorial search problems using DNA, in terms of space usage. Whereas his work required infeasible counts of molecules to solve very large problem instances, our space requirements are efficient. However, unlike Adleman's approach to DNA computing, our energy-efficient simulation and that of Qian, Soloveichik, and Winfree (2011) have an implicit "single-copy" assumption (Condon *et al.* 2012) where it is expected that only a single molecule is initially present for certain species types. These simulations fail to be logically reversible when this assumption is violated (Thachuk 2013).

We conclude with two open questions.

BALANCE VERSUS SPACE EFFICIENCY?

We say that a logically reversible computation is k-*balanced* if, for all transition types, within every computation prefix, the number of times the transition is executed in the forward direction differs from the number of times the transition is executed in the reverse direction by at most k.

The logically reversible simulation of irreversible space-bounded Turing machines of Lange, McKenzie, and Tapp (2000) is not balanced: just like the standard binary counter CRN, transitions are always executed in the forward direction, because they simulate an Eulerian tour of a configuration graph. Thus, if this simulation were implemented chemically using DSDs, it would incur an exponential space blowup.

In contrast, while Bennett did not explicitly design his recursive simulation of irreversible space-bounded Turing machines by logically reversible Turing machines to be balanced, it seems possible to implement it in a balanced way by exploiting the back-tracking inherent to recursive procedures to follow transitions in the reverse direction. Perhaps Bennett's simulation can be adapted to show that DSPACE $(s(n))$ is contained in BalancedSPACE $(s^2(n))$, thereby improving Theorem 1 by a logarithmic factor.

Showing that there is a balanced simulation of irreversible space-bounded Turing machines that avoids a quadratic increase in space may be more challenging. That is, if BalancedSPACE $(s(n))$ is the class of languages recognizable by $O(1)$-balanced, logically reversible Turing machines, can we show that DSPACE $(s(n)) = $ BalancedSPACE $(s(n))$?

DSDS WITH UNIVERSAL TRANSFORMERS?

Of course, balanced simulations may not be necessary to avoid space blowups in chemical implementations of logically reversible computations. Is there a DSD system (or other molecular system) in which transformers are universal and thus identical in the forward and reverse directions of a reaction? Such a system could simulate computations in which all reactions occur in the forward direction, such as the standard binary counter, without incurring a significant space increase. ✦

Acknowledgments

We thank Erik Winfree, David Soloveichik, Tom Ouldridge, and David Sivak for helpful discussions and the anonymous reviewers for their feedback.

REFERENCES

Adleman, L. M. 1994. "Molecular Computation of Solutions to Combinatorial Problems." *Science* 266 (5187): 1021–1024.

Bennett, C. H. 1973. "Logical Reversibility of Computation." *IBM Journal of Research and Development* 17 (6): 525–532.

———. 1981. "The Thermodynamics of Computation—A Review." *International Journal of Theoretical Physics* 21 (12): 905–940.

———. 1989. "Time/Space Trade-Offs for Reversible Computation." *SIAM Journal on Computing* 18 (4): 766–776.

Cardelli, L. 2013. "Two-Domain DNA Strand Displacement." *Mathematical Structures in Computer Science* 23 (2): 247–271.

Condon, A., A. J. Hu, J. Maňuch, and C. Thachuk. 2012. "Less Haste, Less Waste: On Recycling and Its Limits in Strand Displacement Systems." *Journal of the Royal Society: Interface Focus* 2 (4): 512–521.

Cook, M., D. Soloveichik, E. Winfree, and J. Bruck. 2009. "Programmability of Chemical Reaction Networks." In *Algorithmic Bioprocesses,* 543–584. New York: Springer.

Esposito, M., and C. Van den Broeck. 2011. "Second Law and Landauer Principle Far from Equilibrium." *Europhysics Letters* 95 (4): 40004.

Grochow, J. A., and D. H. Wolpert. 2018. "Beyond Number of Bit Erasures: New Complexity Questions Raised by Recently Discovered Thermodynamic Costs of Computation." *SIGACT News* 49, no. 2 (June): 33–56.

Landauer, R. 1961. "Irreversibility and Heat Generation in the Computing Process." *IBM Journal of Research and Development* 5 (3): 183–191.

Lange, K. J., P. McKenzie, and A. Tapp. 2000. "Reversible Space Equals Deterministic Space." *Journal of Computer Systems Science* 60 (2): 354–367.

Lecerf, Y. 1963. "Logique Mathématique: Machines de Turing réversibles." *Comptes rendus des séances de l'académie des sciences* 257:2597–2600.

Lewis, H. R., and C. H. Papadimitriou. 1982. "Symmetric Space-Bounded Computation." *Theoretical Computer Science* XX:161–187.

Parrondo, J. M. R., J. M. Horowitz, and T. Sagawa. 2015. "Thermodynamics of Information." *Nature Physics* 11 (2): 131–139.

Qian, L., D. Soloveichik, and E. Winfree. 2011. "Efficient Turing-Universal Computation with DNA Polymers [extended abstract]." In *Proceedings of the 16th Annual Conference on DNA Computing,* 123–140. Berlin, Heidelberg: Springer-Verlag.

Sagawa, T. 2014. "Thermodynamic and Logical Reversibilities Revisited." *Journal of Statistical Mechanics: Theory and Experiment* 2014 (3): P03025.

Savage, C. 1997. "A Survey of Combinatorial Gray Codes." *SIAM Review* 39 (4): 605–629.

Soloveichik, D., M. Cook, E. Winfree, and J. Bruck. 2008. "Computation with Finite Stochastic Chemical Reaction Networks." *Natural Computing* 7 (4): 615–633.

Soloveichik, D., G. Seelig, and E. Winfree. 2010. "DNA as a Universal Substrate for Chemical Kinetics." *Proceedings of the National Academy of Sciences of the United States of America* 107 (12): 5393–5398.

Thachuk, C. 2013. "Space and Energy Efficient Molecular Programming and Space Efficient Text Indexing Methods for Sequence Alignment." PhD diss., University of British Columbia.

Thachuk, C., and A. Condon. 2012. "Space and Energy Efficient Computation with DNA Strand Displacement Systems." *Lecture Notes in Computer Science* 7433:135–149.

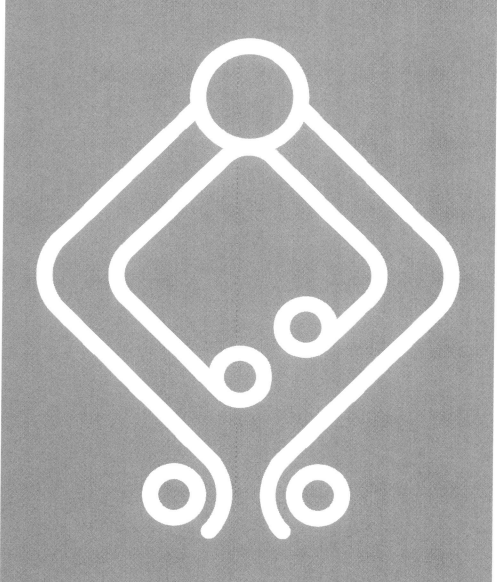

BEYOND NUMBER OF BIT ERASURES: COMPUTER SCIENCE THEORY OF THE THERMODYNAMICS OF COMPUTATION

Joshua A. Grochow, University of Colorado Boulder
David H. Wolpert, Santa Fe Institute

Introduction

With the ever-increasing demands for computing resources—at home, in supercomputers, and in massive data centers—the issue of how much energy is used to run computers is becoming a major challenge for modern society.[1] At the broadest scale, it is estimated that 2.2% to 5% of all US energy is already expended on computing, resulting in a major component of our carbon footprint.[2] At a smaller scale, a supercomputer powerful enough to accurately model the future of the Earth's climate would need to be one hundred times faster than today's fastest supercomputers, but such an "exascale computer would consume electricity equivalent to 200,000 homes and might cost $20 million or more annually to operate" (Markoff 2015). As a final example, well known to everyone who uses a mobile phone or a laptop, one of the major limitations constraining the use of mobile computing is how to store enough energy to run such computers.

---

[1] See Grochow and Wolpert (2018) for an earlier version of the material in this chapter.

[2] See the preface of this volume for references and a discussion on the difficulties in nailing down the precise percentage.

By conservation of energy, so long as all configurations of the bits in a computer have the same energy level, all of the energy used by that computer is ultimately dumped into the world as heat. (Indeed, one of the major challenges in constructing exascale computers is how to keep them from melting.) In other words, computers are energy transducers, converting one kind of energy into another kind, processing information at the same time. The science of energy transduction—the science that governs how much energy is needed to run a computer—is statistical physics. More specifically, it is *nonequilibrium* statistical physics, since computational systems are far from thermal equilibrium.

Because the energy transduction in computation is so intimately coupled to the processing of information, to fully understand the thermodynamics of computation, we need to exploit the relation between information theory and statistical physics. Landauer (1961, 1991, 1996a, 1996b), following the work of Brillouin (1953, 1962) and Szilard (1964), highlighted the connection between Shannon's information-theoretic entropy and thermodynamic entropy, arguing that erasing a bit necessarily pumps $kT \ln 2$ bits of *physical* entropy—essentially, heat—into the environment. Following this line of reasoning, Bennett (1973) began studying how one might design algorithms that needed less energy to run by making more of their steps reversible. This was followed by a series of papers over several decades clarifying and improving the various time-space trade-offs in making computations reversible (Bennett 1982; Fredkin and Toffoli 1982; Bennett 1989; Levine and Sherman 1990; Shizume 1995; Crescenzi and Papadimitriou 1995; Li, Tromp, and Vitányi 1998a, 1998b; Frank, Knight, and Margolus 1998; Lange, McKenzie, and Tapp 2000; Plenio and Vitelli 2001; Buhrman, Tromp, and

Vitányi 2001a, 2001b; Bennett 2003; Sutner 2004; Vitányi 2005; Morita 2008; Tyagi, Lynch, and Demaine 2016; Demaine *et al.* 2016).

However, it was recognized even in the early papers on this subject (Landauer 1961; Bennett 1982) that bit erasure was at most only a small part of the whole story of the thermodynamics of computation. The problem was that when that early work was being done, nonequilibrium statistical physics was still very immature, unable to provide substantial insights beyond those of equilibrium statistical physics. Yet, as mentioned, computers are very far from equilibrium systems, and therefore they cannot be fully analyzed solely using equilibrium statistical physics.

Recently, though, there have been major breakthroughs in nonequilibrium statistical physics, providing mathematical tools far beyond any available when Landauer and others were wrestling with the thermodynamics of bit erasure (Jarzynski 1997; Crooks 1998, 1999; Touchette and Lloyd 2004; Sagawa and Ueda 2009; Hasegawa *et al.* 2010; Takara, Hasegawa, and Driebe 2010; Esposito and Van den Broeck 2010, 2011; Seifert 2012; Sagawa and Ueda 2012; Chejne Janna, Moukalled, and Gómez 2013; Deffner and Jarzynski 2013; Horowitz and Esposito 2014; Pollard 2016; Sagawa 2014; Prokopenko and Einav 2015). These new tools have already resulted in a fully formal analysis of the thermodynamics of bit erasure, in the form of a *generalized Landauer principle* that is both broader in scope and more detailed than the original analysis of bit erasure (Sagawa and Ueda 2009, 2012; Hasegawa *et al.* 2010; Takara, Hasegawa, and Driebe 2010; Sagawa 2014; Parrondo, Horowitz, and Sagawa 2015). These new tools go far further, though, potentially allowing us to analyze issues in computer science theory far more richly than simply counting the number of bit erasures in a computation.

However, reflecting its development, most of the new work in nonequilibrium statistical physics to date has focused on nanoscale physics, without considering the potential implications for computer science. As a result, in the research we and others are currently pursuing on the connections between these new results in nonequilibrium statistical physics and computer science, we are continually stumbling over new connections we had not anticipated. These connections are the topic of this chapter. We emphasize that we fully expect to keep stumbling over many more such connections. As such, this is just a snapshot, consisting primarily of background and questions, which we hope serves as an enticing invitation to researchers in both nonequilibrium statistical physics and computer science theory.

OUTLINE

The primary intended audience for this chapter is computer science theorists, though we hope physicists and others will find the questions raised here interesting and useful as motivation, at the least. Accordingly, we do not review computer science theory here.[3] We presume that many computer science theorists are not well versed in the most recent developments in stochastic thermodynamics (and nonequilibrium statistical physics more generally) and the implications for our understanding of the thermodynamic costs of computation. Accordingly, in the section

---

[3] The reader who wishes a quick summary of some of the relevant definitions in computer science theory is directed to Chapter 2 of this volume, and the reader interested in a more complete introduction to the subject is directed to many excellent texts, such as Arora and Barak (2009), Sipser (2012), Moore and Mertens (2011), Savage (1998), and Hopcroft, Motwani, and Rotwani (2000).

"Quick Introduction to the Modern Thermodynamic Cost of Computation," we review these recent developments. A more detailed review still written for the nonspecialist can be found in Chapter 2 of this volume.

In the section "Thermodynamic Complexity of Circuits," we use the material in "Quick Introduction to the Modern Thermodynamic Cost of Computation" to consider the thermodynamic complexity of circuits. Then, in the section "Uniform Computer Average-Case Thermodynamic Complexity Classes?" we consider average-case *uniform* thermodynamic complexity, that is, the average complexity of machines like Turing machines or more standard algorithms, which are defined for arbitrary-length inputs (in contrast to individual circuits, which are referred to as "nonuniform" computers).

In the section "Randomized Algorithms," we highlight questions about the thermodynamic complexity of randomized algorithms. In the section "Noisy, Approximate, or Inexact Computation," we highlight questions about approximate or inexact computing—computing with inexact hardware—and its thermodynamic implications. Specifically, we ask about the thermodynamic trade-offs of using inexact hardware to implement approximation algorithms and the thermodynamic trade-offs in computing error-correcting codes.

Quick Introduction to the Modern Thermodynamic Cost of Computation

In this section, we review the modern understanding of the thermodynamic cost of computation arising from recent advances in stochastic thermodynamics and in nonequilibrium statistical mechanics more generally (Wolpert and Kolchinsky 2018; Kolchinsky and Wolpert 2016). The computer scientist wishing to get a better understanding of the underlying physics should see Chapter 2 in this volume, and for further details concerning the relationship between logical and thermodynamic reversibility, we recommend starting with Sagawa (2014) and then proceeding to several of the papers cited earlier.

In general, the computational state of a computer could be represented by a *microstate* of a physical system implementing that computer. However, in a typical modern computer, where each location in memory can store one bit—say, the value 0 or the value 1—that one bit is not represented by the state of a single particle (except, perhaps, in quantum computers, where some of these issues go away and others seem to become much more complicated). Rather, a single memory location is represented by the states of a small macroscopic system—too small for us to see but still much larger than a single particle. Thus, when we say a given memory location is in "state 0," we are actually describing a huge number of possibilities for the states of the corresponding particles, which we typically model as follows: the joint state space of all those particles is divided into two regions Z (for "zero") and O (for "one") (or sometimes three regions, indicating an "invalid" value, giving a gap between Z and O). The *macrostate* 0 corresponds, implicitly, to some distribution whose support is Z. This means that our notion of expectation makes sense even for what we would normally consider deterministic computations. When the computation itself is random (as in randomized algorithms) or the input is random (as in average-case analysis), this merely changes the input to a distribution that is not concentrated on any one macrostate, but the thermodynamic formalism is otherwise unchanged.

As is conventional, for simplicity, we assume that at the end of each step by a physical system implementing a given computer, all

Chapter 9: Beyond Number of Bit Erasures

of the computational states of the system have the same energy (or same free energy, if those computational states are macrostates of the underlying physical system rather than microstates; see, e.g., Chapters 2, 13, and 14 in this volume). This has the advantage that it allows us to focus on the thermodynamic implications of the dynamics over those computational states, without having to keep track of the different energy levels of those states.

Following this convention, Landauer argued that the number of bit erasures in a computation is a thermodynamic cost of that computation. This number of erasures is now understood to be a special case of a more generally applicable thermodynamic cost of a computation, sometimes called the *generalized Landauer bound*. This more general cost is the change in entropy (of the distribution over states of a physical computer) as the computer's input is mapped to its output, times kT, where k is Boltzmann's constant and T is the temperature. As its name suggests, this cost is a lower bound on the actual amount of energy needed to run a computation, achieved only by idealized systems.[4]

As an example, suppose that we are given a system with two states and that the initial distribution over those states is uniform. Then, after an erasure, the distribution becomes a delta function. Accordingly, the change in entropy is $\ln[2]$, leading to a generalized Landauer's bound of $kT \ln[2]$. Moreover, in many scenarios, if a computer performs a sequence of n successive bit erasures, then the lower bound on the total amount of energy needed is $nkT \ln[2]$, in agreement with Landauer's conclusion.

The recent advances in nonequilibrium statistical mechanics provide an *exact* decomposition of the full thermodynamic cost of

---

[4]When the system is connected to more than one heat bath, at different temperatures, the thermodynamic implications of a given drop in entropy are more subtle. In this case, the drop in entropy is a lower bound on the "entropy flow" from the system to the heat baths. See Van den Broeck and Esposito (2015), Esposito and Van den Broeck (2010), and Seifert (2012).

a computation on a real computer, going beyond the lower bound provided by the generalized Landauer bound. More precisely, we now have an exact decomposition of the total thermodynamic work done on the system during the computation (given our presumption that all states of the system have the same energy). This work is a sum of three terms:

1. The *generalized Landauer cost*. As described earlier, this is the natural generalization of the traditional cost identified by Landauer. Note that, unlike merely counting the number of bits erased, this generalization naturally handles other functions, such as AND—which, while it loses information, loses *less* information than fully erasing a bit, in that for some values of its output, its input can be recovered. This generalization also handles stochastic or randomized "functions."[5]

   Crucially, the drop in entropy during a computation varies with changes to the distribution of inputs to that computation. For example, a bit erasure has a drop in entropy of $kT \ln 2 - kT \ln 1 = kT \ln 2$ (the Landauer bound) only if the distribution over inputs to a bit-erasing device is uniform; nonuniform distributions will result in a smaller drop in entropy. One particularly important implication of this dependence is that the manner in which the early gates in a circuit modify the input distributions of the downstream gates affects the (summed total over all gates in the circuit of the) Landauer cost of running that circuit.

2. The *cost of doing business* in a physical world. The theoretical minimum of thermodynamic cost given by Landauer's

---
[5] It is worth noting that Landauer (1961) discussed stochastic functions in general terms.

Chapter 9: Beyond Number of Bit Erasures

bound can only be reached if each computational step is performed *quasistatically slowly*; formally, this means so slowly that the system is always at equilibrium and therefore there is no heat exchange with the environment, but, practically, this just means "infinitely slowly." When each computational step takes finite time—as it does in any real computer—an additional cost is incurred, which depends on just how long that step takes. This cost has two terms, both of which are nonnegative:

a) The *mismatch cost*, sometimes alternatively called the "cost due to incorrect priors." Any physical device implementing a given computation is, either implicitly or explicitly, "tuned" to a certain input distribution I_{opt}, for which it incurs a minimal thermodynamic cost. The mismatch cost for using that device with an actual input distribution p is given by the drop in Kullback–Leibler (KL) divergence between p and the optimal distribution I_{opt}, as the computation unfolds (Cover and Thomas 2012). (The fact that this drop in KL divergence is nonnegative is sometimes called the "data processing inequality for KL divergence.") More precisely, if a physical device computing a function f is thermodynamically optimized for the input distribution I_{opt}, but it is run on input distribution I_{actual}, then the mismatch cost is

$$D(I_{\text{actual}}||I_{\text{opt}}) - D(f(I_{\text{actual}})||f(I_{\text{opt}})), \quad (9.1)$$

where $D(p||q)$ denotes the KL divergence $\sum_x p(x) \log \frac{p(x)}{q(x)}$.

b) The *residual entropy production*. This is a linear function of the initial distribution, p.

In the optimal case, the residual EP equals zero for all initial distributions p, and in addition, $I_{\text{actual}} = I_{\text{opt}}$, so that the drop in KL divergence is zero, that is, the mismatch cost is zero. In this best possible case, the cost of doing business is zero, and the total work used to perform the computation is just the Landauer cost. Note, though, that even if residual EP equals 0 for all p, in general, changing I_{actual} from I_{opt} will result in nonzero cost of doing business. In other words, even if a physical computer has been engineered to perform optimally, in general, that optimal behavior will only hold for one particular initial distribution p; changing the initial distribution while not making corresponding changes to the physical computer will result in nonzero cost of doing business, due to the nonzero mismatch cost.

There is a physical distinction between the Landauer cost and the other two thermodynamic costs, namely, the Landauer cost is the energy (more precisely, thermodynamic work) that can always in principle be recovered from the environment. In contrast, the other two thermodynamic costs are the expected *dissipated work*—essentially, how much work is irreversibly lost to the environment during the computation. Throughout this chapter, we will ignore this distinction, for two reasons: (1) while, in principle, the Landauer cost can be recovered from the environment, in practice, this seems horrendously difficult, if not effectively impossible from an engineering perspective, and (2) we are trying to present as simple a model as possible that captures the relevant phenomenon, in the hope of attracting computer science theorists to work on it!

From now on, when we refer simply to the "thermodynamic cost" of a computation, we mean the sum of these three terms:

> generalized Landauer cost + mismatch cost + residual EP.

Chapter 9: Beyond Number of Bit Erasures

(The sum of the second two terms, i.e., the cost of doing business, is also called *(irreversible) entropy production* (EP) in the literature.) While the decomposition into generalized Landauer cost plus cost of doing business has been known for around a decade, the decomposition of the cost of doing business into terms (2a) and (2b) is a more recent result. Calculating (2b) is mostly a question of the physics implementation of the computer, and so we set it aside for the remainder of this chapter. However, cost (2a), which was only discovered in the past two years (Kolchinsky and Wolpert 2016), leads to many new interesting questions in computational complexity, which we elaborate on in the following pages.

The thermodynamic costs of running a system are, by definition, expectations.[6] However, recall that the computational states of the system may be encoded as macrostates, defined by a coarse graining of the microstates of the system. As a result, the expectations defining the thermodynamic cost *makes sense even for deterministic, worst-case analysis*. The expectation need not be an expectation over the Boolean inputs—whose binary values can be encoded in the macrostates of a physical system—but will always be an expectation over the probability distribution over *microstates* of the system, that is, the states of all of the $\approx 10^{23}$ physical particles that make up a single computational state of the system.

---

[6]This is not entirely accurate, as there has been extremely important work in stochastic thermodynamics that focuses on the relative probabilities of individual trajectories and their thermodynamic consequences. Indeed, focusing on individual trajectories is crucial for deriving the famous detailed and integral fluctuation theorems, e.g., Crooks's theorem and the Jarzynski equality (Van den Broeck *et al.* 2013; Crooks 1998; Jarzynski 1997). Such a focus is also the basis for so-called one-shot thermodynamics (see, e.g., Müller and Scharlau 2014; Halpern *et al.* 2015). However, as far as we are aware, none of this work has yet been scrutinized for its potential implications for the thermodynamics of computation, so we won't explore these issues further.

A CAVEAT ABOUT CIRCUITS

We often think of a two-bit gate such as binary AND as having two inputs and one output. Several different physical systems are consistent with such an interpretation. In one of them, an AND gate is a physical device hooked up to three wires, two of which we think of as "inputs" and one as the "output," and therefore the AND gate is actually a function taking three bits to three bits. If this function is $(x, y, z) \mapsto (x, y, z \oplus (x \wedge y))$, we get the reversible Toffoli gate. If it is instead $(x, y, z) \mapsto (x, y, x \wedge y)$, we get a different (irreversible) gate. Alternatively, we might consider an AND "gate" as merely a convenient representation of some time-varying Hamiltonian that alters the state of two bits, perhaps transforming the pair (x, y) into the pair $(x \wedge y, y)$; yet, alternatively still, we might consider it as the function $(x, y) \mapsto (x \wedge y, 0)$. One might argue that this latter interpretation is closest to the "two-input, one-output" interpretation, in that its Landauer cost matches the natural interpretation of an AND gate. So we will use this one throughout the following analysis, unless otherwise specified. As one might imagine, the computation of thermodynamic cost will depend, sometimes mildly and sometimes more significantly, on which interpretation we choose, so it is important to consider this issue carefully in any calculation.[7]

ON REVERSIBLE COMPUTATION

Note that if the function f being computed is logically invertible, then both the Landauer cost and the mismatch cost

---

[7] Multiple subtleties arise when interpreting gates as subsystems of circuits, due to the need to wire all those gates together, and due to the common (albeit often implicit) requirement that the costs of running a circuit must be independent of how often we have already run it (see Wolpert and Kolchinsky 2018). However, none of those subtleties are crucial for the following analyses.

Chapter 9: Beyond Number of Bit Erasures

become zero, regardless of what input distribution we use (since the KL divergence doesn't change when applying f). In this sense, reversible simulation, in which we embed some desired irreversible computation f on n bits into a reversible computation over more than n bits (an idea going back to Bennett 1982) still has a desirable effect.

However, for a computer to be reused—or for a program to be used as a subroutine in a larger program—we must take into account the fact that both the input and the *entire* output must be reset at the end of the computation.

Bearing this in mind, consider the now-standard trick of embedding an irreversible function f into the reversible function $f_r(x) = (x, f(x))$. The cost of erasing the second component of the output, $f(x)$, is a "sunk cost," which would arise no matter what kind of computer we used.[8] Moreover, because f_r is logically reversible, there is no Landauer cost incurred when it produces the output $(x, f(x))$. On the other hand, if we had instead used a conventional computer, then in addition to the sunk cost, there would be a nonzero Landauer cost for producing that computer's output $f(x)$ from the input x. So it would seem that by using the reversible function f_r, we have reduced the Landauer cost over what it would be with the conventional computer computing f.

This is misleading, though. In the case of embedding f in a reversible function f_r, in addition to the sunk cost of erasing $f(x)$, we face an additional cost of erasing the other

---

[8] Typically in the literature, this cost is hidden: one copies $f(x)$ to an external system *that has been initialized to be all 0s*, erasing the original version of the output on the computer in the process. This precise operation has zero direct Landauer cost, because no information is lost. However, there is still a nonzero Landauer cost when we consider the total energy budget over all time—it arose when we initialized the external system to receive the copy of $f(x)$.

component of the output, x. (This cost does not apply for the conventional computer, because its output does not contain a copy of x.) It is straightforward to show that the cost of doing this is precisely the Landauer cost of simply using the conventional computer to implement f directly, without the complication of embedding it as f_r and then running that. Ultimately, whether f_r is considered to have a lower Landauer cost than f depends on the computational machinery in which the circuit is embedded that feeds inputs into the circuit and reads outputs out of it, on how that machinery couples with the circuit, and on how we do the accounting of ascribing costs to either the circuit itself or that embedding machinery.

In light of this, here we focus almost entirely on implementing circuits with AND, OR, NOT, and XOR gates, for several reasons. First, as just discussed, the single stipulation that we use only reversible gates doesn't eliminate thermodynamic cost. More practically, whatever advantages they may or may not have (depending on the embedding computational machinery), reversible gates are not yet in widespread use, despite more than forty years of arguing for their thermodynamic efficiency, and it does not seem likely that reversible gates will completely replace standard CMOS transistors any time soon. Third, even when reversible gates are more readily available—perhaps due to increasing economic pressure coming from the end of Moore's law—there may be advantages to using computations that are only *partially* reversible (Demaine *et al.* 2016; Tyagi, Lynch, and Demaine 2016). And finally, our initial investigations suggest that just over the horizon is a rich and beautiful theory to be developed of the thermodynamic cost of circuits and computations built using AND, OR, NOT, and (sometimes) XOR, beyond just the fact that erasing a bit has a Landauer cost of 1, and we hope to convince the reader of this.

The interested reader should read the last section of Chapter 2 of this volume for a more elaborate and formal analysis of these issues.

Thermodynamic Complexity of Circuits

For nonuniform (circuit) complexity, it is simple and natural (though not necessarily easy!) to consider the average case as being the uniform distribution over $\{0,1\}^n$, though of course we are also free to consider other distributions. For example, other distributions arise naturally when the input describes a graph coming from some stochastic model network growth; see, e.g., Newman, Barabási, and Watts (2006, chapter 4) and Chung and Lu (2006).

Accordingly, we presume one has a box full of gates, each optimized for the uniform distribution, and cost-free wires to connect them, and we wish to know the best way to do so to compute a given Boolean function. Although we feel well motivated to restrict our attention to AND, OR, and NOT gates, doing so is anathema to the general theme of considering measures of complexity that are robust to some large and natural class of changes in the underlying model and runs the risk that some future technological change would invalidate the ensuing results. So instead of using only $\{\wedge, \vee, \neg\}$, we allow any gate set that is "input–output conservatively equivalent" to $\{\wedge, \vee, \neg\}$, in the following sense.

Definition. *(Input–output conservative simulation). One gate set* $G = \{g_1, \ldots, g_k, \ldots\}$ *IO-conservatively* $s(n)$*-simulates another gate set* $H = \{h_1, \ldots, h_k, \ldots\}$ *if, for all i, if h_i has arity a_i, then there exists a G-circuit C_i with at most $s(a_i)$ gates,* *and possibly some of its inputs fixed to constants, that computes the same function as h_i and has the same number of inputs as h_i (excluding the constant inputs) and the same number of outputs as h_i.*

G and H are IO-equivalent *if each IO-conservatively O(1)-simulates the other.*

We won't explicitly use this definition until we consider uniform complexity classes (see the section "Uniform Computer Average-Case Thermodynamic Complexity Classes?"). Nonetheless, it is implicitly present throughout the rest of this section.

Note that the gate sets we consider may be infinite, as in the case of the standard AC^0 gate set, which contains AND gates of arbitrary arity. Note also that while we allow fixing some of the inputs to constants, we do not allow discarding outputs in this definition; "discarding an output" must be accounted for thermodynamically. For example, $\{\wedge_2, \vee_2, \neg\}$ and $\{\wedge_2, \oplus_2\}$ (we use subscripts to denote fan-in or arity) are IO-conservatively equivalent, but $\{\wedge_2, \vee_2, \neg\}$ is not IO-conservatively equivalent to the singleton gate set consisting of the arity-3 Toffoli gate, as any Toffoli-only circuit has at least three outputs.

For a given Boolean function f, we may calculate the difference in entropy between its input and output, giving us the Landauer cost of f. We use the term *Landauer cost of implementing f* to mean the drop in entropy of a given initial distribution that is evolved according to f. This is the minimal thermodynamic cost that could be achieved by an *unconstrained* physical system, that is, by an *all-at-once* (AO) device that directly implements f with a single gate (see for a more formal definition of AO devices Wolpert and Kolchinsky 2018). But when Boolean functions are implemented in practice, they are typically implemented on a device more like a circuit composed of gates from our bag than by a monolithic, AO device. Requiring the physical system to allow only the couplings among its variables specified by the "wiring diagram" of such a circuit introduces extra constraints, which do not restrict an AO device. These extra constraints on circuits that do not apply to AO devices can raise the minimal thermodynamic cost that can be achieved by such a circuit above the minimal cost that can be achieved by an AO device, even though both implement the same function f. Another way of seeing this, illustrated later, is to notice that when implementing f with a circuit whose gates come from a fixed gate set, running *each gate* in the circuit incurs a Landauer cost, and the sum of these costs over all the gates may be more than the Landauer cost of f. Indeed, for a fixed set of gates, this may be true for *any* circuit composed of those gates that implements a given f (see the related discussion in Wolpert and Kolchinsky (2018) and Boyd (2018); see also Chapter 2 in this volume). Minimizing this additional Landauer cost that arises due to using circuits rather than AO devices is a new question in circuit complexity, which appears to pose numerous rich research issues.

The same general phenomenon can be true not just of the Landauer cost but also of the mismatch cost; that cost is also dependent on circuit architecture. Moreover, the change in mismatch cost (due to using a circuit rather than an AO device) has a different dependence on the circuit than does the generalized Landauer cost. (Indeed, as shown subsequently, the mismatch cost may *shrink* when using a circuit rather than an equivalent AO device.) Accordingly, choosing a thermodynamically optimal design for a circuit involves trading off these two costs.

All of these phenomena are elaborated in full formal rigor in Wolpert and Kolchinsky (2018). However, here our interest is more pedagogical, to point to the kinds of questions that arise when computer science theory is expanded to incorporate thermodynamic costs, while eliding the subtleties of the underlying device physics. Accordingly, here we do not present the fully rigorous equations governing these phenomena (and the background machinery needed to understand the associated equations). Rather, in the next subsection, we give some illustrative examples.

The general question that the reader should bear in mind while going through those examples is the following:

Question 1. *Fix a universal gate set. For a given function f, what is the minimum over circuits C computing f of the worst-case thermodynamic cost of C? Is this problem* NP-*hard when f is given as a truth table? Does it reduce to the Minimum Circuit Size Problem?*

ILLUSTRATIVE EXAMPLES

As a matter of notation, except where explicitly indicated otherwise, all logarithms are base 2, with the multiplicative factor of ln[2] being implicitly absorbed into Boltzmann's constant by our choice of units. The reader should also recall standard computer science theory notation, for example, that $\oplus$ is the XOR, which is also the two-bit parity operation. Also, we use the standard notation that $\mathbb{F}_2$ is the group of bits where the binary operation is addition modulo 2, and so $\mathbb{F}_2^n$ is the group given by n-tuples of bits where the binary operation is simultaneous n-fold addition modulo 2. Finally, recall that a distribution over n-tuples is *pairwise independent* if each of its two-variable marginals is a product distribution. For the case of triples of binary variables, every pairwise independent distribution can be described as follows: there is some value p such that each string of even parity has probability $p/4$ and each string of odd parity has probability $(1-p)/4$. We will denote this distribution as d_p.

In our first example, using a limited gate set doesn't significantly change the Landauer cost but has a significant effect on the mismatch cost.

Example 1. (Landauer Cost of an AND Tree). *The traditional transformation of an unbounded fan-in AND gate into a tree of binary AND gates increases Landauer cost, by an amount that goes to zero in the limit that the number of input bits, k, goes to infinity.*

Proof. To see this, first note that the Landauer cost of a fan-in k AND gate is just $k - H(2^{-k})$, where we use $H(p)$ to denote $-p \log p - (1-p) \log(1-p)$. Now suppose we replace such a fan-in k AND gate with a tree of fan-in 2 AND gates; for simplicity, suppose that k is a power of 2, say, $k = 2^\ell$. So we have a perfect binary tree of AND gates of depth $\ell - 1$.

The gates at the input actually receive uniformly random inputs, so each of these incurs a Landauer cost of $2 - H(\frac{1}{4}) \approx 1.189$, for a total cost (among all gates at the first level) of $\frac{k}{2}(2 - H(\frac{1}{4}))$. However, each output of one of the first-layer AND gates has a $(\frac{1}{4}, \frac{3}{4})$ distribution. The entropy of the input of a second-layer AND gate is then $2H(\frac{1}{4}) \approx 1.623$, and the output distribution is then $(\frac{1}{16}, \frac{15}{16})$, with entropy $H(1/16)$, so the Landauer cost of a second-layer AND gate is $2H(\frac{1}{4}) - H(\frac{1}{16}) \approx 1.285$. As there are $k/4$ gates in this second level, the gates at the second level have a total Landauer cost of $(k/4)(2H(\frac{1}{4}) - H(\frac{1}{16})) = \frac{k}{4}H(\frac{1}{4}) - \frac{k}{4}H(\frac{1}{16})$.

Noting that the $(k/2)H(\frac{1}{4})$ cancels the negative term from the first layer, it is easily verified that this pattern continues and we get a telescoping sum whose result is $k - H(4^{-(\ell-1)})$; for large k, this is asymptotically equal to the cost of a single k-input AND gate. □

Example 2. (Cost due to Incorrect Priors for an AND Tree). *Continuing the preceding example, suppose all of our AND gates with fan-in 2 are thermodynamically optimized for uniformly random inputs; that is, on uniformly random inputs, the thermodynamic cost of any one of these AND gates is precisely the (generalized) Landauer cost, with zero mismatch cost. However, the input distributions to the gates in the AND tree will not all be uniform, even if the input distribution to the first layer of gates is uniform. Indeed, for such a distribution of inputs, using the AND tree rather than a single AO device increases the mismatch cost by*

$$= \sum_{i=1}^{\ell} 2^{\ell-i} \left(\frac{2^{2i} - 1}{2^{2i}} \log(3) + \frac{2^i - 1}{2^i} \log(2^i - 1) \right)$$

$$\approx 2^{\ell}(\log(3) + 2) - \ell \approx 3.6k - \log k = \Theta(k),$$

where we define $k = 2^{\ell}$.

Proof. Under our hypothesis, the first-layer AND gates receive the uniform distribution, which is what they are optimized for, so the mismatch cost for these gates is zero.

But second-layer AND gates get a $(\frac{1}{4}, \frac{3}{4})$ distribution, so they also incur a cost because they are optimized for a different input distribution than they receive. The mismatch cost for each such second-layer AND gate is then

$$\frac{6}{16} \log(3/4) + \frac{9}{16} \log(9/4) - \left(\frac{15}{16} \log(5/4) \right) \approx 0.201.$$

Iterating through the entire circuit, and with abuse of notation, we find that the mismatch cost of all the gates is

$$\sum_{i=1}^{\ell} 2^{\ell-i} \left((2H(2^{-i}) - H(2^{-2i})) \right)$$
$$+ KL(2^{-i} \times 2^{-i} || 1/2 \times 1/2)$$
$$- KL(2^{-2i} || 1/4))$$

$$= \sum_{i=1}^{\ell} 2^{\ell-i} \left(\frac{2^{2i}-1}{2^{2i}} \log(3) + \frac{2^{i}-1}{2^{i}} \log(2^{i}-1) \right)$$

$$\approx 2^{\ell}(\log(3)+2) - \ell \approx 3.6k - \log k = \Theta(k),$$

as claimed. □

In contrast to the previous example, the mismatch cost of implementing a function with a circuit can also be *less than* the mismatch cost of an equivalent AO device—if both those gates and the AO device are optimized for a distribution q that differs from the actual distribution p. This is illustrated in the next example.

Example 3. *Consider the three-bit function $f(xyz) = x1z$, which maps y to 1 while leaving x and z unchanged. Suppose we run this function on the input distribution $p(000) = p(111) = 1/2$. Suppose as well that we can implement f using either an AO device that is thermodynamically optimized for the input distribution $q(000) = q(010) = q(101) = q(111) = 1/4$, or a set of three unary gates that are separately optimized for the three marginal distributions of q (one marginal for each bit). So the three unary gates are each optimized for a uniform distribution over a single bit.*

First, note that, using the AO device, we find that the mismatch cost is infinite, as we are taking the KL divergence

$D(p||q)$ but the support of q is strictly larger than that of p. (We can avoid infinities by changing all zero-valued probabilities to be some small ε, but the mismatch cost is still arbitrarily large.)

In contrast, note that the input distribution p marginalizes to the uniform distribution over each bit, so if we use the unary devices to implement f, the mismatch cost vanishes, as the unary gates are being run on the distribution for which they were optimized. (The Landauer cost differs in the opposite direction— 0 for the AO case versus 1 in the unary case—but this small constant is outweighed by the arbitrarily large difference in the mismatch cost.)

Although the preceding example is obviously artificial, we could attempt to make it less so by restricting the minimum probability assigned to any n-bit string to be, say, $\Omega(2^{-n})$.

These examples suggest the following question, which illustrates some of the range of research issues in the thermodynamics of circuits.

Question 2. *Is there an infinite family of functions $f_n \colon \{0,1\}^n \to \{0,1\}^m$ such that the thermodynamic mismatch cost of implementing f_n with gates of bounded fan-in on uniformly random inputs is asymptotically smaller than the mismatch cost of implementing f_n in an AO device on uniformly random inputs?*

Here are two examples to provide some more food for thought:

Example 4. *Consider the function $g \colon \mathbb{F}_2^3 \to \mathbb{F}_2^3$ defined by $g(x,y,z) = (x \wedge y, x \wedge z, y \wedge z)$. Let u be the uniform distribution over all of $\mathbb{F}_2^3$, and suppose we have a device that computes g "in one shot," with an AO device, and that is thermodynamically optimized for u. However, suppose that the* *actual input distribution is the pairwise uniform distribution d_p (see beginning of this subsection). The Landauer cost of g on d_p is not the nicest expression in terms of p. Neither is the mismatch cost. However, when we add the two, many of the terms cancel, and we are left with $(3/2) - p$ as the total thermodynamic cost for an AO device that implements g when the actual distribution is d_p but the device is optimized for u (see calculation 1 in the appendix).*

In contrast, if we implement g in the obvious way with three binary AND gates, and assume that those binary AND gates are each optimized for the uniform distribution on two bits, then the mismatch cost becomes zero, *while the Landauer cost becomes three times that of a single AND gate, or approximately* 3.566. *So while implementing g with binary gates significantly reduces the mismatch cost (because each AND gate only sees two bits at a time), and the two-bit marginals of d_p are equal to those of u, implementing the function with a circuit significantly* increases *the Landauer cost, roughly speaking, because no single AND gate could see sufficiently much information concerning the three-bit output. Moreover, that increase in Landauer cost overwhelms the decrease in the mismatch cost.*

Example 5. *Consider the not-all-equal function*

$$NAE_k \colon \mathbb{F}_2^k \to \mathbb{F}_2,$$

which is zero precisely when all k input bits are equal. The Landauer cost of an AO device implementing this function on uniformly random inputs is $k - H(2^{1-k})$, for example, the Landauer cost of an AO device implementing NAE_3 on such a distribution is $3 - H(\frac{1}{4})$.

Now consider the implementation of NAE_3 as a circuit of three binary gates, implementing the operation $(x \oplus y) \vee (x \oplus z)$.

Each of the three gates in this circuit receives uniformly random inputs, so there is zero mismatch cost (one of the advantages of using XOR gates!). Therefore the total thermodynamic cost is twice that of a single XOR gate plus that of an OR gate, namely, $2 \times (2-1) + (2 - H(\frac{1}{4})) = 4 - H(\frac{1}{4})$. So it is exactly one bit more than the Landauer cost of an AO device that implements NAE_3.

In addition, it is not hard to see that if the actual distribution over the three input bits is the pairwise-independent input distribution d_p (as in the previous example), then the mismatch cost remains zero both for the AO device that implements NAE_3 and for the preceding circuit that implements it. The Landauer cost of the preceding circuit on input d_p also doesn't change. However, the Landauer cost of the AO device implementing NAE_3 reduces to $2 + H(p) - H(\frac{1}{4})$.

Note that just because these examples involve two-input gates that are optimized for uniform input distributions does not mean that using an actual input distribution that is pairwise-independent distribution automatically makes the mismatch cost zero. The reason is that any gates beyond the first layer could still receive inputs that are not uniformly distributed, essentially capturing correlations in the results of the partial computations along the way. We invite the reader to consider using circuits made from the gates $\{\vee, \wedge, \neg\}$ for implementing NAE_3, their Landauer cost, and the mismatch cost on such circuits when the input from $\mathbb{F}_2^3$ is drawn from d_p, where this effect can be seen.

LINEAR FUNCTIONS

The class of linear functions is one of the simplest nontrivial classes of functions. Both as a test case in complexity theory and for practical purposes, for a given linear function $A: \mathbb{F}_2^n \to \mathbb{F}_2^m$, it is natural to ask about the smallest *linear circuit* that computes A, that is, the circuit consisting entirely of XOR gates that computes A.

Although much is obviously known about linear functions, much less is known about their complexity when implemented with linear circuits. For example, a simple counting argument shows that with high probability, a randomly chosen linear function cannot be computed by a linear circuit of fewer than $\omega(n^2/\log n)$ XOR gates. Nonetheless, there is no known *explicit* family of linear functions $A_n\colon \mathbb{F}_2^n \to \mathbb{F}_2^n$ that requires even $\omega(n)$ XOR gates to compute. Valiant (1976) famously introduced the notion of matrix rigidity and showed that a family of sufficiently rigid linear functions could not be computed by linear circuits that were simultaneously of $O(n)$ size and $O(\log n)$ depth; see Codenotti (2000) and Lokam (2000) for surveys and the introduction of Alman and Williams (2016) for more recent citations. In this section, we begin exploring the thermodynamics of linear circuits.

Unlike the case of the AND tree, note that if a binary XOR gate is given the uniform distribution over two bits as input, then its output is a uniformly random bit. This means that the mismatch cost is the same for an AO device implementing an n-bit XOR as it is for an n-bit tree of binary XORs (assuming that the prior distribution q is uniform for all gates and for the AO device).

Furthermore, given a linear circuit over $\mathbb{F}_2$, when provided uniformly random input u, the distribution of pairs of bits (x, y) on *any* pair of wires satisfies one of the following: it is uniform over $\mathbb{F}_2^2$, it is uniform subject to $x = y$, it is uniform subject to $x \neq y$, or at least one of the bits takes a fixed value. Now, consider the two bits (x, y) coming in to some XOR gate in the middle of the circuit. If they are in any of the latter three cases, then that XOR gate is redundant: if the two bits are always equal, or always unequal, then the output of the XOR gate is fixed and the XOR gate can be replaced by a constant; if one of the inputs is fixed, then the XOR gate acts either as the identity map on the other input or as negation, and negations can be safely pushed to the output gate of the circuit (using the fact that $z \oplus \neg y = z \oplus (1 \oplus y) = 1 \oplus (z \oplus y)$), where they must all cancel, since the map A itself is linear, that is, it has no constant term. Thus, if the inputs are uniformly distributed, then every XOR gate in the circuit gets as input the uniform distribution over $\mathbb{F}_2^2$, and the mismatch cost is zero.

Next, given any linear map $A\colon \mathbb{F}_2^n \to \mathbb{F}_2^m$, if u is chosen uniformly from $\mathbb{F}_2^n$, then Au is uniformly distributed over the image of A. Thus, in this case, the entropy of the input is precisely n bits, and the entropy of the output is rkA bits, so the Landauer cost of an AO device is the nullity of A, $n - rkA = \dim \ker A$. If A is invertible, then this is zero, as expected from the classical analysis of reversible computations.

However, when we implement such a linear map with a linear circuit of XOR gates, there is now a nontrivial Landauer cost. In particular, the Landauer cost of a single binary XOR gate with uniformly random input is 1 bit; thus, for uniform inputs, we essentially return to the hard problem with which we began this section:

Proposition 1. *Given a linear map $A\colon \mathbb{F}_2^n \to \mathbb{F}_2^m$, the minimum Landauer cost of a linear XOR circuit computing A is precisely equal to the minimal number of gates in such a linear circuit.*

As computing the minimum linear circuit is NP-hard (Boyar, Matthews, and Peralta 2013), we get the following corollary.

Corollary 1. *Computing the minimum thermodynamic cost of linear circuits with uniform input distribution is NP-hard.*

However, if the input is no longer uniformly distributed, the picture can change significantly. In particular, if we still assume that our XOR gates are fixed to optimize their thermodynamic cost for the uniform distribution, then we must suddenly take into account the mismatch cost, and this cost can change from gate to gate within a circuit. However, we can then drop this assumption and ask about the complexity.

Given a distribution μ over $\mathbb{F}_2^2$, for brevity, let us say the μ-*thermodynamic linear complexity* of a given $\mathbb{F}_2$-linear map A is the minimum expected thermodynamic work over all linear circuits composed of XOR gates that are thermodynamically optimized for inputs drawn from μ. In addition, recall that a rational distribution d on Σ^* is called polynomial-time computable if, given $x \in \Sigma^*$, $d(x)$ can be computed in poly($|x|$) time.

Question 3. *Given a polynomial-time computable distribution on $\mathbb{F}_2^n$ and a matrix $A \in \mathbb{F}_2^{n \times m}$, is it NP-hard to compute the distribution μ over $\mathbb{F}_2^2$ that minimizes the μ-thermodynamic linear complexity of A? Given μ and A, is it NP-hard to compute the linear circuit that minimizes the μ-thermodynamic linear complexity?*[9]

Because a linear circuit need never have more than $O(n^2)$ gates, it is clear that the decision version of this problem—that is, "given μ and $A \in \mathbb{F}_2^{n \times m}$, and $t \in \mathbb{Q}$, is there a linear circuit computing A whose μ-thermodynamic linear complexity is $\leq t$?"—is in NP. Indeed, if we generalize from linear circuits and ask if a given function has a circuit of size s with μ-thermodynamic cost at most t (i.e., a total thermodynamic cost for

---

[9]The reason we require polynomial-time computability in this question is that the distribution on $\mathbb{F}_2^n$ must be *somehow* succinctly specified. After all, if it required 2^n bits to specify, then we would easily be able to solve the preceding problem in polynomial time in its input size, because the input size itself would be exponentially large in n.

initial distribution $P = \mu$ of t), then the corresponding decision problem is also in NP. We don't expect any of these problems to be in P; they seem closely related to the Minimum Circuit Size Problem (for a recent reference, see, e.g., Hirahara, Oliveira, and Santhanam 2018).

Uniform Computer Average-Case Thermodynamic Complexity Classes?

In this section, we focus on "uniform computers," that is, computational machines like Turing machines (TMs), which, in contrast to individual circuits, are defined for arbitrary size inputs (see Arora and Barak (2009); see also Chapter 2 in this volume). Despite its potential utility, in general, much less is known about the average-case complexity of running uniform computers rather than about their worst-case complexity, where "complexity" is quantified, for example, as the number of iterations needed by the computer to perform a given computation. Indeed, the rigorous study of average-case complexity really only began with Levin (1986); see Bogdanov and Trevisan (2006) for a relatively recent survey that, despite being ten years old, is not particularly out of date. In this section, we consider the average-case *thermodynamic* complexity of uniform computers, run on inputs drawn from some distribution. However, before attempting to define uniform complexity classes for average-case thermodynamic complexity, we begin, in the next subsection, with a review of some of the relevant issues from computer science theory.

REVIEW OF UNIFORM AVERAGE-CASE COMPLEXITY

First, we recall the standard average-case complexity classes. A *distributional problem* consists of a decision problem $L \subseteq \{0, 1\}^*$ and a probability distribution μ over $\{0, 1\}^*$. The class AvgP consists of those distributional problems (L, μ) such that there exists an algorithm A for L whose running time t_A is polynomial on μ-average, in the sense that there is some $\varepsilon > 0$ such that $E_{x \sim \mu}[t_A(x)^\varepsilon] < \infty$.

However, for uniform computers like TMs, we have to be careful about what distributions μ over the inputs we consider. In the case of circuit complexity, it was natural to use the uniform probability distribution on $\{0, 1\}^n$. But when we wish to consider uniform computers like TMs, we need distributions over the set $\{0, 1\}^*$ of all strings, on which there is no natural "uniform" distribution (which assigns equal probability to every string). One approach is to consider the distribution given by the universal semi-computable semi-measure (Levin 1974), as in algorithmic information theory; this leads to some interesting theory but, as with Kolmogorov complexity itself, is highly uncomputable. Another natural approach, more widely applicable, would be to allow arbitrary distributions over $\{0, 1\}^*$. But then we can get pathological behavior, and the results that we hope would be true seem impossible to prove, if not outright false.

On the other hand, an argument can be made that all distributions that occur "in nature" should be polynomial-time samplable, in the following standard sense (see, e.g., Bogdanov and Trevisan 2006).

Definition. *A distribution D on $\{0, 1\}^*$ is* polynomial-time samplable *or* p-samplable *if there is a polynomial p and a randomized algorithm A such that $A(n) \in \{0, 1\}^n$ for every n, $A(n)$ runs in at most $p(n)$ steps regardless of its coin tosses, and for all $n \in \mathbb{N}$ and $x \in \{0, 1\}^n$, we have $Pr[A(n) = x] = D(x)$.*

Another more restrictive but still useful definition covers the case when the cumulative distribution can be computed in polynomial time.

Definition. *A distribution D on $\{0,1\}^*$ is* polynomial-time computable *or* p-computable *if there is a polynomial p and a deterministic algorithm A such that, for every x, $A(x)$ runs in at most $p(|x|)$ steps, and $A(x) = Pr_{y \sim D}[y \preceq_{lex} x]$, where $\preceq_{lex}$ denotes length-lexicographic order.*

The class DistNP consists of those distributional problems (L, μ) such that $L \in$ NP and μ is p-computable. DistNP has complete problems (Levin 1986), though this is much less obvious than the case for NP. If AvgP = DistNP, then EXP = NEXP (Ben-David et al. 1992). (While there is also a class SampNP, where μ is only p-samplable, the most commonly used average-case analogs of P and NP are AvgP and DistNP, respectively.)

POSSIBLE DEFINITIONS OF THERMODYNAMIC AVERAGE-CASE COMPLEXITY CLASSES

One natural definition of average-case thermodynamic complexity classes to investigate would be the classes of distributional problems (L, μ) that can be solved by a TM such that the thermodynamic cost of running that TM with an AO device, t, satisfies $E_{x \sim \mu}[c^{t(x)}] < \infty$ for some $c > 1$. In keeping with the notation introduced earlier, we might write this class as

$$\text{AvgTHERMO}(\log n).$$

It is similarly natural to investigate $\text{SampTHERMO}(t(n))$ and $\text{DistTHERMO}(t(n))$, defined in the obvious way.

At least two other complexity classes might be worth investigating. First, we could consider time–space–energy classes $\text{AvgTISPE}(t(n), s(n), e(n))$, bounding the time, space, and energy (thermodynamic cost) simultaneously. Note that, when we restrict time and space, it becomes possible that restricting the thermodynamic cost to be at most n, or even at most n^2, n^3, and so on, is a nontrivial restriction (unlike the case of the THERMO class discussed earlier). This is a natural setting for studying the three-way trade-off between time, space, and thermodynamic cost.

Second, we could attempt to modify THERMO($t(n)$) to rule out reversible simulation using the notion of IO-conservation (def. 9) by only considering uniform circuit complexity classes built on gate sets that are IO-conservatively equivalent to $\{\wedge, \vee, \neg\}$. A prototype of such a class is distributional problems (L, μ) such that L is in DLOGTIME-uniform NC^1, μ is p-computable, and the μ-expected thermodynamic cost of some DLOGTIME-uniform NC^1 family of circuits on inputs of length n is poly($\log n$).

Of course, these are just prototypes; the true measure of how interesting a complexity class is is whether it sheds light on natural and interesting problems. As difficult as average-case circuit complexity is, general results in uniform average-case complexity seem even harder to come by (see, e.g., the survey in Bogdanov and Trevisan 2006). We mention the preceding ideas as we believe there are further definitions to be developed and results to be mined in this direction.

Randomized Algorithms

Randomized algorithms use random bits to speed up or simplify the computation of a deterministic function (Mitzenmacher and Upfal 2005; Motwani and Raghavan 1995). Perhaps best known for their use in Monte Carlo integration and cryptography, they underlie many other algorithms and computing technologies, ranging from computational algebra packages like MATLAB (MathWorks, Inc. 2017) and Mathematica (Wolfram 1999) to linear programming (Motwani and Raghavan 1995, section 9.10) and computational geometry (Mulmuley 1993) to distributed computation (Lynch 1996), Boolean satisfiability solvers (Een, Mishchenko, and Sörensson 2007; Selman, Kautz, and Cohen

1993), and the foundations of nearly all database software (Bloom 1970).

If we view randomizing a bit ("flipping a coin," as we often say) as a map that takes in a 0 and outputs a uniform distribution over 0 and 1, then the change in entropy between the input and the output is *negative* $kT \ln 2$. So doing this can (in principle) refrigerate the surrounding environment, while raising the energy level of the attached computer's battery (Landauer 1961; Bennett 1982). This might lead one to think that an algorithm that uses more randomness could actually be used to refrigerate the environment. The issue here is that those random bits need to be reset at the end of the computation for the computer to be reused. If this is performed by resetting the bits to 0, then the generalized Landauer cost of the erasure exactly cancels out any thermodynamic benefit gained from randomization in the first place. However, this still leaves us with the fact that the overall generalized Landauer cost of using randomness in an algorithm is zero. At the very least, this motivates the study of partially reversible simulations of randomized algorithms, akin to those undertaken by Demaine *et al.* (2016) for some deterministic algorithms.

Some modern computers do contain hardware-based random number generators (see, e.g., the list by Wikipedia Contributors 2017), many of which use thermal noise as part or all of their source of randomness. However, the goal of nearly all current hardware random number generators is to harness a physical source of randomness to generate *secure* random bits quickly, ignoring any potential thermodynamic benefits. To truly harness these thermodynamic benefits, practical issues suggest another theoretical question. Namely, in many existing physical random number generators, the entropy rate in the physical source—or, more precisely, the entropy rate of the bit sequence output by the measuring device—may decay over time, without providing any warning to the user. Therefore such devices should be put through periodic statistical tests of randomness, to determine when this decay happens and the device needs to be replaced or updated.

This leads to consideration of the following scenario and associated question. Consider a device that harnesses a physical source of randomness as a stream of random bits whose entropy rate decays over time (where the decay itself may occur according to a stochastic process). Each such source of randomness may entail bespoke statistical tests to determine when its entropy rate has fallen below levels acceptable to the user.

Question 4. *For a given source of physical randomness—modeled as earlier—what is the trade-off between the quality of the statistical tests and the thermodynamic cost of executing them?*

Noisy, Approximate, or Inexact Computation

Approximate or *inexact* computing (Kugler 2015; Baek and Chilimbi 2010; St. Amant *et al.* 2014; Diethelm 2016; Palem *et al.* 2013; Palmer, Düben, and McNamara 2014; Leyffer *et al.* 2016) is the practice whereby certain parts of a circuit or program are only computed approximately, to reduce the energy they use or the heat they generate. At the moment, this is primarily an engineering practice, though a glimmer of theory has begun (see, e.g., Augustine, Palem, and Parishkrati 2017).

The theoretical models of inexact computing are very similar to those of noisy computing—which have been studied since von Neumann (1956)—in which certain gates can err with some specified probability. One major difference is that, as discussed previously, we now have the tools to calculate the thermodynamic *benefits* of using noisy gates. This suggests a broad range of research questions, in which the computational costs of using noisy gates are traded off the thermodynamic benefits of doing so.

THERMODYNAMICS OF ERROR-CORRECTING CODES

Von Neumann (1956) showed how to simulate reliable computation using unreliable, or noisy, gates. To simplify things a bit, the solution is to use an error-correcting code. In inexact computing, one typically starts from a reliable computer and makes as many components inexact (/unreliable/noisy) as possible without sacrificing too much accuracy in the output. But we may also ask about the trade-off coming from the other direction, namely, suppose we start with noisy gates—which have lower thermodynamic cost—and want to construct a reliable computer using error correction. What is the trade-off in terms of generalized Landauer cost or total thermodynamic cost? For generalized Landauer cost, the first of these questions is essentially asking to extend the analysis of Demaine *et al.* (2016) to error-correcting codes.

Question 5. *What is the thermodynamic cost of computing (encoding and/or decoding) an error-correcting code?*

Question 6. *For producing reliable computers out of noisy components, what is the trade-off between reliability and the thermodynamic cost of the corresponding error correction?*

APPROXIMATE COMPUTATION FOR APPROXIMATION ALGORITHMS?

Because many optimization problems of practical interest are NP-hard, researchers have pursued approximation algorithms (Vazirani 2001) for these problems, which can be either deterministic or randomized algorithms that output *some* solution to the problem, with the proven guarantee that the output is within some factor of the optimum. (This body of research should not be confused with the work on "approximate computing" described just before.) However, at present, when they are implemented on real systems, such approximation algorithms use the typical exact, reliable computing components. This begs the question,

> Can we use approximate computing to get thermodynamically efficient approximation algorithms?

Because many approximation algorithms boil down to a linear programming or semidefinite programming relaxation of the original problem (Balas, Ceria, and Cornuéjols 1993; Lovász and Schrijver 1991; Sherali and Adams 1990; Lasserre 2001a, 2001b, 2010, 2015), we ask: What is the trade-off between reducing thermodynamic cost and the distance-from-optimum achieved when using approximate computations to solve linear or semidefinite programs?

For the specific case of linear programming computations, it makes sense to ask this question not just in terms of polynomial-time algorithms (Karmarkar 1984; Khačijan 1979) but also in terms of the widely used simplex algorithm (Dantzig 1990; Cormen *et al.* 2009; Schrijver 1986; Chvátal 1983). More generally, we can turn the question on its head to guide the development of new, more thermodynamically efficient algorithms!

Of course, this question can be asked for any specific optimization problem, but we believe the preceding question may have a good balance of specificity (so it might actually be possible to solve) and generality (its solution would be broadly applicable).

Conclusion

In this chapter, we've raised and highlighted many questions about the thermodynamics of computation, revealed by the new understanding coming out of recent advances in nonequilibrium statistical mechanics. This has covered everything from general questions about the thermodynamic complexity of circuits to the average-case thermodynamic complexity of uniform algorithms to more specific questions such as the thermodynamic costs of error-correcting codes and of linear circuits. Much exciting work is in progress and remains to be done.

We close by mentioning one more line of research motivated by these considerations that may be of interest to the complexity community, which in a certain sense is even more fundamental than thermodynamics. At the underlying physical level, essentially all (semiclassical) processes are modeled as (time-inhomogeneous) continuous-time Markov chains (CTMCs). One can ask, given a conditional distribution π over a finite state space X, is there *any* CTMC that implements π in a given time interval? Perhaps surprisingly, this turns out to be an open problem, sometimes called the problem of "embedding" a discrete-time Markov chain in a continuous-time one. However, recent work has shown that the answer is always yes for any π—so long as we expand the state space of the CTMC from X to some larger space Y, with π implemented only over the subspace X.

This leads to an obvious question: How many additional "hidden" states are required to compute any given π by a CTMC? Some preliminary results have recently been derived concerning this question (Owen, Kolchinsky, and Wolpert 2019; Wolpert, Kolchinsky, and Owen 2017), especially for the special case of single-valued π, that is, deterministic maps (Wolpert, Kolchinsky, and Owen 2017). Interestingly, this analysis has uncovered a natural notion of "time steps" for a CTMC, where each time step of a CTMC is delineated by the moments when the support of the CTMC changes: a nonzero transition probability becomes zero, or vice versa. Moreover, it turns out that there is a trade-off between the number of hidden states used to embed a given π and the minimal number of time steps needed by any CTMC that embeds π using that number of hidden states.

Although not precisely concerning circuits, this leads to questions closely analogous to those considered in circuit complexity. For example, what is the precise trade-off between the number of additional states used and the number of time steps needed to compute a given function? For deterministic functions π, results from semi-group theory have been used to answer this question (Wolpert, Kolchinsky, and Owen 2017) *for the case where we allow arbitrary freedom in how the CTMC operates*. However, essentially nothing is known about the question, even for deterministic functions, when we impose very natural restrictions on π. Moreover, for stochastic functions, the situation is more complicated; at present, the best upper bound we have on the number of states needed to implement a given stochastic function is one less than its nonnegative rank, a familiar quantity from communication complexity and the complexity of linear programming (Yannakakis 1991; Kol *et al.* 2014).

Appendix

Calculation 1. *(Calculation for Example 4). Let $g(x, y, z) = (x \wedge y, x \wedge z, y \wedge z)$. Let u be the uniform distribution over all of $\mathbb{F}_2^3$, and let d_p be the pairwise-independent distribution that assigns weight $p/4$ to the four strings with an even number of 1s in them and weight $(1 - p)/4$ to the four strings with an odd number of 1s. We suppose that all our operations are optimized for u but are run on d_p.*

First we calculate the Landauer cost the function g itself (computed "all-at-once"), on the distribution u for which it is thermodynamically optimized by assumption, and on the distribution d_p. $H(u) = 3$ and $H(g(u)) = 1/2 \log 2 + 4/8 \log 8 = 2$, so the Landauer cost on the uniform distribution is 1 bit. $H(d_p) = 2 + H(p)$, where $H(p) = -p \log p - (1-p) \log(1-p)$ is the entropy of a binary distribution with bias p. For the output, given input distribution d_p, we have $H(g(d_p)) = (3/4 - p/2) \log(1/(3/4 - p/2)) + \frac{3p}{4} \log(4/p) + \frac{1-p}{4} \log(4/(1-p))$, and therefore the Landauer cost with input distribution d_p is

$$3/4 - p/2 + \frac{3}{4} H(p) + (3/4 - p/2) \log(3/2 - p) + \frac{p}{2} \log p. \quad (9.2)$$

One may verify that when $p = 1/2$, this is indeed 1 bit, in agreement with the preceding calculation of the Landauer cost on the uniform distribution.

The mismatch cost of running g on d_p when it was optimized for u is

$$D(d_p || u) - D(g(d_p) || g(u)).$$

The first term, $D(d_p || u)$, is easily calculated as $4 \cdot \frac{p}{4} \log(\frac{p}{4} \cdot 8) + 4 \cdot \frac{1-p}{4} \log(\frac{1-p}{4} \cdot 8) = 1 - H(p)$. The second term is a bit more involved, so we write it out carefully: $D(g(d_p) || g(u))$

$$\begin{aligned}
&= \left(\frac{3}{4} - \frac{p}{2}\right) \log\left(\left(\frac{3}{4} - \frac{p}{2}\right) \cdot 2\right) + 3 \cdot \frac{p}{4} \log\left(\frac{p}{4} \cdot 8\right) + \frac{1-p}{4} \log\left(\frac{1-p}{4} \cdot 8\right) \\
&= \left(\frac{3}{4} - \frac{p}{2}\right) \log\left(\left(\frac{3}{4} - \frac{p}{2}\right) \cdot 2\right) + \frac{3p}{4} \log(2p) + \frac{1-p}{4} \log(2(1-p)) \\
&= \left(\frac{3}{4} - \frac{p}{2}\right) \log\left(\frac{3}{2} - p\right) + \frac{3p}{4} \log(p) + \frac{1-p}{4} \log((1-p)) + \frac{1+2p}{4} \\
&= \left(\frac{3}{4} - \frac{p}{2}\right) \log\left(\frac{3}{2} - p\right) - \frac{1}{4} H(p) + \frac{p}{2} \log p + \frac{1+2p}{4}.
\end{aligned}$$

Thus the mismatch cost is $D(d_p \| u) - D(g(d_p) \| g(u))$

$$\begin{aligned}
&= 1 - H(p) - \left[\left(\frac{3}{4} - \frac{p}{2}\right) \log\left(\frac{3}{2} - p\right) - \frac{1}{4} H(p) + \frac{p}{2} \log p + \frac{1+2p}{4}\right] \\
&= \left(\frac{3}{4} - \frac{p}{2}\right) - \frac{3}{4} H(p) - \left(\frac{3}{4} - \frac{p}{2}\right) \log(3/2 - p) - \frac{p}{2} \log p. \quad (9.3)
\end{aligned}$$

Interestingly, when we add up the Landauer cost (9.2) of running g on d_p and the mismatch cost (9.3), nearly all the terms cancel, and we are left with a total thermodynamic cost of

$$\frac{3}{2} - p.$$

Now, suppose we implement g with the obvious three AND gates, each of which is thermodynamically optimized for the uniform distribution over $\mathbb{F}_2^2$. Because the distribution d_p is pairwise independent, each AND gate in fact sees the distribution for which it was optimized, so the mismatch cost becomes zero. The Landauer cost $L_\wedge$ of a single binary AND gate with uniform distribution as input is easily calculated, and the total Landauer cost of this implementation of g is thus $3L_\wedge \approx 3.566$.

REFERENCES

Alman, J., and R. Williams. 2016. *Probabilistic Rank and Matrix Rigidity.* arXiv:1611.05558 [cs.CC]. 49th ACM Symposium on the Theory of Computing, STOC '17.

Arora, S., and B. Barak. 2009. *Computational Complexity: A Modern Approach.* Cambridge University Press.

Augustine, J., K. Palem, and Parishkrati. 2017. *Sustaining Moore's Law through Inexactness.* arXiv:1705.01497 [cs.CC].

Baek, W., and T. M. Chilimbi. 2010. "Green: A Framework for Supporting Energy-Conscious Programming Using Controlled Approximation." In *Proceedings of the 31st ACM SIGPLAN Conference on Programming Language Design and Implementation,* 198–209. PLDI '10. Toronto, Ontario, Canada: ACM.

Balas, E., S. Ceria, and G. Cornuéjols. 1993. "A Lift-and-Project Cutting Plane Algorithm for Mixed 0–1 Programs." *Math. Programming* 58 (3, Ser. A): 295–324.

Ben-David, S., B. Chor, O. Goldreich, and M. Luby. 1992. "On the Theory of Average Case Complexity." *J. Comput. System Sci.* 44 (2): 193–219.

Bennett, C. H. 1973. "Logical Reversibility of Computation." *IBM Journal of Research and Development* 17 (6): 525–532.

———. 1982. "The Thermodynamics of Computation—A Review." *Internat. J. Theoret. Phys.* 21 (12): 905–940.

———. 1989. "Time/Space Trade-Offs for Reversible Computation." *SIAM Journal on Computing* 18 (4): 766–776.

———. 2003. "Notes on Landauer's Principle, Reversible Computation, and Maxwell's Demon." *Studies In History and Philosophy of Science Part B: Studies In History and Philosophy of Modern Physics* 34 (3): 501–510.

Bloom, B. H. 1970. "Space/Time Trade-Offs in Hash Coding with Allowable Errors." *Commun. ACM* (New York, NY, USA) 13, no. 7 (July): 422–426.

Bogdanov, A., and L. Trevisan. 2006. "Average-Case Complexity." *Found. Trends Theor. Comput. Sci.* 2 (1): 1–106.

Boyar, J., P. Matthews, and R. Peralta. 2013. "Logic Minimization Techniques with Applications to Cryptology." *Journal of Cryptology* 26 (2): 280–312.

Boyd, A. B. 2018. "Thermodynamics of Modularity: Structural Costs Beyond the Landauer Bound." *Physical Review X* 8 (3).

Brillouin, L. 1953. "Negentropy Principle of Information." *Journal of Applied Physics* 24:1152–1163.

———. 1962. *Science and Information Theory.* Academic Press.

Buhrman, H., J. Tromp, and P. Vitányi. 2001a. "Time and Space Bounds for Reversible Simulation." In *Automata, Languages and Programming: 28th International Colloquium, ICALP 2001 Proceedings,* edited by F. Orejas, P. G. Spirakis, and J. van Leeuwen, 1017–1027. Berlin, Heidelberg: Springer Berlin Heidelberg.

———. 2001b. "Time and Space Bounds for Reversible Simulation." *J. Phys. A* 34 (35): 6821–6830.

Chejne Janna, F., F. Moukalled, and C. A. Gómez. 2013. "A Simple Derivation of Crooks Relation." *Internat. J. Thermodyn.* 16 (3): 97–101.

Chung, F., and L. Lu. 2006. *Complex Graphs and Networks.* 107:viii+264. CBMS Regional Conference Series in Mathematics. Published for the Conference Board of the Mathematical Sciences, Washington, DC; by the American Mathematical Society, Providence, RI.

Chvátal, V. 1983. *Linear Programming.* A Series of Books in the Mathematical Sciences. W. H. Freeman / Company, New York.

Codenotti, B. 2000. "Matrix Rigidity." *Linear Algebra Appl.* 304 (1-3): 181–192.

Cormen, T. H., C. E. Leiserson, R. L. Rivest, and C. Stein. 2009. *Introduction to Algorithms.* Third. MIT Press, Cambridge, MA.

Cover, T. M., and J. A. Thomas. 2012. *Elements of Information Theory.* John Wiley & Sons. Accessed January 11, 2014.

Crescenzi, P., and C. H. Papadimitriou. 1995. "Reversible Simulation of Space-Bounded Computations." *Theoret. Comput. Sci.* 143 (1): 159–165.

Crooks, G. E. 1998. "Nonequilibrium Measurements of Free Energy Differences for Microscopically Reversible Markovian Systems." *J. Stat. Phys.* 90 (5--6): 1481–1487.

———. 1999. "Entropy Production Fluctuation Theorem and the Nonequilibrium Work Relation for Free Energy Differences." *Phys. Rev. E* 60 (3): 2721.

Dantzig, G. B. 1990. "Origins of the Simplex Method." In *A History of Scientific Computing,* edited by S. G. Nash, 141–151. New York, NY, USA: ACM.

Deffner, S., and C. Jarzynski. 2013. "Information Processing and the Second Law of Thermodynamics: An Inclusive, Hamiltonian Approach." *Physical Review X* 3 (4): 041003.

Demaine, E. D., J. Lynch, G. J. Mirano, and N. Tyagi. 2016. "Energy-Efficient Algorithms." In *Proceedings of the 2016 ACM Conference on Innovations in Theoretical Computer Science,* 321–332. ITCS '16. Cambridge, Massachusetts, USA: ACM.

Diethelm, K. 2016. "Tools for Assessing and Optimizing the Energy Requirements of High Performance Scientific Computing Software." *PAMM* 16 (1): 837–838.

Een, N., A. Mishchenko, and N. Sörensson. 2007. "Applying Logic Synthesis for Speeding Up SAT." In *Theory and Applications of Satisfiability Testing— SAT 2007: 10th International Conference, Lisbon, Portugal, May 28-31, 2007. Proceedings,* edited by J. Marques-Silva and K. A. Sakallah, 272–286. Berlin, Heidelberg: Springer Berlin Heidelberg.

Esposito, M., and C. Van den Broeck. 2010. "Three Faces of the Second Law. I. Master Equation Formulation." *Physical Review E* 82 (1): 011143.

———. 2011. "Second Law and Landauer Principle Far from Equilibrium." *EPL (Europhysics Letters)* 95 (4): 40004.

Frank, M., T. Knight, and N. Margolus. 1998. "Reversibility in Optimally Scalable Computer Architectures." In *Unconventional Models of Computation (Auckland, 1998),* 165–182. Springer Ser. Discrete Math. Theor. Comput. Sci. Springer, Singapore.

Fredkin, E., and T. Toffoli. 1982. "Conservative Logic." *Internat. J. Theoret. Phys.* 21 (3): 219–253.

Grochow, J. A., and D. H. Wolpert. 2018. "Beyond Number of Bit Erasures: New Complexity Questions Raised by Recently Discovered Thermodynamic Costs of Computation." *SIGACT News* 54 (2).

Halpern, N. Y., A. J. P. Garner, O. C. O. Dahlsten, and V. Vedral. 2015. "Introducing One-Shot Work into Fluctuation Relations." Focus Issue on Quantum Thermodynamics. Preprint available as arXiv:1409.3878 [cond-mat.stat-mech], *New J. Phys.* 17 (095003).

Hasegawa, H.-H., J. Ishikawa, K. Takara, and D. J. Driebe. 2010. "Generalization of the Second Law for a Nonequilibrium Initial State." *Physics Letters A* 374 (8): 1001–1004.

Hirahara, S., I. C. Oliveira, and R. Santhanam. 2018. "NP-Hardness of Minimum Circuit Size Problem for OR-AND-MOD Circuits." In *33rd Computational Complexity Conference, CCC 2018, June 22-24, 2018, San Diego, CA, USA,* 5:1–5:31.

Hopcroft, J. E., R. Motwani, and U. Rotwani. 2000. *JD: Introduction to Automata Theory, Languages and Computability.*

Horowitz, J. M., and M. Esposito. 2014. "Thermodynamics with Continuous Information Flow." *Phys. Rev. X* 4 (3): 031015.

Jarzynski, C. 1997. "Nonequilibrium Equality for Free Energy Differences." *Physical Review Letters* 78 (14): 2690.

Karmarkar, N. 1984. "A New Polynomial-Time Algorithm for Linear Programming." *Combinatorica* 4 (4): 373–395.

Khačijan, L. G. 1979. "A Polynomial Algorithm in Linear Programming." *Dokl. Akad. Nauk SSSR* 244 (5): 1093–1096.

Kol, G., S. Moran, A. Shpilka, and A. Yehudayoff. 2014. "Approximate Nonnegative Rank Is Equivalent to the Smooth Rectangle Bound." In *Automata, Languages, and Programming (ICALP 2014)*, 701–712. Springer.

Kolchinsky, A., and D. H. Wolpert. 2016. *Dependence of Dissipation on the Initial Distribution over States*. Preprint available as arXiv:1607.00956 [cond-mat.stat-mech].

Kugler, L. 2015. "Is 'Good Enough' Computing Good Enough?" *Commun. ACM* 58, no. 5 (April): 12–14. doi:10.1145/2742482.

Landauer, R. 1961. "Irreversibility and Heat Generation in the Computing Process." *IBM Journal of Research and Development* 5 (3): 183–191.

———. 1991. "Information is Physical." *Physics Today* 44:23.

———. 1996a. "Minimal Energy Requirements in Communication." *Science* 272 (5270): 1914–1918.

———. 1996b. "The Physical Nature of Information." *Physics Letters A* 217 (4): 188–193.

Lange, K.-J., P. McKenzie, and A. Tapp. 2000. "Reversible Space Equals Deterministic Space." Twelfth Annual IEEE Conference on Computational Complexity (Ulm, 1997), *J. Comput. System Sci.* 60 (2, part 2): 354–367.

Lasserre, J. B. 2001a. "An Explicit Exact SDP Relaxation for Nonlinear 0–1 Programs." In *Integer Programming and Combinatorial Optimization (Utrecht, 2001)*, 2081:293–303. Lecture Notes in Comput. Sci. Springer, Berlin. doi:10.1007/3-540-45535-3_23.

———. 2001b. "Global Optimization with Polynomials and the Problem of Moments." *SIAM J. Optim.* 11 (3): 796–817. doi:10.1137/S1052623400366802.

———. 2010. *Moments, Positive Polynomials and their Applications.* 1:xxii+361. Imperial College Press Optimization Series. Imperial College Press, London.

Lasserre, J. B. 2015. *An Introduction to Polynomial and Semi-Algebraic Optimization.* xiv+339. Cambridge Texts in Applied Mathematics. Cambridge University Press, Cambridge.

Levin, L. A. 1974. "Laws on the Conservation (Zero Increase) of Information, and Questions on the Foundations of Probability Theory." *Problemy Peredači Informacii* 10 (3): 30–35.

———. 1986. "Average Case Complete Problems." *SIAM J. Comput.* 15 (1): 285–286.

Levine, R. Y., and A. T. Sherman. 1990. "A Note on Bennett's Time–Space Tradeoff for Reversible Computation." *SIAM J. Comput.* 19 (4): 673–677.

Leyffer, S., S. M. Wild, M. Fagan, M. Snir, K. Palem, K. Yoshii, and H. Finkel. 2016. *Doing Moore with Less—Leapfrogging Moore's Law with Inexactness for Supercomputing.* arXiv:1610.02606 [cs.OH].

Li, M., J. Tromp, and P. Vitányi. 1998a. "Reversible Simulation of Irreversible Computation." In *Proceedings of the Fourth Workshop on Physics and Computation,* 168–176. PhysComp96. Boston, Massachusetts: Elsevier Science Publishers B. V.

———. 1998b. "Reversible Simulation of Irreversible Computation." *Phys. D* 120, nos. 1-2 (September): 168–176.

Lokam, S. V. 2000. "On the Rigidity of Vandermonde Matrices." *Theoret. Comput. Sci.* 237 (1–2): 477–483.

Lovász, L., and A. Schrijver. 1991. "Cones of Matrices and Set-Functions and 0–1 Optimization." *SIAM J. Optim.* 1 (2): 166–190.

Lynch, N. A. 1996. *Distributed Algorithms.* Morgan Kaufmann, San Francisco, CA.

Markoff, J. 2015. "A Climate-Modeling Strategy That Won't Hurt the Climate." *The New York Times* (May).

MathWorks, Inc. 2017. *MATLAB.* Natick, MA, USA.

Mitzenmacher, M., and E. Upfal. 2005. *Probability and Computing.* Randomized Algorithms and Probabilistic Analysis. Cambridge University Press, Cambridge.

Moore, C., and S. Mertens. 2011. *The Nature of Computation.* Oxford University Press.

Morita, Kenichi. 2008. "Reversible computing and cellular automata---a survey." *Theoret. Comput. Sci.* 395 (1): 101–131. http://dx.doi.org/10.1016/j.tcs.2008.01.041.

Motwani, R., and P. Raghavan. 1995. *Randomized Algorithms*. Cambridge University Press, Cambridge.

Müller, M., and J. Scharlau. 2014. *Single-Shot Quantum Thermodynamics—Lecture Notes*. Course materials available at http://www.mpmueller.net/lecture2.html.

Mulmuley, K. D. 1993. *Computational Geometry: An Introduction through Randomized Algorithms*. Pearson.

Newman, M., A.-L. Barabási, and D. J. Watts. 2006. *The Structure and Dynamics of Networks*. Princeton Studies in Complexity. Princeton University Press.

Owen, J., A. Kolchinsky, and D. H. Wolpert. 2019. *Number of Hidden States Needed to Physically Implement a Given Conditional Distribution*, January.

Palem, K., A. Lingamneni, C. Enz, and C. Piguet. 2013. "Why Design Reliable Chips when Faulty Ones are Even Better." In *2013 Proceedings of the ESSCIRC*. IEEE.

Palmer, T., P. Düben, and H. McNamara. 2014. "Stochastic Modelling and Energy-Efficient Computing for Weather and Climate Prediction." *Phil. Trans. R. Soc. A* 372:20140118.

Parrondo, J. M. R., J. M. Horowitz, and T. Sagawa. 2015. "Thermodynamics of Information." *Nature Physics* 11 (2): 131–139.

Plenio, M. B., and V. Vitelli. 2001. "The Physics of Forgetting: Landauer's Erasure Principle and Information Theory." *Contemp. Phys.* 42 (1): 25–60.

Pollard, B. S. 2016. "A Second Law for Open Markov Processes." Preprint available as arXiv:1410.6531 [cond-mat.stat-mech], *Open Syst. Inf. Dyn.* 23:1650006.

Prokopenko, M., and I. Einav. 2015. "Information Thermodynamics of Near-Equilibrium Computation." *Phys. Rev. E* 91 (6): 062143.

Sagawa, T. 2014. "Thermodynamic and Logical Reversibilities Revisited." *Journal of Statistical Mechanics: Theory and Experiment* 2014 (3): P03025.

Sagawa, T., and M. Ueda. 2009. "Minimal Energy Cost for Thermodynamic Information Processing: Measurement and Information Erasure." *Phys. Rev. Lett.* 102 (25): 250602.

———. 2012. "Fluctuation Theorem with Information Exchange: Role of Correlations in Stochastic Thermodynamics." *Phys. Rev. Lett.* 109 (18): 180602.

Savage, J. E. 1998. *Models of Computation*. Vol. 136. Addison-Wesley.

Schrijver, A. 1986. *Theory of Linear and Integer Programming*. Wiley-Interscience Series in Discrete Mathematics. A Wiley-Interscience Publication. John Wiley & Sons, Ltd., Chichester.

Seifert, U. 2012. "Stochastic Thermodynamics, Fluctuation Theorems and Molecular Machines." *Rep. Progress Phys.* 75 (12): 126001.

Selman, B., H. A. Kautz, and B. Cohen. 1993. "Local Search Strategies for Satisfiability Testing." In *Cliques, Coloring, and Satisfiability, Proceedings of a DIMACS Workshop, New Brunswick, New Jersey, USA, October 11–13, 1993*, 521–532.

Sherali, H. D., and W. P. Adams. 1990. "A Hierarchy of Relaxations between the Continuous and Convex Hull Representations for Zero–One Programming Problems." *SIAM J. Discrete Math.* 3 (3): 411–430.

Shizume, K. 1995. "Heat Generation Required by Information Erasure." *Phys. Rev. E* 52 (4): 3495.

Sipser, M. 2012. *Introduction to the Theory of Computation.* 3rd. Cengage Learning.

St. Amant, R., A. Yazdanbakhsh, J. Park, B. Thwaites, H. Esmaeilzadeh, A. Hassibi, L. Ceze, and D. Burger. 2014. "General-Purpose Code Acceleration with Limited-Precision Analog Computation." In *Proceeding of the 41st Annual International Symposium on Computer Architecuture,* 505–516. ISCA '14. Minneapolis, Minnesota, USA: IEEE Press.

Sutner, K. 2004. "The Complexity of Reversible Cellular Automata." *Theoret. Comput. Sci.* 325 (2): 317–328.

Szilard, L. 1964. "On the Decrease of Entropy in a Thermodynamic System by the Intervention of Intelligent Beings." *Behavioral Science* 9 (4): 301–310.

Takara, K., H.-H. Hasegawa, and D. J. Driebe. 2010. "Generalization of the second law for a transition between nonequilibrium states." *Phys. Lett. A* 375 (2): 88–92.

Touchette, H., and S. Lloyd. 2004. "Information-Theoretic Approach to the Study of Control Systems." *Physica A* 331 (1): 140–172.

Tyagi, N., J. Lynch, and E. D. Demaine. 2016. "Toward an Energy Efficient Language and Compiler for (Partially) Reversible Algorithms." In *Reversible Computation,* 9720:121–136. Lecture Notes in Comput. Sci. Springer.

Valiant, L. G. 1976. "Graph-Theoretic Properties in Computational Complexity." Working papers presented at the ACM-SIGACT Symposium on the Theory of Computing (Albuquerque, NM, 1975), *J. Comput. System Sci.* 13 (3): 278–285.

Van den Broeck, C., *et al.* 2013. "Stochastic Thermodynamics: A Brief Introduction." *Phys. Complex Colloids* 184:155–193.

Van den Broeck, C., and M. Esposito. 2015. "Ensemble and Trajectory Thermodynamics: A Brief Introduction." *Physica A: Statistical Mechanics and its Applications* 418:6–16.

Vazirani, V. V. 2001. *Approximation Algorithms*. Springer-Verlag, Berlin.

Vitányi, P. 2005. "Time, Space, and Energy in Reversible Computing." In *Proceedings of the 2nd Conference on Computing Frontiers*, 435–444. CF '05. Ischia, Italy: ACM.

von Neumann, J. 1956. "Probabilistic Logics and the Synthesis of Reliable Organisms from Unreliable Components." In *Automata Studies*, 43–98. Annals of Mathematics Studies, no. 34. Princeton University Press, Princeton, N. J.

Wikipedia Contributors. 2017. *List of random number generators*. Wikipedia, The Free Encyclopedia. https://en.wikipedia.org/wiki/List_of_random_number_generators.

Wolfram, S. 1999. *The Mathematica® Book*. Fourth. Wolfram Media, Inc., Champaign, IL; Cambridge University Press, Cambridge.

Wolpert, D. H., and A. Kolchinsky. 2018. "The Entropic Costs of Straight-Line Circuits." *arXiv preprint arXiv:1806.04103*.

Wolpert, D. H., A. Kolchinsky, and J. Owen. 2017. *The Minimal Hidden Computer Needed to Implement a Visible Computation*. arXiv:1708.08494.

Yannakakis, M. 1991. "Expressing Combinatorial Optimization Problems by Linear Programs." *J. Comput. System Sci.* 43 (3): 441–466.

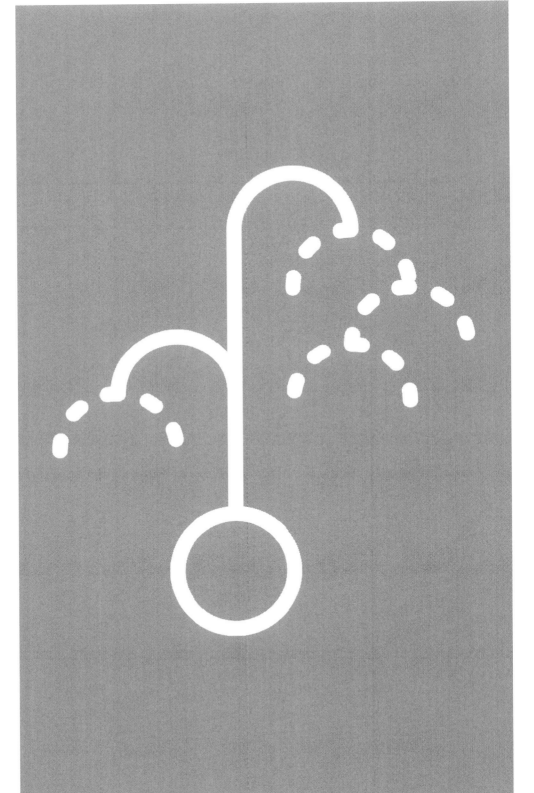

AUTOMATICALLY REDUCING ENERGY CONSUMPTION OF SOFTWARE

Jeremy Lacomis, Carnegie Mellon University
Jonathan Dorn, GrammaTech, Inc.
Westley Weimer, University of Michigan
Stephanie Forrest, Arizona State University

Introduction

As computation continues to migrate from personal computers to large-scale data centers, the energy required to run computers has become a significant economic and environmental concern. For example, between 2005 and 2010, data center electricity consumption grew by 24%. By 2014, data centers accounted for 1.8% of US energy consumption (Shehabi *et al.* 2016). Although this large energy footprint has led to some mitigation efforts, energy consumption in data centers continues to rise. Current estimates project that US energy use will increase a further 4% from 2014 to 2020. In 2016, in anticipation of the possible environmental impact of its growing energy demand, Google announced a $2.5 billion commitment to the purchase of energy from renewable sources.[1]

Computer hardware efficiency directly affects data center energy consumption, and this effect is multiplied by the support systems required for deployment. Mechanical and electrical systems, such as lighting, cooling, air circulation,

---

[1] https://sustainability.google.

and uninterruptible power supplies, can quadruple the power required by the computational hardware itself (Hoelzle and Barroso 2009).

The software running in the center can further multiply energy consumption. For example, data centers must be provisioned with sufficient hardware to run the desired algorithms in a timely fashion. Algorithmic inefficiencies in software implementations can increase run times, leading to greater emphasis on parallelism to compensate. Contention for resources like networks, disks, memory, or caches leads to overprovisioning hardware (Mars *et al.* 2012). At sufficient scales, hardware reliability concerns require implementation of redundant resources and computations; for example, Microsoft implements redundancy for all customer data in Azure storage accounts to meet the uptime guaranteed by its service-level agreements.[2] Because the load on the support systems scales with computational load, improving computational efficiency could significantly reduce overall energy costs of data centers.

The problem of computational energy consumption has typically been addressed by optimizing hardware (Douglis, Krishnan, and Bershad 1995; Delaluz *et al.* 2001; Nowka *et al.* 2002), compilers (Lee *et al.* 1997; Hsu and Kremer 2003; Reda and Nowroz 2012), or cluster scheduling (Mars *et al.* 2012), leaving open the question of how to write energy-efficient applications. These optimizations are largely independent, and energy reductions from different perspectives are composable to achieve greater savings. Reducing software energy use is challenging because it is often unclear how implementation decisions impact energy consumption, making it difficult for

---

[2] https://docs.microsoft.com/en-us/azure/storage/common/storage-redundancy.

developers to write programs that minimize energy use (Manotas, Pollock, and Clause 2014).

In this chapter, we focus on emerging techniques for *automatically* reducing energy consumption in existing computer programs. First, we give an overview of specialized approaches that leverage knowledge about specific properties of some software. Next, we consider new, more general approaches that use insights from evolutionary computation (Schulte *et al.* 2014; Dorn *et al.* 2017). By modifying software and measuring the difference in energy consumption, these approaches are able to automate the "change, observe, iterate" loop that a developer might use when she finds it difficult or impossible to reason about how changes to the software will impact performance. These generic techniques are applicable to many types of software and sometimes reveal new generalizable methods for optimizing specific types of software.

Reducing Software Energy Consumption

Reducing the energy consumption of a program requires that its execution be changed in some way. There are three main approaches used today: semantics-preserving techniques, approximate computing techniques, and techniques based on stochastic search.

SEMANTICS-PRESERVING TECHNIQUES

These techniques require that all transformations to the program preserve input/output behavior identically to the original program. For example, standard compiler optimizations for reducing run-time guarantee that program semantics are not changed. These techniques generate programs that are correct by construction.

INSTRUCTION SCHEDULING TECHNIQUES

Modern CPU internals are primarily composed of transistors used as switches. CPUs combine these transistors into small units that perform simple operations (e.g., addition, division, logical operations), select memory to operate on, or do basic control and input/output. When a computer is running, the CPU is given *instructions* that direct the CPU to combine these units and perform the desired computation.

The transistors inside the CPU account for a majority of its energy use. There are two types of energy consumption in transistors: *static* and *dynamic* (Niu and Quan 2004). Static energy consumption is the energy required to hold a transistor at a steady state, while dynamic energy consumption is the energy dissipated as heat when a transistor is switched from one state to another. Although static energy consumption in modern processors is important, the majority of energy consumption is dynamic (Sarwar 1997). To minimize dynamic energy consumption, researchers use *instruction scheduling* (Lee et al. 1997).

Instruction scheduling techniques reduce the number of times that transistors change states, thus reducing dynamic energy consumption. This is accomplished by optimizing the ordering of independent instructions—instructions that produce the same computation if executed in a different order—to minimize the number of times that transistors change states. For example, if a transistor T is used in the independent instructions I_1, I_2, and I_3, where the values of T are T_{off}, T_{on}, and T_{off}, respectively, the instruction ordering $I_1 \to I_2 \to I_3$ requires T to switch twice ($T_{off} \to T_{on} \to T_{off}$), while the instruction ordering $I_1 \to I_3 \to I_2$ only requires T to switch once ($T_{off} \to T_{off} \to T_{on}$). This ordering consumes less dynamic energy while executing the same instructions and performing the same computation.

A limitation of instruction scheduling is that any reordered instructions must be independent of one another, which is often not the case. It also requires an available energy model for each instruction (as different transistors may have different dynamic power consumption). As a result, its use has typically been limited to applications where minimizing energy usage is paramount, such as in ultra-low-power mobile devices.

SUPEROPTIMIZATION

Superoptimization (Massalin 1987; Schkufza, Sharma, and Aiken 2013) is similar to instruction scheduling in that both reorder low-level instructions. Superoptimization, however, is more general: instead of considering specific properties (e.g., reduced switching), it reorders instructions in any way that preserves semantics, and then performance is measured empirically. Assuming an effective mechanism to verify functionality, superoptimization can identify the best sequence of instructions by exhaustive enumeration. However, the large size of modern instruction sets means that exhaustive search is infeasible for more than a few instructions; this strategy is thus limited to very short sequences. Stochastic search enables these techniques to scale to longer sequences that compute more complex functions but cannot guarantee that the best sequence has been found. Stochastic superoptimization approaches remain constrained by the verification mechanism, as complex functions are generally difficult or impossible to verify formally (Rice 1953).

APPROXIMATE COMPUTING

Approximate computing relaxes the requirement to preserve exact semantics, allowing a trade-off between computational accuracy and run-time or energy consumption (Palem 2014). This trade-off is analogous to the way that lossy compression formats, such as MP3 or MPEG4, achieve smaller file sizes than lossless formats in exchange for reduced output fidelity. Approximate computing techniques exist for both hardware

(Lu 2004; Gupta *et al.* 2013; Palem and Lingamneni 2013; Yang, Han, and Lombardi 2015) and software (Han and Orshansky 2013; Venkataramani *et al.* 2015). We discuss three approximate computing approaches that are suitable for software implementation: *precision scaling*, *task skipping*, and *loop perforation*.

PRECISION SCALING

Precision scaling reduces computational cost by modifying the precision of floating-point variables used to represent real numbers in arithmetic (Sarbishei and Radecka 2010; Tian *et al.* 2015). "Precision" refers to the number of bits that are used to represent a number in memory; higher-precision representations use more bits and are more accurate. However, more bits imply more space in memory, which can affect energy consumption. In certain cases, it is possible to adjust the precision of variables to reduce energy. For example, reducing precision in hardware can reduce the size of the circuit required for computation, thereby reducing its power consumption; in software, changing the precision of a variable can change memory layouts, reducing the time to look up values[3] and decreasing energy usage. However, these effects are difficult to predict, and scaling may have no effect at all.

TASK SKIPPING

Task skipping (Rinard 2006, 2007) reduces energy consumption or run-time by halting or skipping the execution of tasks in a program when the results are unneeded. The strategy uses a model

---

[3] Frequently used values are often stored in a *cache*, a special type of extremely fast memory located inside the CPU itself. Although cache memory is fast, it is expensive, and its size is limited. Larger data might not fit into the cache and can also introduce problems with *alignment*, a topic beyond the scope of this chapter.

to characterize the trade-off between accuracy and performance to decide if an execution should be skipped. Before task skipping can be applied, the program must be manually decomposed into tasks by a developer. This requires programmers with domain-specific knowledge to perform a nontrivial amount of manual work, and sometimes this type of decomposition is not possible (Tilevich and Smaragdakis 2002). This greatly limits the applicability of the technique.

LOOP PERFORATION

Like task skipping, loop perforation reduces run-time and energy consumption by skipping unnecessary computation (Hoffmann *et al.* 2009; Sidiroglou-Douskos *et al.* 2011). In loop perforation, individual iterations of loops are skipped during a calculation. This approach is effective for algorithms that iteratively improve the accuracy of a computation, such as spigot algorithms that iteratively compute the digits of π (Rabinowitz and Wagon 1995) or iterative approximations of integrals (Kammerer and Nashed 1972). Skipping iterations of these loops produces an approximate answer with less energy than a fully precise answer.

Loop perforation has two main advantages over task skipping: loops do not have to be manually specified by domain experts because they can be identified easily and automatically; and loops are ubiquitous in software, so there are many more applicable programs for loop perforation. However, not all loops are improved with perforation, and, in certain cases, loop perforation actually *decreases* performance. For example, a loop might be used to filter a list before it is passed to an expensive processing step (Hoffmann *et al.* 2009), so skipping iterations of the filter loop can actually increase energy consumption.

ENERGY REDUCTION USING GENETIC IMPROVEMENT

Although these earlier approaches to software energy reduction can be effective, they either use very limited transformations (e.g., instruction scheduling and superoptimization) or require programs to have specific properties (e.g., task skipping). Generally, it is difficult to predict the impact of a given transformation on the behavior and energy consumption of a program. This difficulty motivates the use of stochastic optimization methods, such as simulated annealing (Kirkpatrick, Gelatt, and Vecchi 1983), ant colony optimization (Dorigo and Birattari 2011), or *genetic algorithms* (GAs) (Holland 1992).

In the following, we focus on GAs and related methods, which are known collectively as *evolutionary computation* (EC), because they have been applied successfully to software modification. A GA takes as input a representation and a fitness function. The representation specifies a set of properties that can be assembled into a *chromosome* to form a candidate solution (called an *individual*). The fitness function computes the goodness, or *fitness*, of each individual, returning a numerical rating. Execution of a GA begins with an initial *population* of individuals, either generated randomly or provided by some other means. Each individual's fitness is evaluated and used stochastically to *select* individuals, which are then mutated and recombined with other individuals to form the next generation. Although the details of these transformations vary according to the specific implementation, there are two main techniques for creating new individuals from the previous generation: *mutation* and *crossover*. Mutation randomly modifies an individual (e.g., through bit flips), whereas crossover recombines the chromosomes from two or more parents, analogous to crossing over in biology. These processes of fitness evaluation, selection, and variation are iterated for many generations, *evolving* an improved solution to a problem.

An important subfield of EC is genetic programming (GP) (Koza 1992), where programs are evolved to approximate the input/output behavior of a hidden function (a form of function approximation). GP methods have been applied to several problems in software engineering, such as repairing bugs (Le Goues *et al.* 2012), obfuscating code (Petke 2016), and implementing new functionality (Harman, Jia, and Langdon 2014). In these examples, GP is used to improve extant software, and the term *genetic improvement* refers to this class of applications (Langdon 2015).

Energy optimization is, of course, another form of software improvement, and there are several recent efforts along these lines (Schulte *et al.* 2014; Bruce, Petke, and Harman 2015; Linares-Vásquez *et al.* 2015; Bruce *et al.* 2018). In these applications, fitness corresponds to energy reduction, which can be directly measured or modeled. An extension of this work uses multiobjective optimization to trade off energy and accuracy (a form of approximate computing). For example, our work on POWERGAUGE (Dorn *et al.* 2017) relaxes the requirement of strict preservation of semantics to achieve greater energy reductions. indexenergy!reduction Note that the optimization of software along two dimensions does not return one "best" implementation but a set of *Pareto optimal* (Zitzler *et al.* 2003) programs. A program that is Pareto optimal is one for which no other observed program has lower energy consumption without a corresponding increase in error, nor a lower error without an increase in energy consumption. The fitnesses of these Pareto optimal programs lie on a *Pareto frontier*, which can be visualized on an error versus energy reduction graph, as shown in figure 10.1.

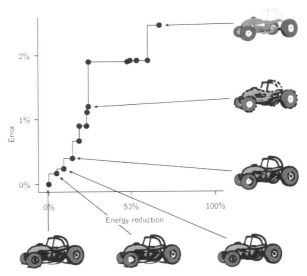

Figure 10.1. Pareto frontier for the **Blender** benchmark. The x-axis indicates the percentage energy reduction (greater is better), and the y-axis indicates error (lower is better). Each point corresponds to a program generated by PowerGAUGE. Example output images are provided for some programs. The image in the lower left has no error and corresponds to a 1% energy savings. The image in the lower right has a small amount of error but corresponds to a 10% energy savings.

The PowerGAUGE Energy Reduction Algorithm

The PowerGAUGE energy reduction algorithm uses a multi-objective GA to optimize multidimensional fitness functions. We use the NSGA-II implementation (Deb *et al.* 2002) to optimize software with respect to both energy consumption and error. PowerGAUGE takes as input a program to be optimized and an n-dimensional fitness function. In our case, $n = 2$, and the fitness function computes the energy consumption of the program and error in the program output. Note that PowerGAUGE is not restricted to optimizing energy consumption and output error; it could optimize with respect to any properties that can be represented by a fitness function, such as program file size or memory usage.

Chapter 10: Automatically Reducing Energy Consumption of Software

Recall that one of the fundamental concepts in EC is the choice of how to represent individuals in the population. Although we start with programs written in the C programming language, we represent them in *assembly language*, a low-level representation of the instructions to be executed by the CPU, which is generated from high-level code by a *compiler*.[4] There are several advantages to representing programs at this level:

- Compilers optimize the performance of software by applying semantics-preserving transformations to the code. By manipulating the output of the compiler, we can incorporate these optimizations.

- Optimization passes consume the majority of time spent compiling. Thus the transformation from high-level code to low-level assembly takes much more time than the transformation from low-level assembly into a binary format suitable for execution on a CPU. By modifying the assembly directly, we avoid repeating these costs for each fitness evaluation.

- High-level code has many syntactic restrictions that must be respected when making modifications. Assembly is much less strict with many more sequences of instructions considered to be legal. This flexibility enhances expressiveness and the creation of programs that could not be generated from source code.

- When targeting a particular machine, compilers for different high-level languages (e.g., C, Pascal, Fortran) all use identical assembly language. This feature implies

---

[4] Assembly code can also be handwritten, but it is typically verbose and difficult to understand, so this practice is unusual.

that PowerGAUGE can operate on programs generated from multiple source-level languages.

One challenge of applying EC to programs, especially assembly programs, is the large number of instructions and high redundancy of code across individuals of a population. To cope with this, we adopt the *patch representation* developed in earlier work on automated program repair (Le Goues et al. 2012). Instead of representing an individual as a complete program, each individual is represented as a list of modifications (edits) to the original program. In our case, we treat the original assembly code as a sequence of instructions and then represent each individual as a sequence of modifications in which instructions are *deleted*, *swapped* (i.e., exchanging the positions of two instructions), or *copied* (i.e., duplicating an instruction and inserting it at a random location). Although other types of code transformations are possible, for example, synthesizing new code (Mechtaev, Yi, and Roychoudhury 2016) or template-based edits (Kim et al. 2013), we found these particular mutations to be both effective and computationally inexpensive.

The second input to a genetic algorithm is its fitness function. In PowerGAUGE, fitness is two-dimensional: one dimension represents the level of error in the program's output, and the other corresponds to the amount of energy consumed while running the program. Creating a fitness function for energy consumption is conceptually straightforward—simply run the program and measure or model the energy consumed. In practice, this is surprisingly difficult for some applications. The fitness function for error is more domain specific, however. For the purposes of this chapter, we consider a simple example: the image produced by Blender, an open-source 3D rendering application. The fitness function measures error as the "distance" from the image output

Chapter 10: Automatically Reducing Energy Consumption of Software

by the program to a reference image. There are natural distance metrics for images, but for other applications it may not be feasible to quantify the output error (i.e., for an email client). This may seem like a limitation, but in these cases PowerGAUGE can still optimize for energy consumption alone.

Figure 10.1 shows the Pareto frontier of different versions of `Blender` generated during an actual PowerGAUGE run. The x-axis indicates the percentage energy reduction (greater is better), whereas the y-axis indicates the percentage error (lower is better). The "ideal" program would fall in the lower right of this graph, generating output identical to the original program while consuming zero energy. To give some insight into what these errors look like, several of the Pareto programs are annotated with their output. The program in the lower left generates an image with no error and uses 1% less energy than the original program. As we tolerate more error and move along the Pareto frontier, energy savings increase, until we reach the upper right of the figure. At this point, error is high, but there is a 67% energy savings over the original program. Users of PowerGAUGE could generate these Pareto optimal programs and select the one that has an acceptable level of error for their applications to maximize energy savings.

Open Questions and Future Directions

Although PowerGAUGE and related approaches show promise for reducing the energy budgets of software, there are still many open questions and future directions to explore. Here we discuss two: the open problem of effective energy measurement and the practical application of PowerGAUGE to security.

ENERGY MEASUREMENT

Any empirically based search method for energy reduction requires a technique for measuring or estimating energy consumption of candidate programs. Modeling is one approach that avoids the expense and communication overhead of additional hardware. However, in our preliminary studies, we found that energy models were inaccurate to the point of interfering with the search, and physical measurements using specialized hardware became an attractive option. This observation may not apply to all energy models or all search methods, but others have made similar observations (Haraldsson and Woodward 2015), which focused our attention on energy measurement methods.

Modern processors often include internal features, such as running average power limit (RAPL) (David *et al.* 2010), that report consumption measurements, but in our experiments we found these to be unreliable, and their documentation inadequate. In addition, measuring only internal CPU energy ignores the energy consumption of other system components that might be affected by a change to software (e.g., memory modules, hard disks, graphics cards). An ideal technique would capture the energy consumption of the entire system. Commercial power monitoring devices exist, but they tend either to be prohibitively expensive, to apply only to mobile devices, or to have insufficient resolution with respect to time and/or energy. In our case, the simplest solution was to construct our own device to measure energy accurately and cost-effectively to use in our search process.

Although we used open-source hardware and provide the source code for the measurement device, better built-in support for energy monitoring would enable anyone to run a search algorithm more easily, without having to construct his or her own hardware. For example, monitoring circuits with a software interface could be included in server power supplies, or CPU manufacturers could provide more detailed energy consumption measurements.

Chapter 10: Automatically Reducing Energy Consumption of Software

SIDE-CHANNEL ATTACK MITIGATION

A common security concern is the leaking of sensitive information through *side-channels*. A side-channel is a source of information related to the effects of a computation on its environment rather than weaknesses in the algorithm itself. An example of this is a timing attack in the Unix `login` command. Early versions of `login` only ran the expensive hash function required to check password validity if the provided login name was a known system user. This allowed an attacker to determine easily whether a username was legal: if the login failed quickly, the username did not exist, whereas if the login took a long time to fail, the username did exist, and the system was checking the validity of the password. Once a legitimate username was discovered, the attacker could move on to trying common passwords and often succeeded in gaining entry. This type of side-channel attack was also a key component of the recent Spectre and Meltdown vulnerabilities found in x86 processors (Kocher *et al.* 2018; Lipp *et al.* 2018).

Similarly, differences in energy consumption can leak information about computations. Search-based techniques could be used to mitigate these effects by normalizing the energy consumption of commands. In the case of the `login` program, the side-channel attack was mitigated by adding a random delay when a login failed, purposefully increasing the run-time to improve security. Although in this chapter we focus primarily on the reduction of energy consumption, search-based techniques could also potentially be used to *increase* the energy consumption of software or even target a specific level of energy used by a computation.

Conclusion

As computation migrates to large-scale data centers, concern about the environmental and economic impact of software is growing. Despite interest in limiting the energy footprint of software, few general-purpose techniques have specifically targeted the creation of low-energy programs. This chapter describes a search-based approach to this problem. Using methods from evolutionary computation combined with sufficiently accurate energy measurements or models, PowerGAUGE automatically modifies programs and identifies those that consume less energy while preserving required functionality. Genetic improvement methods like PowerGAUGE do not require human guidance or prior knowledge of how a particular software modification will impact program behavior. Although the research described here is still in its early stages, we are optimistic that effective and developer-friendly techniques for improving the energy efficiency of software will ultimately play a role in reducing energy consumption.

REFERENCES

Bruce, B. R., J. Petke, and M. Harman. 2015. "Reducing Energy Consumption Using Genetic Improvement." In *Genetic and Evolutionary Computation Conference*, 1327–1334.

Bruce, B. R., J. Petke, M. Harman, and E. Barr. 2018. "Approximate Oracles and Synergy in Software Energy Search Spaces." *Transactions on Software Engineering* PP, no. 99 (April): 1–1.

David, H., E. Gorbatov, U. R. Hanebutte, R. Khanna, and C. Le. 2010. "RAPL: Memory Power Estimation and Capping." In *International Symposium on Low-Power Electronics and Design*, 189–194.

Deb, K., A. Pratap, S. Agarwal, and T. Meyarivan. 2002. "A Fast and Elitist Multiobjective Genetic Algorithm: NSGA-II." *Transactions on Evolutionary Computation* 6, no. 2 (April): 182–197.

Delaluz, V., M. Kandemir, N. Vijaykrishnan, A. Sivasubramaniam, and M. J. Irwin. 2001. "DRAM Energy Management Using Software and Hardware Directed Power Mode Control." In *International Symposium on High-Performance Computer Architecture*, 159–169.

Dorigo, M., and B. Birattari. 2011. "Ant Colony Optimization." In *Encyclopedia of Machine Learning*, 36–39. New York: Springer.

Dorn, J., J. Lacomis, W. Weimer, and S. Forrest. 2017. "Automatically Exploring Tradeoffs between Software Output Fidelity and Energy Costs." *Transactions on Software Engineering* PP, no. 99 (November): 1–1.

Douglas, F., P. Krishnan, and B. Bershad. 1995. "Adaptive Disk Spin-Down Policies for Mobile Computers." In *Symposium on Mobile and Location-Independent Computing*, 121–137. April.

Gupta, V., D. Mohapatra, A. Raghunathan, and K. Roy. 2013. "Low-Power Digital Signal Processing Using Approximate Adders." *Transactions on Computer-Aided Design of Integrated Circuits and Systems* 32, no. 1 (January): 124–137.

Han, J., and M. Orshansky. 2013. "Approximate Computing: An Emerging Paradigm for Energy-Efficient Design." In *European Test Symposium*, 1–6.

Haraldsson, S. O., and J. R. Woodward. 2015. "Genetic Improvement of Energy Usage Is Only as Reliable as the Measurements Are Accurate." In *Genetic and Evolutionary Computation Conference*, 821–822.

Harman, M., Y. Jia, and W. B. Langdon. 2014. "Babel Pidgin: SBSE Can Grow and Graft Entirely New Functionality into a Real World System." In *International Symposium on Search-Based Software Engineering*, 247–252.

Hoelzle, U., and L. A. Barroso. 2009. *The Datacenter as a Computer: An Introduction to the Design of Warehouse-Scale Machines*. 1st ed. Morgan / Claypool.

Hoffmann, H., S. Misailovic, S. Sidiroglou, A. Agarwal, and M. Rinard. 2009. *Using Code Perforation to Improve Performance, Reduce Energy Consumption, and Respond to Failures*. Technical Report MIT-CSAIL-TR-2009-042. Cambridge, MA: MIT.

Holland, J. H. 1992. *Adaptation in Natural and Artificial Systems*. 2nd ed. Cambridge, MA: MIT Press.

Hsu, C.-H., and U. Kremer. 2003. "The Design, Implementation, and Evaluation of a Compiler Algorithm for CPU Energy Reduction." In *Programming Language Design and Implementation*, 38–48. June.

Kammerer, W. J., and M. Z. Nashed. 1972. "Iterative Methods for Best Approximate Solutions of Linear Integral Equations of the First and Second Kinds." *Journal of Mathematical Analysis and Applications* 40, no. 3 (December): 547–573.

Kim, D., J. Nam, J. Song, and S. Kim. 2013. "Automatic Patch Generation Learned from Human-Written Patches." In *International Conference on Software Engineering*, 802–811. May.

Kirkpatrick, S., C. D. Gelatt, and M. P. Vecchi. 1983. "Optimization by Simulated Annealing." *Science* 220 (4598): 671–680.

Kocher, P., D. Genkin, D. Gruss, W. Haas, M. Hamburg, M. Lipp, S. Mangard, T. Prescher, M. Schwarz, and Y. Yarom. 2018. "Spectre Attacks: Exploiting Speculative Execution." ArXiv:1801.01203.

Koza, J. R. 1992. *Genetic Programming: On the Programming of Computers by Means of Natural Selection*. Cambridge, MA: MIT Press.

Langdon, W. B. 2015. "Genetically Improved Software." In *Handbook of Genetic Programming Applications,* edited by Amir H. Gandomi, Amir H. Alavi, and Conor Ryan. New York: Springer.

Le Goues, C., M. Dewey-Vogt, S. Forrest, and W. Weimer. 2012. "A Systematic Study of Automated Program Repair: Fixing 55 out of 105 Bugs for $8 Each." In *International Conference on Software Engineering*, 3–13.

Lee, M. T.-C., V. Tiwari, S. Malik, and M. Fujita. 1997. "Power Analysis and Minimization Techniques for Embedded DSP Software." *Transactions on Very Large Scale Integration Systems* 5 (1): 123–135.

Linares-Vásquez, M., G. Bavota, C. E. B. Cárdenas, R. Oliveto, M. Di Penta, and D. Poshyvanyk. 2015. "Optimizing Energy Consumption of GUIs in Android Apps: A Multi-objective Approach." In *Joint Meeting of the European Software Engineering Conference and the Symposium on the Foundations of Software Engineering,* 143–154.

Lipp, M., M. Schwarz, D. Gruss, T. Prescher, W. Haas, S. Mangard, P. Kocher, D. Genkin, Y. Yarom, and M. Hamburg. 2018. "Meltdown." ArXiv:1801.01207.

Lu, S.-L. 2004. "Speeding Up Processing with Approximation Circuits." *IEEE Computer* 37 (3): 67–73.

Manotas, I., L. Pollock, and J. Clause. 2014. "SEEDS: A Software Engineer's Energy-Optimization Decision Support Framework." In *International Conference on Software Engineering,* 503–514.

Mars, J., L. Tang, R. Hundt, K. Skadron, and M. L. Soffa. 2012. "Increasing Utilization in Modern Warehouse-Scale Computers Using Bubble-Up." *IEEE Micro* 32, no. 3 (May): 88–99.

Massalin, H. 1987. "Superoptimizer: A Look at the Smallest Program." *ACM SIGARCH Computer Architecture News* 15 (5): 122–126.

Mechtaev, S., J. Yi, and A. Roychoudhury. 2016. "Angelix: Scalable Multiline Program Patch Synthesis via Symbolic Analysis." In *International Conference on Software Engineering*, 691–701.

Niu, L., and G. Quan. 2004. "Reducing Both Dynamic and Leakage Energy Consumption for Hard Real-Time Systems." In *International Conference on Compilers, Architecture, and Synthesis for Embedded Systems*, 140–148.

Nowka, K. J., G. D. Carpenter, E. W. MacDonald, H. C. Ngo, B. C. Brock, K. I. Ishii, T. Y. Nguyen, and J. L. Burns. 2002. "A 32-bit PowerPC System-on-a-Chip with Support for Dynamic Voltage Scaling and Dynamic Frequency Scaling." *IEEE Journal of Solid-State Circuits* 37 (11): 1441–1447.

Palem, K. V. 2014. "Inexactness and a Future of Computing." *Philosophical Transactions of the Royal Society, Series A* 372, no. 2018 (May): 20130281.

Palem, K. V., and A. Lingamneni. 2013. "Ten Years of Building Broken Chips: The Physics and Engineering of Inexact Computing." *Transactions on Embedded Computing Systems* 12, no. 2s (May): 87:1–87:23.

Petke, J. 2016. "Genetic Improvement for Code Obfuscation." In *Genetic and Evolutionary Computation Conference Companion*, 1135–1136.

Rabinowitz, S., and S. Wagon. 1995. "A Spigot Algorithm for the Digits of π." *The American Mathematical Monthly* 102, no. 3 (March): 195–203.

Reda, S., and A. N. Nowroz. 2012. "Power Modeling and Characterization of Computing Devices: A Survey." *Foundations and Trends in Electronic Design Automation* 6 (2): 121–216.

Rice, H. G. 1953. "Classes of Recursively Enumerable Sets and Their Decision Problems." *Transactions of the American Mathematical Society* 74:358–366.

Rinard, M. 2006. "Probabilistic Accuracy Bounds for Fault-Tolerant Computations That Discard Tasks." In *International Conference on Supercomputing*, 324–334.

———. 2007. "Using Early Phase Termination to Eliminate Load Imbalances at Barrier Synchronization Points." In *International Conference on Object-Oriented Programming, Systems, Languages, and Applications*, 369–386.

Sarbishei, O., and K. Radecka. 2010. "Analysis of Precision for Scaling the Intermediate Variables in Fixed-Point Arithmetic Circuits." In *International Conference on Computer-Aided Design*, 739–745.

Sarwar, A. 1997. *CMOS Power Consumption and C_{pd} Calculation*. Technical report. Texas Instruments, June.

Schkufza, E., R. Sharma, and A. Aiken. 2013. "Stochastic Superoptimization." In *International Conference on Architectural Support for Programming Languages and Operating Systems*, 305–316.

Schulte, E., J. Dorn, S. Harding, S. Forrest, and W. Weimer. 2014. "Post-Compiler Software Optimization for Reducing Energy." In *International Conference on Architectural Support for Programming Languages and Operating Systems*, 639–652.

Shehabi, A., S. Smith, D. Sartor, R. Brown, M. Herrlin, J. Koomey, E. Masanet, N. Horner, I. Azevedo, and W. Linter. 2016. *United States Data Center Energy Usage Report*. Technical report. Berkeley, CA: Ernest Orlando Lawrence Berkeley National Laboratory.

Sidiroglou-Douskos, S., S. Misailovic, H. Hoffmann, and M. Rinard. 2011. "Managing Performance vs. Accuracy Trade-Offs with Loop Perforation." In *Joint Meeting of the European Software Engineering Conference and the Symposium on the Foundations of Software Engineering*, 124–134.

Tian, Y., Q. Zhang, T. Wang, F. Yuan, and Q. Xu. 2015. "ApproxMA: Approximate Memory Access for Dynamic Precision Scaling." In *Great Lakes Symposium on VLSI*, 337–342.

Tilevich, E., and Y. Smaragdakis. 2002. "J-Orchestra: Automatic Java Application Partitioning." In *European Conference on Object-Oriented Programming*, 178–204.

Venkataramani, S., S. T. Chakradhar, K. Roy, and A. Raghunathan. 2015. "Approximate Computing and the Quest for Computing Efficiency." In *Design Automation Conference*, 120:1–120:6.

Yang, Z., J. Han, and F. Lombardi. 2015. "Transmission Gate-Based Approximate Adders for Inexact Computing." In *International Symposium on Nanoscale Architectures*, 145–150.

Zitzler, E., L. Thiele, M. Laumanns, C. M. Fonseca, and V. Grunert da Fonseca. 2003. "Performance Assessment of Multiobjective Optimizers: An Analysis and Review." *Transactions on Evolutionary Computation* 7 (2): 117–132.

TRADE-OFFS BETWEEN COST AND PRECISION AND THEIR POSSIBLE IMPACT ON AGING

Hildegard Meyer-Ortmanns, Jacobs University Bremen

Introduction

A current vision in the development of antiaging measures is to consider aging as a disease that can be cured like a disease (De Magalhães 2014; Werfel, Ingber, and Bar-Yam 2015). Typical manifestations of biological aging are accumulating deleterious mutations in older age. Here we associate with aging an increasing accumulation of "errors" in the genetic and cellular "code" that steers fundamental biological processes. Errors may be wrong signaling pathways, insufficient accuracy in reproduction events, imprecise copying processes, or imprecise cellular sensing of concentrations outside the cells, to name a few. If errors in the code exceed a certain threshold, their accumulation will compromise the function of the biological unit.

Our approach is to explore the conjecture that aging of living organisms might be intrinsically unavoidable owing to fundamental principles from physics so that aging can be delayed but not completely reversed or healed like a disease. A possible rationale could proceed along the following line of argument:

- Living organisms are information processing.

- Owing to fundamental trade-offs between cost for the use of resources and precision in its performance, this information processing is inherently error prone and defective.

- As a result, errors will accumulate, either upon the interaction of many defective subsystems or in the reproduction of defective structures.

- Admittedly, different sources of stochastic fluctuations need not automatically add up to a noisy output of reactions if the sources are anticorrelated. In addition, nature has invented a repertoire of repair mechanisms to counteract a rapid accumulation of errors. However, repair mechanisms are information processing, and thus defective, themselves.

- Because "information" in biological systems is physically implemented, its processing costs energy (or other kinds of resources, such as space or time); the higher the required precision, the higher the costs.

- Although the size of external reservoirs is practically infinite for any organism, only a finite amount is accessible within a finite period of time.

- Given these circumstances, one may wonder for how long the accessible energy suffices to pay the price for the required precision and maintaining the errors at a finite but tolerably low fraction such that the functioning of the whole organism is guaranteed.

We summarize results of a simple model in which we studied the time evolution of errors in a reproduction process of bitstrings under constraints on the achievable precision and the accessible energy (Voit and Meyer-Ortmanns 2018, 2019). The bitstring is recursively reproduced at step n from step $n-1$, starting from a correct string at step one. The assumed fundamental trade-off is not modeled explicitly but is effectively absorbed in a cost–precision function that diverges for 100% precision and takes a finite value for random reproduction. To avoid an immediate accumulation of errors, we implement repair and correct all errors beyond a certain threshold with a probability ≤ 1. Importantly, the required energy for repair is taken from the same reservoir that supports the reproduction process. It is finite but can be replenished after a sequence of repair and reproduction processes (for simplicity, always with the same amount of energy as was initially available). Therefore repair goes on the price of accuracy of the subsequent reproduction, as less energy is available. This suggests an increasing need for repair to overcome the decreasing precision and the increasing number of errors. At this point, one may be tempted to jump to conclusions and expect an accumulation of errors to be the unavoidable fate of the system.

In a bifurcation analysis of the mean field equations of this model, we found regions of fixed points in phase space. These are fixed points of the discrete map between error fractions at succeeding steps in the reproduction process. This means that, in a certain region of phase space, it is possible to maintain a low and tolerable error fraction at finite energetic costs. The extension of the regions in phase space depend on the very choice of the cost–precision function. The parameters are the costs for repair relative to the costs for reproduction as well as the probability for successful repair. Depending on the choice of parameters and the initial error fraction, errors either become saturated at the desired threshold or accumulate beyond the tolerated ratio (or, equivalently to the latter case, all energy has to be spent for repair). In a first approach, Voit and Meyer-Ortmanns (2018) neglected costs for maintaining the probability for successful repair at a constant (high) level. In Voit and Meyer-Ortmanns (2019) these costs are taken into account, and their implications are discussed.

Returning to our initial question of whether aging in the form of accumulating errors is an unavoidable fate of information-processing systems under constrained resources, the answer our model offers is that it depends on how much of the resources remain for the essential processes (like reproduction in our model) when an increasing amount has to be spent on repair to maintain proper function. In the next section, we summarize some background from biophysics and stochastic thermodynamics along the lines of the described rationale; then, we present further details of our model.

Background from Biophysics and Stochastic Thermodynamics

BIOLOGICAL SYSTEMS AS SYSTEMS OF INFORMATION PROCESSING

Biological systems—such as genetic networks, signaling proteins, and cells—sense, transduce, and process internal and external signals from the environment to optimize their behavior in order to exploit the limited resources for their own benefit. These processes may be summarized as information processing and biological computation. Nowadays, the use of information theory and computational science in biology extends far beyond mere metaphor for pointing out superficial analogies. Tkačik and Bialek (2016) show how biological information flow in different realizations of biological processes under physical constraints can be formalized in the language of information theory.

An example for biological computation is provided by the so-called chromatin-computer (Prohaska, Stadler, and Krakauer 2010; Arnold, Stadler, and Prohaska 2013). Eukaryotic genomes are typically organized as chromatin, which is the complex of DNA and proteins that forms chromosomes within the cell's nucleus. Chromatin provides a powerful cellular memory device that is capable of storing and processing a large amount of information, including logical and arithmetic operations in a massively parallel architecture (Arnold, Stadler, and Prohaska 2013), with rules for reading and writing and analogies to amorphous computing (Prohaska, Stadler, and Krakauer 2010).

FUNDAMENTAL TRADE-OFFS BETWEEN COST AND PRECISION

Thus it is natural to assess biological processes in terms of accuracy, energetic efficiency, and speed of information processing. From a comparison between cellular copy protocols with canonical copy protocols in computational science, it becomes clear (ten Wolde *et al.* 2016) why cellular sensing systems can never reach the Landauer limit (Landauer 1996; Bennett 2003) on the optimal trade-off between accuracy and energetic cost. According to Landauer (1996) and Bennett (2003), the minimal amount of work for a perfect copy cycle is $k_B T \ln 2$, with k_B the Boltzmann constant and T the temperature of the system. Cellular systems have to dissipate more than thermodynamic processes that run according to ideal quasistatic protocols. While quasistatic protocols make it possible to perform repeated copies with 100% precision at finite energy, cellular copy protocols can only reach 100% precision for diverging costs (Ouldridge and ten Wolde 2017; Ouldridge, Govern, and ten Wolde 2017). When living cells measure low chemical concentrations, three factors together impose a fundamental limit on the precision of sensing (ten

Wolde *et al.* 2016): the number of receptors, the receptor correlation time, and the effective integration time set by the downstream networks. As discussed by ten Wolde *et al.* (2016), this precision is remarkably high but finite, and, as mentioned before, even in an optimized trade-off, the precision cannot achieve the value set by the Landauer limit.

COUNTERMEASURES TO AN ACCUMULATION OF ERRORS

Before we focus on the impact of limited precision, some remarks are in order first to appreciate the high precision that can be achieved despite the versatile sources of stochastic fluctuations in biological systems. Inherent stochastic elements, usually termed "noise," are due to fluctuations in basic chemical reactions, fluctuations in the numbers of copied molecules, or the asynchronous occurrence of synthesis and degradation events. Nature has developed a repertoire of mechanisms to prevent an immediate accumulation of errors and to suppress or amplify signals as needed. In particular, negative feedback can both amplify and suppress noise (Bruggeman, Blüthgen, and Westerhoff 2009); however, noise in one part of the feedback loop is reduced at the expense of an increased level of noise in another part of the loop.

Moreover, correlations play an important role in noise propagation through the different layers of a network, from the genetic to the cellular level. Anticorrelations between extrinsic noise from the environment and intrinsic noise inside the cells can improve the robustness of biochemical networks (Tănase-Nicola, Warren, and ten Wolde 2006). Thus different sources of noise need not add up to an overall noisy output, as one may naïvely expect.

Beyond these simple means of noise suppression, more sophisticated countermeasures to the deleterious impact of errors in information processing have evolved, such as *kinetic proofreading* (Hopfield 1974), which is an important repair mechanism in biochemical processes. Yet here we can also find a trade-off between the speed of the process and the remaining error rate. The process itself costs time and energy; with more resources consumed, fewer errors remain. However, when enzyme-substrate reactions are slowed too significantly by a careful proofreading process, it affects their fitness, and their growth and reproduction slow down as well. A stochastic implementation of kinetic proofreading can lead to an increased level of noise in gene expression as a trade-off for improved specificity (Cepeda, Rieckh, and Tkačik 2015).

THERMODYNAMIC UNCERTAINTY RELATIONS

So far, we have given some examples of different required resources, such as energy and time (and, in general, also space), and different manifestations of trade-offs that affect the precision of information processing. It is natural to wonder whether these are just different manifestations of a universal and fundamental kind of trade-off relation, leading us on a short excursion to thermodynamics in and out of equilibrium. Fluctuations near thermodynamic equilibrium are characterized by the fluctuation-dissipation theorem (Kubo, Toda, and Hashitsume 2012). Biological processes proceed in general out of equilibrium. Fluctuations out of equilibrium are less universal, but among the few universal nonequilibrium principles is a thermodynamic uncertainty relation (Barato and Seifert 2015). This relation expresses a trade-off between the square of the relative uncertainty of an observable and the rate of entropy production as dissipation of energy. So, also, in out-of-equilibrium systems, it is dissipation that continues to regulate fluctuations: small ones, as considered by Barato and Seifert (2015), and even large ones, as proven by Gingrich *et al.* (2016) based on large-deviation inequalities.

Applied to biological systems, the thermodynamic uncertainty relation means that controlling current fluctuations by reducing their relative uncertainty costs a minimal amount of dissipation. Thus a certain level of precision requires a minimal energetic cost. The precision can be associated with a random variable, such as the output of a chemical reaction, the number of consumed or produced molecules, or the number of steps of a molecular motor (Pietzonka, Barato, and Seifert 2016; Pietzonka, Ritort, and Seifert 2017; Seifert 2017). This trade-off applies to any biomolecular process, as long as it is a stationary nonequilibrium process. In this sense, it is universal.

FINITE RESOURCES

In conclusion, from the thermodynamic type of fundamental uncertainty, we expect biological information processing to be in general and inherently error prone. The higher the required level of precision, the greater the energy needed. Even if an almost infinite amount of energy (or other resource) is stored in external reservoirs, biological systems can access only a finite amount in a finite period of time. Is the accessible energy then sufficient to pay for the required precision and to maintain a finite (> 0) but tolerably low rate of error? For how long is it possible to maintain the rate at a low level? In the next section, we report on results obtained from a simple model proposed by Voit and Meyer-Ortmanns (2018, 2019) to address these questions.

Time Evolution of Error Fractions under Constraints on Precision and Energy

In abstraction from concrete biological realizations, we consider dynamical systems with two kinds of processes. The first type are out-of-equilibrium processes, running at low entropy and supported by external energy sources. In general, these include transcription, translation, signaling, reproduction, and other information-processing mechanisms in biological networks, as well as manufacturing in factories of artificial systems. The processes are assumed to be repeated regularly and to require a minimum amount of precision in their performance. The second type should guarantee the maintenance of the dynamical systems, ranging from kinetic proofreading to error correction and repair during the "fabrication" of products.

As representative of the first kind, we choose the repeated reproduction or copying of a bitstring of N bits with finite accuracy $p < 1$. More precisely, rather than reproducing a single template in n independent copies, the nth copy results from a reproduction process of the $(n-1)$th copy, starting from an initial error-free bitstring template.

The second type, the repair process, is represented by the correction of erroneously copied bits in the string under the following assumptions. The correction process starts when a certain error threshold ζ is exceeded; ζ is assumed to be tolerable. The correction is successful with probability $\rho \leq 1$, but even if an attempt at correction is not successful, that attempt costs energy. The trade-off between cost and precision is implemented via a limited accuracy by a class of cost–precision functions $c(p)$. These functions are chosen to impose diverging costs for the copy of a single bit with 100% precision, while a copy without any guarantee of its correctness requires a minimal cost, set to 1. For $c(p)$, we choose

$$c(p) = \frac{1}{(1-p^{n_1})^{n_2}} \qquad (11.1)$$

for $n_1 = 1$, $n_2 = 1$, or 2, or $n_1 = 2$, $n_2 = 1$, as well as a logarithmic dependence according to

$$c(p) = 1 - \log(1-p). \qquad (11.2)$$

The choice of these functions according to equations (11.1) and (11.2) should interpolate between the boundary values; otherwise, the choice is arbitrary in our model (in stark contrast to a concrete biological system, for which it would be challenging to derive it).

For the energy costs per copy of a single bit at highest precision p_0, we set $c(p_0) = f/N > 1$, where f denotes the cost of copying the whole string and p_0 the initial precision. Thus the actual value of f depends on the choice of the cost–precision function and on the value of the initial precision. For simplicity, the cost should represent just one type of resource, called energy, without any quantitative relation to the actual energy that a copying process would cost apart from the upper bound (which is, however, considered to be essential).

To prevent an immediate accumulation of errors, we include repair of erroneously copied bits with $R \cdot c(p_0)$ cost per bit and $R \geq 0$, independently of whether the repair is successful. The success of repair is guaranteed with probability ρ with $0 \leq \rho \leq 1$. It is assumed to set in beyond a certain error threshold ζ that is supposed to be tolerable for an efficient performance. Throughout the calculations we choose $\zeta = 0.1$, which is much larger than the tolerated error rate in DNA, for example, but our results do not sensitively depend on this choice. The main distinction in the implementation of repair is between a constant and a decaying success rate ρ over time. Once the copying leads to an error fraction beyond the tolerated threshold, repair of all errors exceeding the threshold begins, with a success rate depending on ρ, followed by a new copying process of the whole string. We call this sequence of repair and copying processes of N bits "one sweep."

Repair is at the expense of the accuracy of the subsequent copying, as follows. We assume that, for each sweep, the energy reservoir can be replenished by a fixed amount, given by $f = N \cdot c(p_0)$, as long as not all energy has been used for repair in the previous step. If no energy remains after repair, the copying process terminates; otherwise, it continues. It should be noted that this quantity of energy corresponds to a minimal choice in our model, as it is only sufficient to copy the whole string once at maximal accuracy p_0. (Additional energy in the reservoir would delay the subsequent time evolution, and fluctuations in the size of the reservoir may change the monotonous behavior, but we expect neither would impact our overall conclusions.)

Importantly, now, once the reservoir gets replenished and some energy has been spent on repairs, the remaining energy is only sufficient for copying the whole string if the copies are made at a reduced accuracy $p < p_0$ (in a stochastic realization, it is on average reduced), where $p = c^{-1}(c(p_0)(1-R(x-\zeta)))$, with c^{-1} the inverse copying function and $x > \zeta$ the error fraction at time step n. Thus, repair is at the expense of copying accuracy so that the subsequent copying process must be expected to generate more errors on average than the preceding one, leading to an increased number of required corrections afterward.

Typical choices are $c(p) = 1/(1-p)$, $p_0 = 0.9$, and $\zeta = 0.1$, while the parameters R and ρ are used as bifurcation parameters. Once we assume the repair success probability ρ decays with time, the difference between success rates (ρ) in subsequent sweeps is d. In a further extension in Voit and Meyer-Ortmanns (2019), we let ρ dynamically decrease, taking into account the costs for maintaining the success of repair attempts at a certain level.

If we approximate the stochastic implementation of these rules in a mean-field approach and replace the probabilities for errors and their corrections by average values that would be approached in the limit of large string size N, we obtain the following discrete map for the time evolution of error fraction x_n from time n to $n+1$, starting from $x_n \geq \zeta$ and with $\rho \leq 1$:

$$x_{n+1} = \zeta + (x_n - \zeta)(1 - \rho)$$
$$+ (1 - \zeta - (x_n - \zeta)(1 - \rho))\frac{1 - p_0}{(1 - (x_n - \zeta)R)}. \quad (11.3)$$

This recursive map holds for a cost–precision function chosen as $c(p) = \frac{1}{1-p}$. Here $(x_n - \zeta)(1 - \rho)$ is the fraction of errors repaired without success, while a fraction $1 - p = \frac{1-p_0}{(1-(x_n-\zeta)R)}$ of the remaining bits gets erroneously copied.

From the physics point of view, we are interested in the possibility of maintaining a copying performance at constant error fraction, so we determine the fixed points of this map. Real fixed points are obtained from $x_{n+1} = x_n$ under the condition that the square root is real, leading to

$$x_\pm^* = \frac{1}{2R\rho}(1 + p_0(\rho - 1) + 2R\rho\zeta)$$
$$\pm \sqrt{1 + p_0^2(\rho - 1)^2 - 2p_0(\rho - 1 - 2R\rho(\zeta - 1)) + 4R\rho(\zeta - 1)}. \quad (11.4)$$

The stability analysis of these fixed points shows that x_+^* is unstable, while x_-^* is stable. Figure 11.1A displays the location of both types of fixed points as a function of the repair cost factor R for $\rho = 1$.

When the repair factor R is varied, we obtain a line of stable and unstable fixed points that terminate in a fold bifurcation (single black dot, linked to the corresponding R_{crit} via the dotted line), from which on no real solution for $x_n = x_{n+1}$ exists. For all initial conditions on x from the shaded area, the error fraction converges to a stable fixed point

Chapter 11: Trade-Offs between Cost and Precision

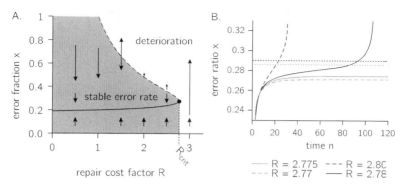

Figure 11.1. (A) Bifurcation diagram for parameter values $p_0 = 0.9$, $\zeta = 0.1$, and fixed $\rho = 1$, corresponding to deterministic repair. Plotted are the locations of the stable (solid line) and unstable (dashed line) fixed points against the size of the repair-cost factor R. States within the shaded region to the left of and below the unstable branch are stable in the sense that the error fraction approaches a constant value, given by the stable fixed point that is approached. For $R > R_{\text{crit}}$, the system always deteriorates. (B) Mean-field error fraction x_n as a function of time for different repair costs R, below and above $R_{\text{crit}} = 2.7$.

on the solid line, depending on the choice of R. Note that the values in figure 11.1A are above the so-called tolerable error threshold of $\zeta = 0.1$, the error fraction after defective copying becomes saturated at some larger value. In a numerical implementation, the larger value is observed when measuring the error fraction after completion of each sweep, and so after copying. Measuring it immediately after the repair, the fraction stays at the desired threshold value of ζ, but after the subsequent, less precise, copying, the error fraction increases because of the reduced accuracy, and the value is the fixed-point value of the map, if it exists. All energy is then consumed for the copying of the whole string at the reduced accuracy. The reservoir is depleted and gets replenished with the fixed amount.

If no fixed points exist, the error fraction from the previous copying increases toward 1 if the reservoir still contains enough energy for copying. The information coded in the initial string would then be completely corrupted in such a copy at a later time. For our choice of reservoir size, the repair absorbs all energy from the reservoir before the fraction approaches 1, leaving nothing for a further copying process; copying thus terminates. Both fates of the system are called "deterioration." In crossing the critical value of R in figure 11.1A, indicated as R_{crit}, or the dashed line of the unstable fixed points, the system "deteriorates" in this sense. The arrows indicate possible time evolutions of the error fraction.

The actual time evolution of the error fraction is plotted in figure 11.1B with a stable evolution for $R = 2.77 < R_{\text{crit}}$; a slow increase at $R = 2.78$ slightly above $R_{\text{crit}} = 2.\bar{7}$, where the flow of error fractions is still reminiscent of the flow in the vicinity of the stable fixed point; and a fast increase for larger R. In a stochastic realization of this system, a fluctuation can easily kick it from the shaded area across the line of unstable fixed points toward a phase of error accumulation. Dotted lines mark the location of the unstable fixed points (which exist only for $R < R_{\text{crit}}$).

If we determine stable and unstable fixed points also for an error repair success rate $\rho < 1$, we obtain the bifurcation diagram of figure 11.2 as a function of both bifurcation parameters ρ and R. The single point at which the fold bifurcation occurs in figure 11.1A becomes the solid line in figure 11.2, continued by the dashed black line. For ρ and R from the shaded area in dark grey, the system stabilizes at a constant error rate. When ρ and R are chosen from the so-called coexistence area (light grey), the fate of the system depends on the initial values for x_n: either the error fraction stabilizes or all energy is spent for repair, because the error fraction would otherwise approach 1 (cf. fig. 11.1, which corresponds to the case of $\rho = 1$). Choosing ρ and R between the solid and black dashed line of fold bifurcations leads to deterioration, while a choice of ρ and R from the small area below the black dashed line corresponds to complex-valued fixed points or fixed points with values larger than 1, so that neither is physically meaningful. The vertical thin dashed line for fixed $R > 1$ in figure 11.2 indicates how a decaying success rate of error correction (indicated by the black arrow) leads from a parameter region with stable error rate to a region in which, sooner or later, all energy must be spent on repair.

Interestingly, the very existence of fixed points may come as a surprise and gives a warning about extrapolating the final fate of the system. Even if the energy for repair is taken from the same reservoir as the energy for copying at the expense of a decreasing accuracy in the subsequent copying process, the errors need not accumulate as one might intuitively expect. There may be fixed points of the map that prevent this evolution.

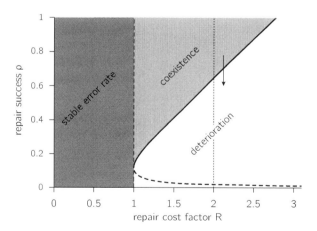

Figure 11.2. Bifurcation diagram for varying repair success factor ρ and repair cost factor R. The solid line marks the critical values $R_{\text{crit}}(\rho)$ of fold bifurcations, continued by the black dotted line. The shaded vertical dotted line indicates a possible path when the repair success factor ρ is decaying. Other parameters are $p_0 = 0.9$ and $\zeta = 0.1$. For further explanation, see the text.

However, in the considerations so far, we have assumed that the success rate of error correction can be maintained at a constant value without charging a cost for this type of maintenance. As soon as repair mechanisms are also subject to some manifestation of the trade-offs (which we assume universally exist), energy should also be spent on maintaining the repair process. Therefore a decreasing repair success rate ($d > 0$ between ρ of subsequent sweeps) amounts to an effective description of a situation where no energy is invested in "repairing" repair. For a dynamically implemented decrease of ρ we refer to Voit and Meyer-Ortmanns (2019).

We have checked the dependence on the choice of the cost–precision function, which turns out to be weak. For the special case of no repair, we have described the accumulation of errors in terms of an increase of entropy toward its maximal value if the initial string had a periodic structure; however, it may happen that an energy reservoir of fixed size gets depleted before the maximal entropy value is reached. For further details, refer to Voit and Meyer-Ortmanns (2018).

Conclusions and Outlook

In real biological organisms, multiple information processing is ongoing in parallel. Different kinds of resources constrain the performance, and themselves fluctuate. The functional form of cost–precision functions depends on the specific process, the organism, the individual, so the complexity of a whole organism is far too rich to predict the time evolution of "errors" on multiple levels.

We think that physics plays a role in the basis of biological aging because trade-offs between cost and precision seem to occur universally, rendering any biological information processing inherently defective, including repair mechanisms themselves.

Unless sufficient energy can be spent on maintaining the quality of repair, the success of repair decreases with time, and errors accumulate over time. It then will still depend on the value of the accumulation point whether such a fate is deleterious for the system or not.

How fast living systems approach such a fate of possible malfunction as a result of error accumulation depends in particular on the cost–precision function and the amount of available resources. These factors can be partially influenced to delay the process of aging by an optimal partition of the costs that are spent for life and for its maintenance. To our understanding, and from our simple model, it remains open whether aging is irreversible in general on a long time scale due to principles of fundamental physics. The answer will depend on the factors mentioned before.

In future work, one may think of isolating suitable subsystems of living organisms that are simple enough to specify the cost of life, the required precision, and the possible cost of repair in order to see whether indications for aging can be traced back to an accumulation of errors within these subsystems. For example, experiments with mice (Farajnia *et al.* 2014) indicate that the circadian clock ages via the suprachiasmatic nucleus in the sense that it gets less adaptive in older age. However, in general, coherent oscillations cannot be sustained for an indefinite time when the period of oscillations fluctuates. According to Barato and Seifert (2016) and Barato and Seifert (2017), the number of coherent oscillations is universally bound by the thermodynamic force that drives the system out of equilibrium and by the topology of the associated biochemical network of states. These results, however, apply to autonomous biochemical oscillators, while the circadian clock is also coupled to external periodic signals of light and darkness. It would be an interesting challenge for future researchers to explore whether the decrease in flexibility of the circadian clock in older age can be traced back to relations between precision and thermodynamics. The precision may then fluctuate on a short time scale, but, over the entire lifespan, it should decrease when facing diverging costs for its repair. ☙

REFERENCES

Arnold, C., P. F. Stadler, and S. J. Prohaska. 2013. "Chromatin Computation: Epigenetic Inheritance as a Pattern Reconstruction Problem." *Journal of Theoretical Biology* 336:61–74.

Barato, A. C., and U. Seifert. 2015. "Thermodynamic Uncertainty Relation for Biomolecular Processes." *Physical Review Letters* 114 (15): 158101.

———. 2016. "Cost and Precision of Brownian Clocks." *Physical Review X* 6 (4): 041053.

———. 2017. "Coherence of Biochemical Oscillations Is Bounded by Driving Force and Network Topology." *Physical Review E* 95 (6): 062409.

Bennett, C. H. 2003. "Notes on Landauer's Principle, Reversible Computation, and Maxwell's Demon." *Studies In History and Philosophy of Science Part B: Studies In History and Philosophy of Modern Physics* 34 (3): 501–510.

Bruggeman, F. J., N. Blüthgen, and H. V. Westerhoff. 2009. "Noise Management by Molecular Networks." *PLoS Computational Biology* 5 (9): e1000506.

Cepeda, S. A., G. Rieckh, and G. Tkačik. 2015. "Stochastic Proofreading Mechanism Alleviates Crosstalk in Transcriptional Regulation." *Physical Review Letters* 115 (24): 248101.

De Magalhães, J. P. 2014. "The Scientific Quest for Lasting Youth: Prospects for Curing Aging." *Rejuvenation Research* 17 (5): 458–467.

Farajnia, S., T. Deboer, J. H. T. Rohling, J. H. Meijer, and S. Michel. 2014. "Aging of the Suprachiasmatic Clock." *The Neuroscientist* (Los Angeles, CA) 20 (1): 44–55.

Gingrich, T. R., J. M. Horowitz, N. Perunov, and J. L. England. 2016. "Dissipation Bounds All Steady-State Current Fluctuations." *Physical Review Letters* 116 (12): 120601.

Hopfield, J. J. 1974. "Kinetic Proofreading: A New Mechanism for Reducing Errors in Biosynthetic Processes Requiring High Specificity." *Proceedings of the National Academy of Sciences of the United States of America* 71 (10): 4135–4139.

Kubo, R., M. Toda, and N. Hashitsume. 2012. *Statistical Physics II: Nonequilibrium Statistical Mechanics*. Vol. 31. New York: Springer Science & Business Media.

Landauer, R. 1996. "The Physical Nature of Information." *Physics Letters A* 217 (4): 188–193.

Ouldridge, T. E., C. C. Govern, and P. R. ten Wolde. 2017. "Thermodynamics of Computational Copying in Biochemical Systems." *Physical Review X* 7 (2): 021004.

Ouldridge, T. E., and P. R. ten Wolde. 2017. "Fundamental Costs in the Production and Destruction of Persistent Polymer Copies." *Physical Review Letters* 118 (15): 158103.

Pietzonka, P., A. C. Barato, and U. Seifert. 2016. "Universal Bound on the Efficiency of Molecular Motors." *Journal of Statistical Mechanics: Theory and Experiment* 2016 (12): 124004.

Pietzonka, P., F. Ritort, and U. Seifert. 2017. "Finite-time generalization of the thermodynamic uncertainty relation." *Physical Review E* 96 (1): 012101.

Prohaska, S. J., P. F. Stadler, and D. C. Krakauer. 2010. "Innovation in Gene Regulation: The Case of Chromatin Computation." *Journal of Theoretical Biology* 265 (1): 27–44.

Seifert, U. 2017. "Stochastic Thermodynamics: From Principles to the Cost of Precision." *Physica A: Statistical Mechanics and Its Applications*.

Tănase-Nicola, S., P. B. Warren, and P. R. ten Wolde. 2006. "Signal Detection, Modularity, and the Correlation between Extrinsic and Intrinsic Noise in Biochemical Networks." *Physical Review Letters* 97 (6): 068102.

ten Wolde, P. R., N. B. Becker, T. E. Ouldridge, and A. Mugler. 2016. "Fundamental Limits to Cellular Sensing." *Journal of Statistical Physics* 162 (5): 1395–1424.

Tkačik, G., and W. Bialek. 2016. "Information Processing in Living Systems." *Annual Review of Condensed Matter Physics* 7:89–117.

Voit, M., and H. Meyer-Ortmanns. 2018. "On the Fate of Dynamical Systems under a Trade-off between Cost and Precision." arXiv:1810.04084v1.

Voit, M., and H. Meyer-Ortmanns. 2019. "How Aging May Be an Unavoidable Fate of Dynamical Systems." *New Journal of Physics* 21 (4): 043045.

Werfel, J., D. E. Ingber, and Y. Bar-Yam. 2015. "Programmed Death Is Favored by Natural Selection in Spatial Systems." *Physical Review Letters* 114 (23): 238103.

16

THE POWER OF BEING EXPLICIT: DEMYSTIFYING WORK, HEAT, AND FREE ENERGY IN THE PHYSICS OF COMPUTATION

Thomas E. Ouldridge, Imperial College London
Rory A. Brittain, Imperial College London
Pieter Rein ten Wolde, FOM Institute AMOLF

Introduction

Interest in the thermodynamics of computation has revived in recent years, driven by developments in science, economics, and technology. Given the consequences of the growing demand for computational power, the idea of reducing the energy cost of computations has gained new importance (DeBenedictis, Mee, and Frank 2017; Frank 2017). Simultaneously, many biological networks are now interpreted as information-processing or computational systems constrained by their underlying thermodynamics (Andrieux and Gaspard 2008; Lan *et al.* 2012; Mehta and Schwab 2012; Ito and Sagawa 2015; Govern and ten Wolde 2014; Barato, Hartich, and Seifert 2014; Mehta, Lang, and Schwab 2016; ten Wolde *et al.* 2016; Ouldridge and ten Wolde 2017; Ouldridge, Govern, and ten Wolde 2017; Barato and Seifert 2017). Indeed, some suggest (Bennett 1982; Adleman 1994; Organick *et al.* 2018) that low-cost, high-density biological systems may help to mitigate the rising demand for computational power and the "end" of Moore's law of exponential growth in the density of transistors (Frank 2017).

In this chapter, we address widespread misconceptions about thermodynamics and the thermodynamics of computation, using biomolecular systems as a conceptual crutch. In particular, we argue against the general perception that a measurement or copy operation can be performed at no cost; against the emphasis placed on the significance of erasure operations; and against the careless discussion of heat and work. Although not universal, these misconceptions are sufficiently prevalent (particularly within interdisciplinary contexts) to warrant a detailed discussion. In the process, we argue that representing fundamental processes explicitly is a useful tool, serving to demystify key concepts.

We give first a brief overview of thermodynamics, then the history of the thermodynamics of computation—particularly in terms of copy and measurement operations inherent to classic thought experiments. Subsequently, we analyze these ideas via an explicit biochemical representation of the entire cycle of Szilard's engine. In doing so, we show that molecular computation is both a promising engineering paradigm and a valuable tool in providing fundamental understanding.

Basic Thermodynamics

Here we briefly recap the key ideas of work, heat, entropy, and free energy. In effect, we introduce several consistent ways to think about the second law of thermodynamics. Experienced readers may wish to skip to the section "A History of Maxwell's Demon and Szilard's Engine;" those wishing for more detail on the background of equilibrium thermodynamics and statistical mechanics may refer to Atkins (2010) and Huang (1987). Note that the quantities we consider here are implicitly statistical averages, rather than quantities defined for individual fluctuating trajectories (Jarzynski 1997; Crooks 1999; Esposito and Van den

Broeck 2011; Seifert 2005), and that we restrict ourselves to the classical regime.

WORK, HEAT, THE SECOND LAW, AND ENTROPY

Energy is conserved, but not all energy is equivalent. Certain forms of energy, like that stored when a mass is raised in a gravitational field, are more useful than others, such as that stored in a hot object. The energy of an ideal mass is stored in one accessible (collective) degree of freedom: the position of the center of mass in a gravitational field. The energy of a hot object, such as a volume of gas, is stored in many degrees of freedom (the momenta of the gas particles). Unlike the mass, it is impossible for an experimenter to couple to these degrees of freedom individually and use the stored energy. This distinction underlies the second law.

Energy is transferred as heat when the transfer occurs through many degrees of freedom that are individually inaccessible. By contrast, work is done when the transfer of energy occurs via the accessible degrees of freedom. We define a heat reservoir as an ideal, arbitrarily large system that can only accept energy transfer via heat (think of a large gas at fixed volume). The heat reservoir is characterized by a temperature T, which quantifies the average energy per degree of freedom. We define an ideal work reservoir as an ideal system, like a mass in a gravitational field, that only stores energy in an accessible degree of freedom. We note that this definition of work—in terms of the transfer of energy via the manipulation of accessible coordinates—is consistent with the definition used in modern stochastic thermodynamics (Horowitz and Jarzynski 2008). The definition of heat, then, is necessarily also consistent.

One of the traditional statements of the second law is that "no process is possible whose sole result is the conversion of heat into work." This statement makes precise the idea that "not all energies are equal." No engine exists that cyclically takes energy from a single heat reservoir and deposits all of it into a work reservoir (fig. 12.1A). The cyclical operation is important here—if the engine underwent a sustained change, then the transfer of energy would not be the "sole result" of the process.

A more typical modern statement of the second law would be that "in a closed system, the entropy is nondecreasing." The entropy quantifies the uncertainty in the precise state of a system. Specifically (Jaynes 1957),

$$\mathcal{S}[p(x)] = -k_\text{B} \sum_x p(x) \ln p(x). \qquad (12.1)$$

Here x denotes the possible states of a system and $p(x)$ the probability that the system will be found in state x. The flatter the distribution $p(x)$ is, the larger the entropy is. If we are completely certain of the state of a system, then all terms in the sum in equation (12.1) are identically zero.

Entropy incorporates the traditional statement of the second law. When energy is transferred to a work reservoir, there is no change in the work reservoir's entropy, because it only has one degree of freedom—its energy uniquely specifies its state. By contrast, when energy is transferred to a heat reservoir, it is shared between the many degrees of freedom. More energy increases the uncertainty, because there are more ways to divide up the energy. Thus the transfer of energy to a heat reservoir is associated with an entropy increase, and converting heat directly into work is forbidden, because it would be an overall decrease in the entropy of a closed system. To be precise, if heat dQ is transferred to a heat reservoir, the reservoir's entropy increase is $d\mathcal{S} = TdQ$. Importantly, the entropic definition of the second law allows us to analyze processes in which systems that are not ideal reservoirs undergo changes, using equation (12.1); this framework allows

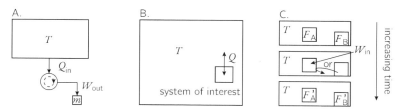

Figure 12.1. (A) Graphical representation of processes forbidden by the second law of thermodynamics: a cyclic engine cannot solely take in heat from a reservoir T and convert it to work to raise a mass m. (B) Illustration of the canonical ensemble: a small system of interest is embedded within a larger system setting the temperature T. Heat Q can be exchanged between system and reservoir. (C) The input of external work, or the coupling to another subsystem, can be used to drive up the free energy of a subsystem A so that $\mathcal{F}'_A > \mathcal{F}_A$. In the absence of the second subsystem, the second law dictates that $\mathcal{W}_{in} \geq \mathcal{F}'_A - \mathcal{F}_A$. In the other case of no external work input, $\mathcal{F}_B - \mathcal{F}'_B \geq \mathcal{F}'_A - \mathcal{F}_A$.

us to analyze noncyclic systems and individual steps of cyclic operations.

EQUILIBRIUM, THE BOLTZMANN DISTRIBUTION, AND FREE ENERGIES

An isolated system with sufficiently many interacting degrees of freedom will evolve toward an equilibrium steady state in which there is no systematic variation of observables with time, or net currents of matter or energy. Because the entropy of an isolated system cannot decrease, the equilibrium distribution $p_{eq}(x)$ maximizes the entropy subject to whatever constraints are relevant (e.g., the overall conservation of energy). It is often convenient to consider a small system of interest coupled to a heat reservoir at temperature T, as illustrated in figure 12.1B. In this case, representing the states of the small system by x, it is possible to show that

$$p_{eq}(x) \propto \exp(-\epsilon(x)/k_B T), \qquad (12.2)$$

where $\epsilon(x)$ is the energy of configuration x (Huang 1987). The distribution $p_{eq}(x)$ maximizes the overall entropy of the combined system and reservoir given a fixed total energy to share between the two. For simplicity, we restrict ourselves to this "canonical ensemble."

Prior to reaching equilibrium, systems have a nonequilibrium distribution, $p(x) \neq p_{\text{eq}}(x)$. The distance from equilibrium can be characterized by the nonequilibrium generalized free energy (Esposito and Van den Broeck 2011; Parrondo, Horowitz, and Sagawa 2015)

$$\mathcal{F}[p(x)] = \sum_x p(x)\epsilon(x) + k_\text{B}T \sum_x p(x) \ln p(x) = \\ \langle \epsilon \rangle - T\mathcal{S}[p(x)] \qquad (12.3)$$

$$= \mathcal{F}[p_{\text{eq}}(x)] + k_\text{B}T \sum_x p(x) \ln \tfrac{p(x)}{p_{\text{eq}}(x)} \geq \mathcal{F}[p_{\text{eq}}(x)].$$

Following a process in which $p(x)$ evolves over time, the change in the total entropy of the system and reservoir is given by $-\Delta\mathcal{F}[p(x)]/T$; $\mathcal{F}[p(x)]$ then *decreases* over time and is *minimal* at equilibrium. This minimum represents a balance between minimization of energy $\langle \epsilon \rangle$ and maximization of entropy $\mathcal{S}[p(x)]$ within the system.

Although nonequilibrium free energies cannot increase spontaneously, they can be driven upward, as in figure 12.1C. Specifically, the change in free energy determines the minimal work that must be applied to move a system A coupled to a heat reservoir between $p(x)$ and $p'(x)$ (Esposito and Van den Broeck 2011): $\mathcal{W}_{\text{in}} \geq \mathcal{F}_A[p'(x)] - \mathcal{F}_A[p(x)]$. As a consequence, the maximum work that can be extracted is $\mathcal{W}_{\text{out}} \leq \mathcal{F}_A[p(x)] - \mathcal{F}_A[p'(x)]$. The system's free energy must change to obtain positive work because work cannot be extracted from the heat reservoir alone.

We could also couple to another system that is not an idealized work reservoir (figure 12.1C). In this case, the second law dictates that the total free energy of the combined system must decrease, but a decrease in the free energy of one subsystem can compensate for an increase in the other. A nonequilibrium system is then fundamentally an exploitable resource. If a system A is out of equilibrium, we can, in principle, couple it to a second system B and use the relaxation of A to drive a change in B. If A is in equilibrium, however, it is not exploitable in this way.

It is important to note that, although the free energy has the dimensions of an energy, and although changes in free energy are bounded by work input, free energies are not energies in a very fundamental sense. Energies obey the first law of thermodynamics; energy can never be destroyed, only converted between different forms. If work is done on a system, either its internal energy must change or energy must be transferred to the environment as heat. Free energies, however, also have an entropic component (e.g., equation 12.3). They are therefore not conserved; it is perfectly possible for the free energy of a system to decrease with no other consequences. We explore this fact in more detail in the subsection "Focusing on Heat and Work Can Be Misleading."

THERMODYNAMIC REVERSIBILITY

Consider applying a time-varying manipulation of a system, driving it from $p(x)$ to $p'(x)$. If and only if this process is thermodynamically reversible, applying the manipulation in a time-reversed manner would convert $p'(x)$ back to $p(x)$, and the intermediate probability distributions obtained during the forward manipulation would be reproduced in the opposite order during the reverse protocol (see Sagawa 2014). Thermodynamically reversible processes necessarily involve no change in the overall entropy of the system and reservoirs to which they are coupled, because a positive entropy increase on the forward manipulation would correspond to a forbidden decrease on the reverse. Reversible processes therefore require work input of exactly $\mathcal{W}_{\text{in}} = F[p'(x)] - F[p(x)]$ (Esposito and Van den Broeck 2011) and constitute an important idealized limit of practical processes.

The Physics of Computation

To perform computation, a physical system—be it silicon-based or biological—must be able to change in response to a range of inputs. Indeed, data and information must have a physical realization simply to exist, let alone to be manipulated during a computation (Landauer 1991). If computation involves changes in the states of physical systems, then these processes can be analyzed thermodynamically, and the resultant thermodynamic costs and requirements can be assessed. This approach is becoming increasingly fashionable, driven by a desire to reduce the consumption of electricity by our increasing demands on computational power (DeBenedictis, Mee, and Frank 2017; Frank 2017).

Although recent years have seen a wider focus (see, e.g., Owen, Kolchinsky, and Wolpert 2011), the study of the thermodynamics of computation has been dominated by the discussion of two processes: erasure and measurement (or copying) of single two-state memories. Partly, this focus follows from the fact that these processes are prototypical computational operations that provide much of the framework for understanding more complex systems. Equally importantly, however, erasure and measurement lie at the heart of a fundamental debate in physics relating to the validity and meaning of the second law of thermodynamics. In the following paragraphs, we summarize this debate.

Chapter 12: The Power of Being Explicit

A HISTORY OF MAXWELL'S DEMON AND SZILARD'S ENGINE

Physicists have long been concerned by the second law. Most famously, James Clerk Maxwell wondered whether an intelligent being (a "demon") could violate the second law by extracting work from random fluctuations of an equilibrium system (Dougal 2016). Maxwell's demon operated a trap door separating two volumes of gas at initially equal temperature. If the demon could determine the velocity of gas particles, it could selectively open the door to allow high-energy particles to pass in one direction and low-energy particles in the other. Over time, a temperature difference would develop, to which a work-extracting heat engine could be coupled. There is no fundamental lower bound on the work needed to open and close a trap door. Therefore it appeared to Maxwell that the demon could use its intelligence to perform measurement and feedback to exploit fluctuations and thereby violate the second law of thermodynamics by extracting work from an initially equilibrated system.

Many authors have attempted to explain why the second law is not violated by Maxwell's demon. Typically, they construct a demon-like system, analyze its behavior, and demonstrate that violation of the second law does not occur. This approach has been criticized (Norton 2005) on the grounds that the accuracy of the second law is generally assumed in setting up the analysis. We, however, believe that it is valid both to test the self-consistency of the second law in this way and to ask the conceptual question of *how* the self-consistency is manifested, as the understanding gained applies beyond the original context.

Although the term *Maxwell's demon* is used quite liberally, we will focus on systems in which a measurement is made, stored, and then used to implement measurement-dependent feedback. In 1929, Leo Szilard described an analytically tractable "engine" with measurement and feedback (Szilard 1964). A version is schematically illustrated in figure 12.2. Here, the demon simply determines whether a single particle is located on the left-hand or right-hand side of a piston in a hollow cylinder and configures the system to exploit this fact. Given knowledge of the particle's position, the piston can be coupled to a weight so that the pressure of the particle can do work to raise the weight. It is clear that measurement and feedback are required to couple the weight correctly. Assuming that this procedure is performed with 100% accuracy, the maximum work extracted per expansion can be calculated by integrating pdV. Here p is the average pressure applied by the particle, and V is the volume in which it is contained, which grows from $V_{\text{cylinder}}/2$ to V_{cylinder}. The result is $k_B T \ln 2$.

Szilard sought to identify a cost that compensated for the apparent ability to extract work from the initial equilibrium system. He realized that the essence of his system could be analyzed without reference to an intelligent being. Instead, he argued that the demon's measurement generates a correlation between two physical degrees of freedom. In figure 12.2, these degrees of freedom are the position of the particle (the data) and the weight's connection to the partition (the memory). Crucially, Szilard emphasized that, post-measurement, the memory must retain its state even if the data bit subsequently changes. In figure 12.2, for example, the weight must stay attached to the same side of the piston even as the particle's accessible volume expands to fill the whole cylinder. Szilard stated that "if the measurement could take place without compensation," then the second law could be violated. He claimed that there must be a minimum "production of

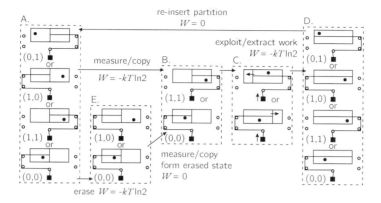

Figure 12.2. A cycle of Szilard's engine, with possible operating states illustrated. Steps are labeled with the minimal work input required for reversible operation (negative values correspond to work extraction). The engine consists of a single particle in a volume, a partition, and a weight that can be attached to either side of the partition via pulleys. (A) Initially, the particle and the weight can both be associated with either side of the volume, and four states are possible (labeled (d, m), where $d = 0[1]$ corresponds to the particle on the left-hand [right-hand] side of the system, and $m = 0[1]$ corresponds to the apparatus on the left [right] side of the partition). (B) A measurement correlates the weight's connection with the particle's position. Two states are possible after a perfectly accurate measurement. (C) The particle is allowed to expand against the partition, raising the weight. Note that the weight's attachment state does not change. Eventually, state (D) is reached, when the particle's position is uncorrelated with the weight's attachment. The system is then restored to its initial condition by reinserting the partition. (E) Alternatively, one might first "erase," setting the pulleys into a guaranteed configuration and leaving two possible states overall, prior to performing the measurement.

entropy" during measurement that compensated for the $k_\text{B}T \ln 2$ of work extracted during the expansion. In the computational sense, he was arguing that there is a cost to copying of data.

In 1951, Brillouin reopened the question. He, too, believed that the demon needed to pay a cost when measuring (Brillouin 1951). He associated this cost with the generation of a nonequilibrium photon that could be used to see the particle. Brillouin did not consider correlating a memory with the measured system but merely the interaction mechanism. His proposed measurement costs are therefore of a fundamentally different character from Szilard's.

A decade later, Landauer (1961) asked whether heat generation was necessary during computation. Although he referred to Brillouin's work, he did not deal with demons. Rather, he argued that computational operations without a single-valued inverse (many-to-one functions) necessarily lead to an increase in the entropy of the environment. Landauer pointed out that such "logically irreversible" functions—for example, "erasing" a bit of unknown value to zero—decrease the entropy of the bit itself. He argued that there must be an entropy increase in the environment to preserve the second law. For a symmetric bit initially in state 0 with 50% probability, the initial entropy is $k_B \ln 2$ (from equation 12.1). After erasure is complete, the entropy of the bit is 0. Thus the entropy of the environment must increase by at least $k_B \ln 2$ to compensate. In a simple system with single heat and work reservoirs, this operation would involve a transfer of at least $k_B T \ln 2$ of energy from the work reservoir to the heat reservoir via the bit.

Eventually, Bennett drew upon Landauer's work to argue that measurement, or copying of a bit, could be performed with no work input, provided that the memory was in a well-defined initial state (say, 0) (Bennett 1982, 2003). In this case, the measurement is logically reversible, because two initial states of the memory and data are mapped to two final states, as shown by the conversion from (E) to (B) in figure 12.2. Such a measurement requires no net increase in the entropy of the environment and hence no work input. However, to operate again, the memory must be reset. If the "data" bit has already been exploited at this point, for example, by a demon extracting work, it no longer retains its previously measured value. The reset cannot then be performed in a logically reversible manner—we must perform a logically irreversible "erase" (fig. 12.2) to reset the memory. This reset must increase the entropy of the environment by Landauer's principle, and hence the work extracted ($\mathcal{W}_{\text{out}} \leq k_B T \ln 2$) is paid for by the resetting of the device ($\mathcal{W}_{\text{in}} \geq k_B T \ln 2$). Bennett argued that this erasure cost explains why the second law is not violated.

Within the wider scientific community, Bennett's argument is generally accepted as the "true" explanation of why the second law survives (e.g., Feynman 1998; Plenio and Vitelli 2001; Bub 2001; Dillenschneider and Lutz 2009; Maruyama, Nori, and Vedral 2009; Bérut *et al*. 2012; Jun, Gavrilov, and Bechhoefer 2014). Although there are exceptions to this perspective (Fahn 1996; Maroney 2005; Sagawa and Ueda 2009; Sagawa 2014; Parrondo, Horowitz, and Sagawa 2015), Szilard's argument that "measurement cannot take place without compensation" has largely been discounted, with measurement viewed as potentially costless and the cost of erasure viewed as fundamental to resolving the paradox. Here we argue against this perspective, leveraging a concrete molecular mechanism for implementing the entire measurement and feedback cycle of Szilard's engine. We use this mechanism to demonstrate that, in fact, Szilard's emphasis on the cost of correlating degrees of freedom, which corresponds to creating a nonequilibrium state, is essentially correct. We will simultaneously highlight fundamental issues that follow from an overemphasis on the significance of erasure and the stages at which work and heat are exchanged with the environment.

Our explicit molecular mechanism overcomes the inherent inscrutability of Maxwell's demon. Previous work has included both physical and thought experiments on systems that perform erasure, measurement, and exploitation (Landauer

1961; Szilard 1964; Bennett 1982; Sagawa and Ueda 2009; Lambson, Carlton, and Bokor 2011; Bérut *et al.* 2012; Jun, Gavrilov, and Bechhoefer 2014; Koski *et al.* 2014; Hong *et al.* 2016; Ouldridge, Govern, and ten Wolde 2017). However, it is extremely rare to study a full cycle of Szilard's engine in which all components and information-processing mechanisms are concrete and explicitly accounted for (we leverage our recent work in Brittain, Jones, and Ouldridge (2018)). Based on our analysis of a molecular Szilard engine, we will make the case for explicit molecular systems as a natural environment in which to embed thermodynamic thought experiments more generally.

Implementing Szilard's Engine with Molecular Reactions

Well-mixed, dilute molecular systems fit naturally into the paradigm of the section "Basic Thermodynamics" (Ouldridge 2018). The state of such systems is given by the number of each molecular species present. Changes in state are chemical reactions, such as $A + B \to C + D$, that happen at rates determined by reactant concentrations. Some of these reactions may correspond to large-scale conformational changes of a single molecule. *Chemical reaction networks* (CRNs) of this kind have been extensively studied in the biophysics, mathematics, and computer science literature (Krishnamurthy and Smith 2017) and provide a realistic model of many biological processes. Furthermore, it has been shown that arbitrary CRNs can be approximated in the laboratory through nucleic acid-based designs (Qian, Soloviechick, and Winfree 2011; Chen *et al.* 2013; Srinivas *et al.* 2017; Plesa 2018).

The discrete nature of the state space at the CRN level introduces a minor modification to the theory of the section "Basic Thermodynamics." Each of these "macrostates" contains an enormous number of microscopic configurations.

Macrostates are then stabilized not only by having low energy but also by having many microscopic configurations (an "entropic" contribution). Thankfully, we can simply work at the level of macrostates and replace the energies of individual states $\epsilon(x)$ appearing in equations (12.2) and (12.3) with chemical free energies of macrostates $f_{\text{chem}}(m)$, which incorporate both energetic and entropic contributions:

$$f_{\text{chem}}(m) = k_\text{B} T \ln \sum_{x \in m} \exp(-\epsilon(x)/k_\text{B} T),$$

$$p_{\text{eq}}(m) \propto \exp(-f_{\text{chem}}(m)/k_\text{B} T),$$

$$\mathcal{F}[p(m)] = \sum_m p(m) f_{\text{chem}}(m)$$
$$+ k_\text{B} T \sum_m p(m) \ln p(m) = \langle f_{\text{chem}} \rangle - T \mathcal{S}[p(m)]$$
$$= \mathcal{F}[p_{\text{eq}}(m)] + k_\text{B} T \sum_m p(m) \ln \frac{p(m)}{p_{\text{eq}}(m)} \geq \mathcal{F}[p_{\text{eq}}(m)]. \tag{12.4}$$

For a detailed discussion of the macrostate perspective, see Ouldridge (2018).

We will consider large baths of "fuel" molecules, or buffers; our systems will mediate chemical reactions that interconvert these fuels (Seifert 2011). To understand the thermodynamic consequences of these reactions, it is helpful to consider the change in chemical free energy of a system when an additional molecule of A is added, the *chemical potential* of A, μ_A. For dilute systems (Ouldridge 2018; Nelson 2004),

$$\mu_A = \mu_A^0 + k_\text{B} T \ln\left([A]/[1\text{M}]\right), \tag{12.5}$$

where μ_A^0 is the (theoretical) value of the chemical potential at a reference concentration of 1M and $[A]$ is its actual concentration. If a reaction consumes a single A and produces a single B, then there is a contribution to the overall chemical free energy change of

$$\Delta f_{\text{chem}}(A \to B) = \mu_B - \mu_A = \mu_B^0 - \mu_A^0 + k_\text{B} T \ln\left([A]/[B]\right). \tag{12.6}$$

This contribution to Δf_{chem} can be used to drive the interconversion of other species. For example, consider the reaction $A+C \rightleftharpoons B+D$. A large, negative value of $\Delta f_{\text{chem}}(A \to B)$ would tend to favor the conversion of C into D, because of the more favorable chemical free energy of B.

BUFFERS AS A SOURCE OF CHEMICAL WORK

We will assume that the buffers are sufficiently large that any reactions involving the system of interest have a negligible effect on the probability distribution of bath macrostates. In this limit, the baths become reservoirs of "chemical work," analogous to the (less concrete) work reservoirs introduced in the subsection "Work, Heat, the Second Law, and Entropy." In this case, we now use m to label the macrostate of only the molecules that are not part of the buffers (the "system of interest"), with $p(m)$ describing the distribution of these macrostates. The change in free energy due to an arbitrary process is then

$$\Delta \mathcal{F} = \Delta \left(\sum_m p(m) f_{\text{chem}}^{\text{sys}}(m) \right)$$
$$+ k_B T \Delta \left(\sum_m p(m) \ln p(m) \right) + \langle \sum_n \Delta f_{\text{chem}}^n \rangle$$
$$= \Delta \langle f_{\text{chem}}^{\text{sys}} \rangle - T \Delta \mathcal{S}^{\text{sys}}[p(m)] + \langle \Delta f_{\text{chem}}^{\text{buffers}} \rangle = \Delta \mathcal{F}^{\text{sys}} - \mathcal{W}_{\text{chem}}, \quad (12.7)$$

where $f_{\text{chem}}^{\text{sys}}(m)$ is the chemical free energy of the system of interest only and $\langle \Delta f_{\text{chem}}^n \rangle$ is the expected change in chemical free energy of buffer n. The chemical work is defined as $-\langle \Delta f_{\text{chem}}^{\text{buffers}} \rangle = -\langle \sum_n \Delta f_{\text{chem}}^n \rangle = \mathcal{W}_{\text{chem}}$; a positive value allows an increase in the free energy of the system of interest while satisfying $\Delta \mathcal{F} \leq 0$ overall.

By varying the chemical potentials of fuel molecules, we can implement a time-varying protocol on the system of interest (Ouldridge, Govern, and ten Wolde 2017; Rao and Esposito 2016; Schmiedl and Seifert 2007; Rao and Esposito

Chapter 12: The Power of Being Explicit

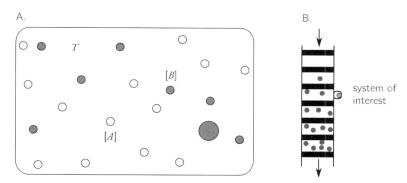

Figure 12.3. Molecular buffers. (A) A single molecule of interest (shown as a larger circle) in a large bath of fuel molecules of two types, A and B, at a temperature T, and concentrations $[A]$ and $[B]$. (B) A system of interest coupled to one of a series of buffers at different concentrations of fuel molecules.

2018). A device for implementing such a protocol is illustrated in figure 12.3B. We have a few molecules of interest that are trapped inside a small reaction volume. These molecules could be tethered to the surface of the volume or trapped by a membrane. Outside of the volume is a large bath of fuel molecules, which can diffuse into the volume, setting the chemical potential and biasing reactions as desired. We implement a time-dependent protocol by connecting a series of buffers with different fuel concentrations to the system of interest—either in a linear fashion, as in figure 12.3B, or in a cycle in order to implement a cyclic protocol.

Despite the explicit apparatus, it is still possible to have implicit decision-making; a demon could move the buffers in a particular fashion following a measurement of the system of interest, for example. We wish to eliminate such behavior and make all information processing and decision-making explicit in the system of interest, thereby avoiding conceptual pitfalls. We therefore demand that any applied protocol incorporates no external feedback based on the state of the system of interest.

One way to view the "no external feedback" criterion is that *a protocol applied to a single system can be trivially applied in parallel to many systems.* This idea marginalizes all costs of applying even complex protocols. In principle, if the buffers are moved slowly enough, there is no lower bound on the mechanical cost of moving them, given pistons of arbitrarily low friction. If we act on many systems in parallel, then any external cost associated with implementing the protocol can be taken as an edge effect. Furthermore, if the differences in fuel concentration in consecutive buffers tend to zero, so does any thermodynamic cost associated with fuel molecules being exchanged between adjacent buffers via the volume of interest. In this limit, therefore, all thermodynamic costs arise from actual chemical reactions involving the system of interest. We now illustrate systems and protocols that implement erasing, copying, and Szilard's engine.

ERASING A BIT

A molecule with two long-lived states is a natural analog of a bit (Ouldridge, Govern, and ten Wolde 2017). Let X and X^* be two states of a molecule, and let us assume that it is possible to switch between X and X^* by coupling to fuel molecules F_1 and F_1^* and a catalyst Y:

$$X + F_1^* + Y \rightleftharpoons X^* + F_1 + Y. \qquad (12.8)$$

The catalyst Y must be present for the reaction to proceed at an appreciable rate, but is not consumed by the reaction. Reactions of this type are extremely common in natural systems (e.g., Y could be a kinase, X its substrate, F_1^*, ATP, and F_1, ADP). They can also be engineered as nucleic acid strand displacement reactions (Qian, Soloviechick, and Winfree 2011; Chen *et al.* 2013; Srinivas *et al.* 2017). The chemical free energy change associated with the forward reaction in equation (12.8) is

$$\Delta f_{\text{chem}}(X + F_1^* + Y \to X^* + F_1 + Y)$$
$$= \Delta\mu_{F_1^* \to F_1} + \Delta\mu^0_{X \to X^*} \qquad (12.9)$$
$$= \mu^0_{F_1} - \mu^0_{F_1^*} + k_\text{B}T \ln \frac{[F_1]}{[F_1^*]} + \mu^0_{X^*} - \mu^0_X.$$

Here $\Delta\mu^0_{X \to X^*}$ is the intrinsic stability difference between the two states (the $\ln[X]/[X^*]$ term is absent because we have a single molecule of X/X^*). For simplicity, we shall assume that $\Delta\mu^0_{X \to X^*} = 0$, equivalent to assuming the bit is symmetric, as is typical. In this case, all changes in chemical free energy arise from the fuel molecules. The fundamental conclusions of this chapter are not affected by this assumption.

We now implement erasure on X/X^* using the apparatus outlined in figure 12.4A (Ouldridge, Govern, and ten Wolde 2017). Let a single Y molecule and a single X/X^* be confined in the volume. Initially, we start with buffers with both $[F_1], [F_1^*] \to 0$, so both X and X^* are essentially stable. We then slowly increase both $[F_1]$ and $[F_1^*]$ to a point where the forward and backward reactions occur at appreciable rates, maintaining a ratio of $[F_1]/[F_1^*]$ that ensures that $\Delta\mu_{F_1^* \to F_1} = 0$. We now see interchanges of X and X^* but do not yet bias the bit's state. In the energy landscape representation shown in figure 12.4, this initial manipulation corresponds to lowering the barrier between macrostates prior to quasistatically destabilizing one with respect to the other. Next, we slowly increase $[F_1]$ while keeping $[F_1^*]$ fixed, thereby increasing $\Delta\mu_{F_1^* \to F_1}$. A large and positive value of $\Delta\mu_{F_1^* \to F_1}$ pushes X^* toward X. When we have sufficiently large $\mu_{F_1^*} - \mu_{F_1} \gg k_\text{B}T$ that the probability of finding X^* is essentially zero, we start to decrease both $[F_1]$ and $[F_1^*]$ toward zero, while maintaining a constant $\mu_{F_1^*} - \mu_{F_1} \gg k_\text{B}T$, raising the barrier and "freezing" our molecular bit in the X state. Finally, we reduce $[F_1]$ further so that $[F_1], [F_1^*] \to 0$ and $\mu_X = \mu_{X^*}$. The original buffer

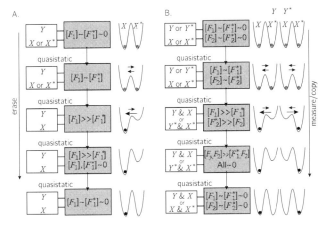

Figure 12.4. Reversible erase and copy protocols can be implemented in molecular systems. (A) Erasure of a molecular bit X/X^* to the X state. The system of interest (far left), containing X/X^* and catalyst Y, is coupled to a series of buffers with varying concentrations of fuel molecules; we indicate key stages in this process. $[F_1] \sim [F_1^*]$ means that $\Delta\mu_{F_1^* \to F_1} = 0$; $[F] \sim 0$ means that the concentration is so low that reactions requiring $[F]$ are negligible. The landscapes on the right-hand side represent the effective free energy landscapes for the interconversion of X and X^* at a given buffer and coupled to the catalyst Y, using an ideal reaction coordinate. Arrows indicate the dominant reaction, if any, and shaded balls indicate the probability of occupying a given state. (B) As in (A), but for copying of a Y/Y^* bit using two pairs of fuel molecules. Landscapes are now shown for both possibilities of the data bit Y/Y^*.

is now reconnected, but the initially uncertain bit has been set to a guaranteed X state.

Overall, the free energy change of the reaction volume itself is $\Delta \mathcal{F}^{\text{sys}} = k_\text{B} T \ln 2$, because both X and X^* have the same chemical free energy, but the entropy $\mathcal{S}^{\text{sys}}[p(m)] = -k_\text{B} \sum_m p(m) \ln p(m)$ decreases from $k_\text{B} T \ln 2$ to 0.

The cost of erasure is supplied by the buffers. Initially, when $[F_1]$ and $[F_1^*]$ are increased simultaneously while maintaining $\mu_{F_1} = \mu_{F_1^*}$, no reactions occur on average. Subsequently, $[F_1]$ is increased relative to $[F_1^*]$, and net reactions do occur, consuming chemical free energy and performing chemical work. If the buffers are updated sufficiently slowly, then the X/X^* system reaches a new buffer-dependent equilibrium at each point (Ouldridge, Govern, and ten Wolde 2017). From equations (12.4) and (12.9), the equilibrium probability of observing X in each buffer is

$$p(X) = \frac{1}{1 + \exp(-\Delta\mu_{F_1^* \to F_1}/k_{\mathrm{B}}T)}. \qquad (12.10)$$

When the system is coupled to the next buffer, there is an infinitesimal change $\mathrm{d}\Delta\mu_{F_1^* \to F_1}$. Consequently, $p(X)$ changes by

$$\begin{aligned}
\mathrm{d}p(X) &= \frac{1}{1+\exp(-(\Delta\mu_{F_1^* \to F_1} + \mathrm{d}\Delta\mu_{F_1^* \to F_1})/k_{\mathrm{B}}T)} \\
&\quad - \frac{1}{1+\exp(-\Delta\mu_{F_1^* \to F_1}/k_{\mathrm{B}}T)} \qquad (12.11) \\
&= \frac{\exp(-\Delta\mu_{F_1^* \to F_1}/k_{\mathrm{B}}T)}{\left(1+\exp(-\Delta\mu_{F_1^* \to F_1}/k_{\mathrm{B}}T)\right)^2} \frac{\mathrm{d}\Delta\mu_{F_1^* \to F_1}}{k_{\mathrm{B}}T}.
\end{aligned}$$

When X^* changes to X, F_1 changes to F_1^* in the buffer, with an associated $\Delta f_{\mathrm{chem}}^{\mathrm{buffers}} = -\Delta\mu_{F_1^* \to F_1}$. The expected drop in free energy, or chemical work done, by the change $\mathrm{d}p(X)$ is

$$\begin{aligned}
\mathrm{d}W_{\mathrm{chem}}^{\mathrm{in}} &= \Delta\mu_{F_1^* \to F_1} \mathrm{d}p(X) \\
&= \Delta\mu_{F_1^* \to F_1} \frac{\exp(-\Delta\mu_{F_1^* \to F_1}/k_{\mathrm{B}}T)}{\left(1+\exp(-\Delta\mu_{F_1^* \to F_1}/k_{\mathrm{B}}T)\right)^2} \frac{\mathrm{d}\Delta\mu_{F_1^* \to F_1}}{k_{\mathrm{B}}T}. \qquad (12.12)
\end{aligned}$$

To find the total expected work done by all of the buffers, we simply integrate over the whole protocol: $\Delta\mu_{F_1^* \to F_1} = 0$ to $\Delta\mu_{F_1^* \to F_1} \to \infty$. Using the substitution $u = \Delta\mu_{F_1^* \to F_1}/k_{\mathrm{B}}T$, we obtain

$$\begin{aligned}
\frac{W_{\mathrm{chem}}^{\mathrm{in}}}{k_{\mathrm{B}}T} &= \int_0^\infty \frac{u \exp(-u)}{(1+\exp(-u))^2} \mathrm{d}u \\
&= \int_0^\infty -\frac{\exp(u)}{1+\exp(u)} + \frac{\mathrm{d}}{\mathrm{d}u}\left(\frac{u \exp(u)}{1+\exp(u)}\right) \mathrm{d}u = \ln 2. \qquad (12.13)
\end{aligned}$$

The chemical work done to erase our molecular bit perfectly is $k_{\mathrm{B}}T \ln 2$, as expected. The process is thermodynamically reversible; applying the buffers in reverse would return the system to the initial state, and the overall change in the nonequilibrium free energy is zero ($W_{\mathrm{chem}}^{\mathrm{in}} = \Delta\mathcal{F}^{\mathrm{sys}}$).

COPYING OR MEASURING A BIT

To allow for copying, we consider two states of the catalyst, Y and Y^*. We label the catalyst as the "data" molecule or data bit. Moreover, we require that Y and Y^* separately catalyze reactions that couple the reactions of the measurement molecule or memory bit $X \rightleftharpoons X^*$ to two distinct fuels (Ouldridge, Govern, and ten Wolde 2017):

$$X + F_1^* + Y \rightleftharpoons X^* + F_1 + Y$$
$$X + F_2^* + Y^* \rightleftharpoons X^* + F_2 + Y^*. \quad (12.14)$$

Such a scheme approximates natural "bifunctional kinases" (Stock, Robinson, and Goudreau 2000) and could easily be engineered with nucleic acid circuitry (Qian, Soloviechick, and Winfree 2011; Chen et al. 2013; Srinivas et al. 2017).

Performing a perfect copy corresponds to correlating the data and measurement molecules. If we knew for certain that the data molecule was in the Y state, we could simply repeat the protocol outlined in the section "Erasing a Bit." However, we also need the system to automatically handle the possibility that the data molecule is in the Y^* state via a single protocol. If we follow the protocol of the section "Erasing a Bit" and figure 12.4A, but with the data molecule in the Y^* state, nothing would happen, because reactions require the presence of either F_1, F_1^* and Y or F_2, F_2^* and Y^*. We can therefore implement two erasures simultaneously as a single protocol: driving the system to X^* if Y^* is present and to X if Y is present. Such a protocol is illustrated in figure 12.4B. The buffer concentrations of $[F_2]$ and $[F_2^*]$ mirror those of $[F_1]$ and $[F_1^*]$, in that $[F_2]$ and $[F_2^*]$ are initially increased at $\Delta\mu_{F_2^* \to F_2} = 0$, but then $[F_2^*]$ is increased further so that $-\Delta\mu_{F_2^* \to F_2} \gg k_B T$, and the system is driven toward the X^* state if Y^* is present.

Calculating the chemical work done during the copy is trivial. We assume for simplicity that there is a 50% chance that Y is present and a 50% chance that Y^* is present (similar arguments can be constructed for any initial bias). In either case, the decrease in chemical free energies of the buffers is $k_B T \ln 2$, and so $\mathcal{W}^{\text{in}}_{\text{chem}} = k_B T \ln 2$. This decrease in free energy exactly matches the increase of $\mathcal{F}^{\text{sys}}$. As in the section "Erasing a Bit," the chemical free energy of the system of interest is unchanged, but $\mathcal{S}[p(m)]$ has decreased from $k_B \ln 4$ to $k_B \ln 2$—we have gone from four equally probable macrostates to just two because of the correlations. We have paid chemical work to decrease the entropy of the data/measurement molecule system, thereby storing free energy.

Crucially, the protocol outlined herein is true measurement in the sense discussed by Szilard. The two molecules' states are correlated without direct interaction and, were the catalyst (the data bit) to undergo a state change, there would be no tendency for the measurement molecule (memory bit) to change. We have previously described this type of copying as *persistent* (Ouldridge, Govern, and ten Wolde 2017).

A MOLECULAR SZILARD ENGINE

Having implemented a molecular copy (or measurement), we now exploit the measurement through feedback. The essential challenge is to allow the data molecule Y/Y^* to evolve so that it decorrelates with the fixed memory X/X^*, releasing the free energy stored in correlations in a controlled way so that chemical work can be extracted. Decorrelation could also be performed by undoing the measurement: here Y/Y^* does not change, but X/X^* is randomized. This process, which doesn't involve feedback, can be achieved reversibly by reversing the copy protocol.

Feedback is normally implicit in Szilard's engine; it is difficult to imagine concrete systems in which one degree of freedom influences the other during an initial period and then this influence is reversed. Fortunately, our chemical system allows this behavior. On top of the reactions used during the copy (equation (12.14)), we consider catalytic interconversion of Y/Y^* by X/X^* using the fuel molecules G_1, G_2, G_1^*, and G_2^*:

$$Y + G_1^* + X \rightleftharpoons Y^* + G_1 + X$$

(12.15)

$$Y + G_2^* + X^* \rightleftharpoons Y^* + G_2 + X^*.$$

The system is now a mutual bifunctional catalyst network. Such a system is unusual but by no means unimaginable.

The overall protocol of our molecular Szilard engine is illustrated in figure 12.5A. The initial measurement, or correlation stage, proceeds exactly as in the section "Copying or Measuring a Bit." The concentration of fuel molecules $[G_1], [G_1^*], [G_2], [G_2^*] \to 0$, preventing the state of the data molecule from changing during this period. This part of the protocol requires $W_{\text{chem}}^{\text{in}} = k_B T \ln 2$.

The subsequent exploitation stage proceeds as follows. First, $[G_1]$ and $[G_2^*]$ are increased while maintaining $[F_1]$, $[F_1^*]$, $[F_2], [F_2^*], [G_1^*], [G_2] \to 0$, until $\Delta\mu_{G_1^* \to G_1}, -\Delta\mu_{G_2^* \to G_2} \gg k_B T$. Because only (Y, X) and (Y^*, X^*) combinations are possible at this point, no reactions occur, and no chemical work is done. Subsequently, $[G_1^*]$ and $[G_2]$ are slowly increased until $\Delta\mu_{G_1^* \to G_1} = \Delta\mu_{G_2^* \to G_2} = 0$. During this period, the data are decorrelated from the measurement, and free energy is transferred back to the buffers. Given a measurement X, the data molecule is in state Y^* with probability

$$p(Y^*|X) = \frac{\exp(-\Delta\mu_{G_1^* \to G_1}/k_B T)}{1 + \exp(-\Delta\mu_{G_1^* \to G_1}/k_B T)}$$

(12.16)

when coupled to a buffer at $\Delta\mu_{G_1^* \to G_1}$. Proceeding as in the subsection "Erasing a Bit," we multiply the increase in

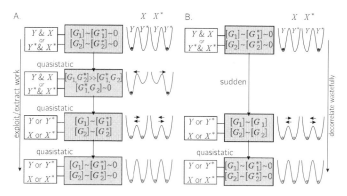

Figure 12.5. Work extraction step of a Szilard engine as implemented in a molecular system. (A) Work extraction step following the copy/measurement step in figure 12.4B. The final state is equivalent to the initial state of the measurement protocol, so combining the two produces a full cyclic operation. The combined data and measurement molecules are exposed to a series of buffers of varying fuel concentrations; key points are illustrated. $[G_1] \sim [G_1^*]$ means that $\Delta\mu_{G_1^* \to G_1} = 0$; $[G] \sim 0$ means that the concentration is so low that reactions requiring $[G]$ are negligible. The landscapes on the right-hand side represent the effective free energy landscapes for the interconversion of Y and Y^* at a given buffer and coupled to either X or X^*, acting as a catalyst. Arrows indicate the dominant reaction, if any, and shaded balls indicate the probability of occupying a given state. (B) Relaxation of the correlated state without careful extraction of chemical work during the decorrelation. Unlike in the scheme of (A), work is not extracted by slowly changing the relative stability of the two states.

$p(Y^*|X)$ associated with a buffer increment of $d\Delta\mu_{G_1^* \to G_1}$ by the resulting free energy change of the buffer for a conversion of Y into Y^*, $\Delta\mu_{G_1^* \to G_1}$, and integrate: $\Delta f_{\text{chem}}^{\text{buffers}}$

$$= -\int_\infty^0 \Delta\mu_{G_1^* \to G_1} \frac{\exp(-\Delta\mu_{G_1^* \to G_1}/k_\text{B}T)}{\left(1 + \exp(-\Delta\mu_{G_1^* \to G_1}/k_\text{B}T)\right)^2} \frac{d\Delta\mu_{G_1^* \to G_1}}{k_\text{B}T}.$$

(12.17)

The integral is identical to equation (12.13), and thus $\Delta f_{\text{chem}}^{\text{buffers}} = k_B T \ln 2$. The same $\Delta f_{\text{chem}}^{\text{buffers}}$ is obtained for a measurement result of X^*, although it is the combined free energy of the G_2/G_2^* fuels that increases. Thus the average increase in the total chemical free energies of the buffers is $-\mathcal{W}_{\text{chem}}^{\text{in}} = k_B T \ln 2$, and so the chemical work done in performing the measurement is extracted. Finally, no chemical work is done as the concentrations of all fuels are taken to zero at fixed $\Delta\mu_{G_1^*\to G_1} = \Delta\mu_{G_2^*\to G_2} = 0$. This final operation corresponds to "raising the barrier" between the metastable states of the bit in the energy landscape analog, as shown in figure 12.5.

Our molecular device is an exact analog of the engine proposed by Szilard, with all components explicit. We perform a binary measurement by creating a record of the data molecule with the measurement molecule. Chemical work of $k_B T \ln 2$ is required by the measurement step to create a low-entropy, high–free energy correlated state. This work is then extracted from these correlations during exploitation, when the measurement molecule itself couples the data molecule to the appropriate fuel reservoir. Importantly, the measurement is persistent in that the X/X^* state does not change during exploitation.

What We Learn from Explicit Biochemical Protocols

ERASURE IS NOT THE MOST COMPELLING EXPLANATION FOR THE SELF-CONSISTENCY OF THE SECOND LAW

There is no erasure in the preceding protocol. In principle, we could add one by including a third molecule Z that couples the transition of $X \rightleftharpoons X^*$ to another pair of fuel molecules H and H^*. By implementing a protocol equivalent to that in the section "Erasing a Bit" after the exploitation step, we could reversibly set the measurement molecule to state X at a cost of $k_B T \ln 2$ of chemical work. Having paid for the erasure, the minimal chemical work required to go from the erased state to the postmeasurement, correlated state would now be zero because $\Delta\mathcal{F}^{\text{sys}} = 0$; in both cases, there are two equally likely macrostates for the measurement/data system $((d, m) = (0, 0)$ OR $(1, 0)$ and $(0, 0)$ OR $(1, 1)$, respectively).

Although cycles involving erasure are possible, the process is not required, and so it seems slightly perverse to attribute the self-consistency of the second law in Szilard's engine to erasure (Maroney 2005). Indeed, in the simplest implementation of measurement with an erased bit, one first "unerases" the measurement molecule to an equal probability of X and X^*, prior even to coupling to Y/Y^* via the F fuels. During this unerase, $k_B T \ln 2$ of chemical work is extracted, giving zero overall chemical work for measurement from an erased state when combined with the $k_B T \ln 2$ required to subsequently create correlations. Erasing, then, appears to be an accounting trick; we add a process to the cycle (erase and unerase) that achieves nothing but then incorporate the unerase into a "measurement from an erased state" process (fig. 12.2) that overall has no cost. We note that Bennett's analysis also includes an "unerase" prior to correlation (Bennett 1982, 2003). It is possible to incorporate erasure in a less trivial manner, but we still find the emphasis on any one step to be misleading. The second law is not violated at *any* stage of the protocol, whether erasure is performed or not (Fahn 1996; Maroney 2005; Sagawa 2014; Parrondo, Horowitz, and Sagawa 2015).

Instead, it is more illuminating to focus on the nonequilibrium free energy. As Szilard originally pointed out, the measured state contains correlations between degrees of freedom that must persist despite an absence of direct interactions. The existence of correlations without direct interactions implies a low-entropy state without a compensating low chemical free energy (Fahn 1996; Sagawa 2014; Parrondo, Horowitz, and Sagawa 2015). In this case, the correlations mean that only two of a total of four equally stable macrostates are possible; the entropy of the post-measurement state is $k_B \ln 2$ rather than $k_B \ln 4$, and an excess free energy of $\mathcal{F}^{sys} - \mathcal{F}^{sys}_{eq} = k_B T \ln 2$ is available. The stored free energy in the measured state makes the exploitation step consistent with the second law, regardless of the details of the cycle. The only difference between protocols with and without erasure is whether we choose to compensate for the creation of the measurement by initially creating an out-of-equilibrium memory bit.

This explanation of Szilard's engine generalizes. In information theory, correlations between two degrees of freedom m_1 and m_2 are described using the mutual information indexinformation (Shannon and Weaver 1949)

$$\mathcal{I}[p(m_1), p(m_2)] = \sum_{m_1, m_2} p(m_1, m_2) \ln \left(\frac{p(m_1, m_2)}{p(m_1) p(m_2)} \right) \geq 0, \quad (12.18)$$

with the equality holding if the two degrees of freedom are independent. It is relatively straightforward to show that if the energy/chemical free energy is additive in m_1 and m_2 (the two degrees of freedom do not interact), then (Esposito and Van den Broeck 2011; Horowitz, Sagawa, and Parrondo 2013; Parrondo, Horowitz, and Sagawa 2015)

$$\mathcal{F}[p(m_1), p(m_2)] = \mathcal{F}[p(m_1)] + \mathcal{F}[p(m_2)] + k_B T \mathcal{I}[p(m_1), p(m_2)]. \quad (12.19)$$

Because m_1 and m_2 must be independent in equilibrium, a nonzero $\mathcal{I}[p(m_1), p(m_2)]$ necessarily represents a store of free energy. This principle has been applied to systems that create and exploit correlations between noninteracting degrees of freedom (Horowitz, Sagawa, and Parrondo 2013; Ouldridge, Govern, and ten Wolde 2017; McGrath *et al.* 2017; Ouldridge and ten Wolde 2017; Brittain, Jones, and Ouldridge 2018;

Stopnitzky *et al.* 2019; Chapman and Miyake 2015; Boyd, Mandal, and Crutchfield 2016).

Bennett's argument that, in a setting with heat and work reservoirs coupled to a pair of bits, measurement can be performed with no work input using an erased memory, and that erasing memories requires work input, is not incorrect. We nonetheless believe that the emphasis on erasure causes unnecessary confusion. Taken at face value, the Bennett explanation might seem to suggest that the second law is transiently violated, because erasure compensates for the work extraction after the fact, when a whole cycle is completed (Landauer 1991). Indeed, the finite size of the demon's memory is sometimes invoked (Feynman 1998; Deffner 2015), suggesting that the second law can be violated until the memory is full and needs to be erased. But if the second law can be violated until memory has to be erased, then what happens if erasure never happens? If, instead, the necessarily nonequilibrium nature of the postmeasurement state is used to support the second law, these confusions evaporate; all substeps are consistent with the second law (Parrondo, Horowitz, and Sagawa 2015; Sagawa 2014). Furthermore, the perceived importance of erasure leads researchers to look for its presence in systems that could perhaps be more fruitfully analyzed in terms of the generation and use of nonequilibrium correlations (Mehta and Schwab 2012; Andrieux and Gaspard 2013).

FOCUSING ON HEAT AND WORK CAN BE MISLEADING
There is a tendency to use heat generation as a shorthand for irreversibility, or the increase in the total entropy. This shorthand, however, can lead to significant confusion. Heat generation does not imply thermodynamic irreversibility.

It is possible, at least in principle, to transfer heat to a system's environment in a thermodynamically reversible way, provided that the increase in the entropy of the environment is compensated by a decrease in the entropy of the system. A common example would be quasistatic compression of a piston. Net heat transfer to the environment only guarantees irreversibility if the system operates cyclically, in which case its entropy cannot change to compensate. Nonetheless, the transfer of heat is frequently described as "dissipation," without clarifying the degree to which the total entropy actually increases (Maroney 2005).

From a technical perspective, the distinction is between entropy transferred and entropy produced. The entropy production is the increase in total entropy. It is the sum of the change in entropy of the system of interest and the entropy transferred to the rest of the environment due to its interaction with the system. In the case where the system of interest is in contact with a single heat reservoir, the entropy transferred is the heat transferred to the reservoir divided by the temperature (Parrondo, Horowitz, and Sagawa 2015); the heat transferred is therefore not directly informative of the total entropy production, even in simple settings with a single heat reservoir.

In Landauer's analysis of erasure, heat generation compensates for the entropy decrease of the bit. Despite the fact that the bound of $k_B T \ln 2$ is explicitly derived for a reversible protocol satisfying $\Delta S^{\text{tot}} = 0$, Landauer himself describes the heat transfer as a necessarily thermodynamically irreversible process (Landauer 1961). Many others have apparently repeated this logic (Plenio and Vitelli 2001; Bub 2001; Ladyman *et al*. 2007; Dillenschneider and Lutz 2009; Bérut *et al*. 2012; Jun, Gavrilov, and Bechhoefer 2014; Hong *et al*. 2016). This confusion contributes to the erroneous narrative that thermodynamic irreversibility of erasure pays for a transient violation of the second law during a cycle of Szilard's engine and leads to statements such as "logical irreversibility is associated with physical irreversibility and requires a minimal heat generation" (Landauer 1961).

We defined thermodynamic reversibility in the subsection "Thermodynamic Reversibility;" it is a statement about whether a particular protocol that converts a system's distribution from $p(x)$ into $p'(x)$ would re-create $p(x)$ when applied in a time-reversed fashion to an initial distribution $p'(x)$. Thermodynamic reversibility depends on the actual physical process undergone to move from $p(x)$ to $p'(x)$. However, exact trajectories $x(t)$ do not need to be reversed; $p(x)$ need only be re-created in the statistical sense. Whether a process is thermodynamically reversible depends on both the protocol and the initial distribution $p(x)$ (Kolchinsky and Wolpert 2017); in general, a protocol that is thermodynamically reversible for one $p(x)$ will not be reversible for another.

By contrast, a process is logically reversible if and only if the initial logical state y_i can be inferred unambiguously from the final logical state y_f—which is not true for erasure, for example (Sagawa 2014). Logical irreversibility is then solely a condition on the overall transition matrix between input and output logical states $T_{y_f y_i}$ and is independent of the initial state to which $T_{y_f y_i}$ is applied; the possibility of multiple physical states x within a logical state y; and the detailed physical mechanism by which $T_{y_f y_i}$ is achieved. Logical reversibility is, however, crucially dependent on the specific input/output map represented by $T_{y_f y_i}$ rather than just the overall statistics. Given these differences between logical and thermodynamic irreversibility, it is not surprising that there is in fact no causal connection between the two; both logically reversible and logically irreversible processes can in principle be implemented in thermodynamically reversible and irreversible ways. Sagawa (2014) has given specific examples to illustrate this fact in general settings.

Moreover, if a process as a whole is thermodynamically irreversible, one cannot immediately associate the part of the process that transfers heat to the environment (or requires work input) as the cause of the irreversibility. For example, irreversibility in certain systems has been ascribed to the need to erase (Mehta and Schwab 2012; Mehta, Lang, and Schwab 2016; Andrieux and Gaspard 2013). But if the erasure considered by Landauer is not necessarily thermodynamically irreversible, then it is a logical non sequitur to attribute thermodynamic irreversibility to the presence of erasure. Landauer's principle in isolation cannot explain an increase in the total entropy of a closed system.

For example, consider the molecular Szilard engine including erasure as outlined in the section "A Molecular Szilard Engine." This device is thermodynamically reversible. We could, however, consider an alternative protocol in which, instead of carefully relaxing the correlated state after the measurement, we simply suddenly expose the system to buffers with high concentrations of $[G_1]$, $[G_1^*]$, $[G_2]$, and $[G_2^*]$ with $\Delta \mu_{G_1^* \to G_1} = \Delta \mu_{G_2^* \to G_2} = 0$, as in figure 12.5B, prior to erasing. The measurement and data molecules would immediately decorrelate, and no chemical work would be extracted. The overall cycle would then require $\mathcal{W}_{\text{chem}} = k_\text{B} T \ln 2$, the total free energy of the buffers would decrease in a cycle, and the total entropy would increase. The same uncontrolled decorrelation would happen if we tried to erase directly from the correlated state using Z and fuels H and H^*, without gradually undoing the correlations first.

It is tempting to associate the irreversibility in this cycle with a step at which we have to expend net chemical work due to Landauer's principle: erasure. In the setting of magnetic bits, this process would require work input and generate heat. However, in either context erasure is thermodynamically reversible. The (chemical) work input is exactly compensated by the increase in $\mathcal{F}^{sys}$. Rather, it is the decorrelation step—which requires no (chemical) work and exchanges no heat with the environment in either the molecular or magnetic context—that is the *cause* of the irreversibility (Fahn 1996; Ouldridge, Govern, and ten Wolde 2017). Reversing this step would not restore the correlated state, and during this step, $\mathcal{F}^{sys}$ decreases by $k_B T \ln 2$ due to the loss of correlations, without a compensating extraction of (chemical) work, implying an increase in the combined entropy of the system and environment.

The foregoing discussion highlights a further issue: the second law does not apply only to the mechanical heat engines that drove the Industrial Revolution; it also applies to the molecular systems that underlie life. In these settings, and others, mechanical work and heat can play a less direct role, with processes relying on other forms of free energy transduction. Indeed, the erasure, copying, and Szilard engine protocols described in the section "Implementing Szilard's Engine with Molecular Reactions" involve no mechanical work at all. This distinction matters because free energy is not an energy; it is not conserved. If *chemical* work W_{chem}^{in} is done on a system during a cyclic evolution (in which $\Delta U = 0$), physical heat $Q^{out} = W_{chem}^{in}$ need not be transferred to the environment to compensate. Nonetheless, heat generation is often used as shorthand for thermodynamic irreversibility, even in molecular contexts (see, e.g., Lan *et al.* 2012; Mehta,

Lang, and Schwab 2016; Govern and ten Wolde 2014; Barato and Seifert 2017), as noted by Seifert (2011). Heat flow is a real energy flux that can be measured; it is misleading to describe entropy generation in a different form in this way. For example, one might incorrectly assume that a living system needs to develop machinery to rid itself of heat exactly equal to $T\dot{S}$, where $\dot{S}$ is its entropy generation rate, to maintain a constant temperature.

THE BENEFITS OF HAVING AN EXPLICIT SUPPLY OF CHEMICAL WORK

Work reservoirs are not usually explicitly modeled like the chemical work buffers in the section "Implementing Szilard's Engine with Molecular Reactions." Using abstract (and interchangeable) work buffers is often conceptually useful, but it can help to make certain results seem overly remarkable. For example, measurement can be performed without net (chemical) work expenditure if the memory is initially in a well-defined state, suggesting that the measurement has no cost. Given that the correlated postmeasurement state stores free energy that can be extracted as work, this lack of a cost seems remarkable. However, there is a cost—the initial nonequilibrium state of the memory is consumed to pay for the measurement. At a fundamental physical level, there is no real difference between using free energy stored in the memory bit or a (chemical) work buffer. Either way, a resource is consumed to create nonequilibrium correlations between memory and data. There is no reason to value resources in a (chemical) work reservoir above sources of free energy explicitly incorporated into the system of interest, which is particularly clear when the work reservoir is made concrete.

An explicit supply of (chemical) work also helps establish what is definitely possible—a so-called "positive result." Work reservoirs are often pictured as masses in a gravitational field, as in figure 12.2. It is often imagined, however, that work can be done by implementing an arbitrary change in the potential energy of a system as a function of its internal coordinates (e.g., Schmiedl and Seifert 2007), despite the fact that a single weight would actually be insufficient even for the simple Szilard engine of figure 12.2. Without an explicit mechanism to achieve the required work coupling, it is arguable whether one can positively prove the possibility of certain operations—even if one can prove the impossibility of others. Perhaps even more challenging than implementing an arbitrary control of a potential is the idea of recovering work back from the system of interest, as is required by Szilard's engine (Deshpande *et al.* 2017).

The operations that can be achieved with the molecular setup outlined in the section "Implementing Szilard's Engine with Molecular Reactions" are clearly limited. However, they are at least definitely possible, albeit in a highly idealized limit. Furthermore, because the interconversion of fuel species directly determines the chemical work done, any chemical work recovered is indeed stored. By contrast, consider recent experiments in which devices such as optical feedback traps are used to implement an arbitrary force on a colloid (Bérut *et al.* 2012; Jun, Gavrilov, and Bechhoefer 2014). Although this force is designed to mimic a conservative, time-dependent potential, the thermodynamic cost of applying the control is orders of magnitude higher than calculated, and work is not actually recovered when the colloid moves against the pseudopotential.

Biochemical Systems as a Platform for Studying Fully Autonomous Systems

Although conceptually useful, the apparatus in the section "Implementing Szilard's Engine with Molecular Reactions" is highly idealized. We assume that molecular states are stable unless enzymes are present and that the fuels only couple to the intended enzymes (in practice, lifetimes need to be long and leak rates need to be slow relative to intended reactions). Furthermore, although we did not consider intelligent intervention or implicit decision-making, it is still somewhat unsatisfying to invoke an external protocol at all. The most intellectually satisfying description would include the device (or demon) implementing any protocol as part of the system itself (Mehta, Lang, and Schwab 2016).

Such systems, which do not involve an externally applied time-varying control, are *autonomous*. Molecular systems are even more suited to exploring autonomous contexts than those with externally applied protocols. Fixed buffers of nonequilibrium fuel concentrations can drive active processes. Crucially, diffusion is automatically present in molecular systems, allowing an arbitrary complex network of interactions at no cost. We have previously shown that a simple, autonomously operating enzymatic network can be mapped directly to a copy process in the sense of Szilard's engine (Ouldridge, Govern, and ten Wolde 2017). The network could not reach the bound set by the nonautonomous, quasistatic protocols considered in the section "Implementing Szilard's Engine with Molecular Reactions." It could come remarkably close, however, despite operating at a finite rate and requiring less stringent assumptions about the slowness of unintended leak reactions.

Conclusion

Basing our discussion around molecular protocols for manipulating bits, we have demonstrated the advantages of being more explicit in our analysis of fundamental thermodynamic and computational systems. We have argued that doing so removes much of the mystery surrounding the physics of computation and measurement, and thermodynamics more generally. We have shown that molecular realizations are a particularly helpful way to understand basic thermodynamic ideas.

We argue that it is time to revisit the generally accepted explanation for the self-consistency of the second law of thermodynamics during the operation of Maxwell's demon and Szilard's engine. Attributing the survival of the second law to one part of the process or another is an accounting exercise that misses the fundamental point: the second law is self-consistent throughout all stages of operation. Nonetheless, Szilard's original contention that creating a correlated state is fundamental to the device, and that creating this correlated state requires "compensation," is essentially correct. Whether this cost is paid by exploiting a work reservoir or a supply of free energy that is part of the explicitly modeled system is moot. We emphasize that Bennett's analysis is not wrong; rather, the significance of erasure in the context of the second law has since been overstated. The lack of a relationship between logical irreversibility and thermodynamic reversibility more generally calls into question the value of pursuing logically reversible computing approaches in an effort to reduce power consumption (Grochow and Wolpert 2018), although tracking the amount of erasure in a human-designed computation is potentially useful for some specific approaches to low-cost computation. We should also not impose a language of erasure on (biological) systems for which it is unnatural.

We are not claiming to advance this view for the first time in this chapter, and we have drawn on a number of works published throughout the last few decades (including Fahn 1996; Maroney 2005; Sagawa and Ueda 2009; Sagawa 2014; Parrondo, Horowitz, and Sagawa 2015). One might ask whether the development of new theories and tools since the middle of the twentieth century has led to this alternative rationalization. On one hand, modern methodology is not really necessary to understand the problem. Nonequilibrium generalized free energies and the language of information theory are a convenient tool that make Szilard's argument more formal, but the essential insight was already provided in 1929, and Landauer's and Bennett's calculations themselves were not incorrect. Really, it is a question of interpretation, and the development and application of stochastic thermodynamics (Jarzynski 1997; Crooks 1999; Esposito and Van den Broeck 2011; Seifert 2005) has facilitated a much deeper general understanding of information and thermodynamics for small, fluctuating systems in recent years. This general understanding has in turn yielded a clearer approach to this particular problem. We note that for many (including ourselves), this insight has been developed while studying concrete molecular systems that accomplish information-processing or computational tasks (Ouldridge, Govern, and ten Wolde 2017).

We additionally make the case for more careful use of the terms *heat* and *work*; *heat* is not a synonym for an increase in the total entropy, and free energy is not an energy. The distinction between free energies and work touches on a further important point: fundamentally, chemical work buffers are reservoirs of free energy due to an entropic component to their deviation from equilibrium. Much has been made of the fact that a low-entropy string of erased bits is a thermodynamic resource (Feynman 1998; Mandal and Jarzynski 2012). However, entropic contributions to stored free energies are natural in molecular contexts. The effective force driving the diffusion of uncharged molecules from a high concentration to a low concentration is entirely entropic. Once the physical object encoding the low entropy state is made explicit, such "information reservoirs" typically seem much less mysterious.

We have also argued that molecular systems are a convenient basis on which to ground explicit (if idealized) representations of processes that we wish to analyze at a fundamental level. Explicit mechanisms can demystify the underlying concepts and, without a method of implementing a protocol, it is difficult to argue for positive rather than negative results. Rao and Esposito (2018) have introduced systematic methods for analyzing more general molecular systems—although care should be taken when interpreting their free energies, which are not defined in the conventional manner discussed here. Furthermore, without a concrete mechanism, it is easy to design systems that are not even thermodynamically well defined—as discussed by Stopnitzky *et al.* (2019). Needless to say, despite our enthusiasm with molecular contexts and their relevance to complex natural systems, alternative paradigms may be more relevant in other settings.

Full autonomy is the natural limit of explicitness. Currently there is a major lack of understanding of the possibilities of autonomous systems (Mehta, Lang, and Schwab 2016). Can they approach the performance of externally manipulated systems, particularly in the limit of finite time operation? How challenging is it to develop systems in which one subsystem effectively generates time-varying conditions for another, and is this effective? Molecular systems are both an important conceptual tool and an experimental testing ground for exploring these ideas, particularly as we move beyond the classic questions of copying and erasing.

Acknowledgments

We gratefully thank Dave Doty, Manoj Gopalkrishnan, and Nick Jones for helpful conversations. T. E. O. acknowledges support from a Royal Society University Research Fellowship, R. A. B. acknowledges support from an Imperial College London AMMP studentship, and for P. R. t. W. this work is part of the research programme of the Netherlands Organisation for Scientific Research (NWO) and was performed at the research institute AMOLF.

REFERENCES

Adleman, L. M. 1994. "Molecular Computation of Solutions to Combinatorial Problems." *Science* 266:1021–1024.

Andrieux, D., and P. Gaspard. 2008. "Nonequilibrium Generation of Information in Copolymerization Processes." *Proceedings of the National Academy of Sciences of the United States of America* 105:9516–9521.

———. 2013. "Information Erasure in Copolymers." *Europhysics Letters* 103:30004.

Atkins, P. 2010. *The Laws of Thermodynamics: A Very Short Introduction*. Oxford: Oxford University Press.

Barato, A. C., D. Hartich, and U. Seifert. 2014. "Efficiency of Cellular Information Processing." *New Journal of Physics* 16:103024.

Barato, A. C., and U. Seifert. 2017. "Coherence of Biochemical Oscillations Is Bounded by Driving Force and Network Topology." *Physical Review E* 95:062409.

Bennett, C. H. 1982. "The Thermodynamics of Computation—A Review." *International Journal of Theoretical Physics* 21:905–940.

———. 2003. "Notes on Landauer's Principle, Reversible Computation, and Maxwell's Demon." *Studies in History and Philosophy of Modern Physics* 34:501–510.

Bérut, A., A. Arakelyan, A. Petrosyan, S. Ciliberto, R. Dillenschneider, and E. Lutz. 2012. "Experimental Verification of Landauer's Principle Linking Information and Thermodynamics." *Nature* 483:187–189.

Boyd, A. B., D. Mandal, and J. P. Crutchfield. 2016. "Identifying Functional Thermodynamics in Autonomous Maxwellian Ratchets." *New Journal of Physics* 18:023049.

Brillouin, L. 1951. "Maxwell's Demon Cannot Operate: Information and Entropy. I." *Journal of Applied Physics* 22:334–337.

Brittain, R. A., N. S. Jones, and T. E. Ouldridge. 2018. *Biochemical Szilard Engines for Memory-Limited Inference.* arXiv: 1812.08401 [cond-mat.stat-mech].

Bub, J. 2001. "Maxwell's Demon and the Thermodynamics of Computation." *Studies in History and Philosophy of Modern Physics* 32:569–579.

Chapman, A., and A. Miyake. 2015. "How an Autonomous Quantum Maxwell Demon Can Harness Correlated Information." *Physical Review E* 92:062125.

Chen, Y.-J., N. Dalchau, N. Srinivas, A. Phillips, L. Cardelli, D. Solveichik, and G. Seelig. 2013. "Programmable Chemical Controllers Made from DNA." *Nature Nanotechnology* 8:755–762.

Crooks, G. E. 1999. "Entropy Production Fluctuation Theorem and the Nonequilibrium Work Relation for Free Energy Differences." *Physics Review E* 60:2721–2726.

DeBenedictis, E. P., J. K. Mee, and M. P. Frank. 2017. "The Opportunities and Controversies of Reversible Computing." *Computer* 50:76–80.

Deffner, S. 2015. "Viewpoint: Exorcising Maxwell's Demon." *Physics* 8:127.

Deshpande, A., M. Gopalkrishnan, T. E. Ouldridge, and N. S. Jones. 2017. "Designing the Optimal Bit: Balancing Energetic Cost, Speed and Reliability." *Proceedings of the Royal Society of London, Series A* 473 (2204).

Dillenschneider, R., and E. Lutz. 2009. "Memory Erasure in Small Systems." *Physical Review Letters* 102:210601.

Dougal, R. C. 2016. "Kelvin, Thermodynamics and the Natural World." In *Kelvin, Maxwell, Clausius and Tait: The Correspondence of James Clerk Maxwell*, edited by M. W. Collins, R. C. Dougal, C. Koenig, and I. S. Ruddock, 135–151. WIT Press.

Esposito, M., and C. Van den Broeck. 2011. "Second Law and Landauer Principle Far from Equilibrium." *Europhysics Letters* 95:40004.

Fahn, P. N. 1996. "Maxwell's Demon and the Entropy Cost of Information." *Foundations of Physics* 26:71–93.

Feynman, R. P. 1998. *Feynman Lectures on Computation*. New York: Addison-Wesley.

Frank, M. P. 2017. "Throwing Computing into Reverse." *IEEE Spectrum* 54:32–37.

Govern, C. C., and P. R. ten Wolde. 2014. "Optimal Resource Allocation in Cellular Sensing Systems." *Proc. Nat. Acad. Sci. USA* 111:17486–17491.

Grochow, J. A., and D. H. Wolpert. 2018. "Beyond Number of Bit Erasures: New Complexity Questions Raised by Recently Discovered Thermodynamic Costs of Computation." *SIGACT News* 49:33–56.

Hong, J., B. Lambson, S. Dhuey, and J. Bokor. 2016. "Experimental Test of Landauer's Principle in Single-Bit Operations on Nanomagnetic Memory Bits." *Science Advances* 2:e1501492.

Horowitz, J. M., T. Sagawa, and J. M. R. Parrondo. 2013. "Imitating Chemical Motors with Optimal Information Motors." *Physical Review Letters* 111:010602.

Horowitz, J., and C. Jarzynski. 2008. "Comment on 'Failure of the Work-Hamiltonian Connection for Free-Energy Calculations'." *Physical Review Letters* 101:098901.

Huang, K. 1987. *Statistical Mechanics*. Second Edition. New York: John Wiley & Sons, Inc.

Ito, S., and T. Sagawa. 2015. "Maxwell's Demon in Biochemical Signal Transduction with Feedback Loop." *Nature Communications* 6:7498.

Jarzynski, C. 1997. "Nonequilibrium Equality for Free Energy Differences." *Phys. Rev. Lett.* 78:2690–2693.

Jaynes, E. T. 1957. "Information Theory and Statistical Mechanics." *Physical Review* 106:620–630.

Jun, Y., M. Gavrilov, and J. Bechhoefer. 2014. "High-Precision Test of Landauer's Principle in a Feedback Trap." *Physical Review Letters* 113:190601.

Kolchinsky, A., and D. H. Wolpert. 2017. "Dependence of Dissipation on the Initial Distribution over States." *Journal of Statistical Mechanics* 2017:083202.

Koski, J. V., V. F. Maisi, T. Sagawa, and J. P. Pekola. 2014. "Experimental Observation of the Role of Mutual Information in the Nonequilibrium Dynamics of a Maxwell Demon." *Physical Review Letters* 113:030601.

Krishnamurthy, S., and E. Smith. 2017. "Solving Moment Hierarchies for Chemical Reaction Networks." *Journal of Physics A* 50:425002.

Ladyman, J., S. Presnell, A. J. Short, and B. Groisman. 2007. "The Connection between Logical and Thermodynamic Irreversibility." *Studies in History and Philosophy of Modern Physics* 38:58–79.

Lambson, B., D. Carlton, and J. Bokor. 2011. "Exploring the Thermodynamic Limits of Computation in Integrated Systems: Magnetic Memory, Nanomagnetic Logic, and the Landauer Limit." *Physical Review Letters* 107:010604.

Lan, G., P. Sartori, S. Neumann, V. Sourjik, and Y. Tu. 2012. "The Energy–Speed–Accuracy Trade-Off in Sensory Adaptation." *Nature Physics* 8:422–428.

Landauer, R. 1961. "Irreversibility and Heat Generation in the Computing Process." *IBM Journal of Research and Development* 5:183–191.

———. 1991. "Information Is Physical." *Physics Today* 44:23–29.

Mandal, D., and C. Jarzynski. 2012. "Work and Information Processing in a Solvable Model of Maxwell's Demon." *Proceedings of the National Acadademy of Sciences of the United States of America* 109:11641–11645.

Maroney, O. J. E. 2005. "The (Absence of a) Relationship between Thermodynamic and Logical Reversibility." *Studies in History and Philosophy of Modern Physics* 36:355–374.

Maruyama, K., F. Nori, and V. Vedral. 2009. "Colloquium: The Physics of Maxwell's Demon and Information." *Reviews of Modern Physics* 81.

McGrath, T., N. S. Jones, P. R. ten Wolde, and T. E. Ouldridge. 2017. "Biochemical Machines for the Interconversion of Mutual Information and Work." *Physics Review Letters* 118:028101.

Mehta, P., A. H. Lang, and D. J. Schwab. 2016. "Landauer in the Age of Synthetic Biology: Energy Consumption and Information Processing in Biochemical Networks." *Journal of Statistical Physics* 162:1153–1166.

Mehta, P., and D. J Schwab. 2012. "Energetic Costs of Cellular Computation." *Proceedings of the National Acadademy of Sciences of the United States of America* 109:17978–17982.

Nelson, P. 2004. *Biological Physics: Energy, Information, Life.* New York: W. H. Freeman.

Norton, J. D. 2005. "Eaters of the Lotus: Landauer's Principle and the Return of Maxwell's Demon." *Stud. Hist. Philos. M. P.* 36:375–411.

Organick, L., S. D. Ang, Y.-J. Chen, R. Lopez, S. Yekhanin, K. Makarychev, M. Z. Racz, et al. 2018. "Random Access in Large-Scale DNA Data Storage." *Nature Biotechnology* 36:242–248.

Ouldridge, T. E. 2018. "The Importance of Thermodynamics for Molecular Systems and the Importance of Molecular Systems for Thermodynamics." *Natural Computing* 17:3–29.

Ouldridge, T. E., C. C. Govern, and P. R. ten Wolde. 2017. "The Thermodynamics of Computational Copying in Biochemical Systems." *Physical Review X* 7:021004.

Ouldridge, T. E., and P. R. ten Wolde. 2017. "Fundamental Costs in the Production and Destruction of Persistent Polymer Copies." *Physical Review Letters* 118:158103.

Owen, J. A., A. Kolchinsky, and D. H. Wolpert. 2011. "Number of Hidden States Needed to Physically Implement a Given Conditional Distribution." *New Journal of Physics* 21.

Parrondo, J. M., J. M. Horowitz, and T. Sagawa. 2015. "Thermodynamics of Information." *Natural Physics* 11:131–139.

Plenio, M. B., and V. Vitelli. 2001. "The Physics of Forgetting: Landauer's Erasure Principle and Information Theory." *Contemporary Physics* 42:25–60.

Plesa, T. 2018. *Stochastic Approximation of High-Molecular by Bi-Molecular Reactions.* arXiv: 1811.02766 [q-bio.MN].

Qian, L., D. Soloviechick, and E. Winfree. 2011. "Efficient Turing-universal computation with DNA polymers." *DNA 16, LNCS* 6518:123–140.

Rao, R., and M. Esposito. 2016. "Nonequilibrium Thermodynamics of Chemical Reaction Networks: Wisdom from Stochastic Thermodynamics." *Physical Review X* 6:041064.

———. 2018. "Conservation Laws and Work Fluctuation Relations in Chemical Reaction Networks." *The Journal of Chemical Physics* 149:245101.

Sagawa, T. 2014. "Thermodynamic and Logical Reversibilities Revisited." *Journal of Statistical Mechanics:* P03025.

Sagawa, T., and M. Ueda. 2009. "Minimal Energy Cost for Thermodynamic Information Processing: Measurement and Information Erasure." *Phys. Rev. Lett.* 102:250602.

Schmiedl, T., and U. Seifert. 2007. "Stochastic Thermodynamics of Chemical Reaction Networks." *Journal of Chemical Physics* 126:044101.

Seifert, U. 2005. "Entropy Production Along a Stochastic Trajectory and an Integral Fluctuation Theorem." *Physical Review Letters* 95:040602.

———. 2011. "Stochastic Thermodynamics of Single Enzymes and Molecular Motors." *European Physical Journal E* 34:26.

Shannon, C. E., and W. Weaver. 1949. *The Mathematical Theory of Communication.* Champaign: University of Illinois Press.

Srinivas, N., J. Parkin, G. Seelig, E. Winfree, and D. Soloveichik. 2017. "Enzyme-Free Nucleic Acid Dynamical Systems." *Science* 358:eaal2052.

Stock, A. M., V. L. Robinson, and P. N. Goudreau. 2000. "Two-Component Signal Transduction." *Annual Review of Biochemistry* 69:183–215.

Stopnitzky, E., S. Still, T. E. Ouldridge, and L. Altenberg. 2019. "Physical Limitations of Work Extraction from Temporal Correlations." In *The Energetics of Computing in Life and Machines,* edited by D. H. Wolpert, C. Kempes, P. F. Stadler, and J. A. Grochow.

Szilard, L. 1964. "On the Decrease of Entropy in a Thermodynamic System by the Intervention of Intelligent Beings." *Behavioral Science* 9:301–310.

ten Wolde, P. R., N. B. Becker, T. E. Ouldridge, and Mugler A. 2016. "Fundamental Limits to Cellular Sensing." *Journal of Statistical Physics* 162:1395–1424.

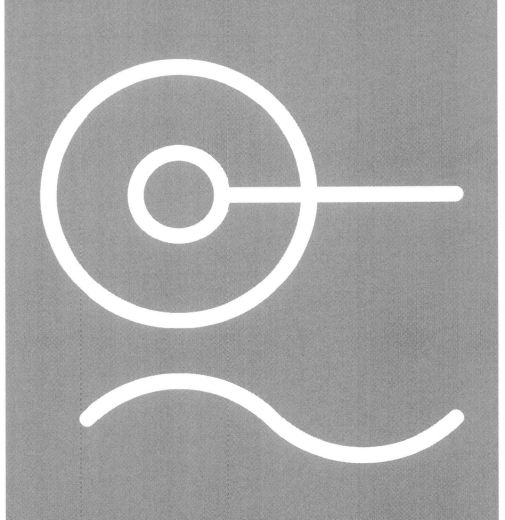

TRANSFORMING METASTABLE MEMORIES: THE NONEQUILIBRIUM THERMODYNAMICS OF COMPUTATION

Paul M. Riechers, University of California, Davis

Modern scientific understanding suggests that computation can be performed without any dissipation at all—a perplexing result, since we still plug in our computers and eat to fuel our brains every day. To reconcile this discrepancy between theoretical possibility and familiar reality requires a thermodynamics of realistic computation, where nonequilibrium distributions, corresponding to metastable memories, are transformed under practical constraints by controlled driving in finite time. From the perspective of nonequilibrium thermodynamics employed here, logical irreversibility is indeed compatible with thermodynamic reversibility if accompanied by a metastable increase in nonequilibrium addition to free energy, which can later be leveraged to reclaim the original work input—hence computation without dissipation. However, our demands for speed and modularity each implies trade-offs that necessitate dissipation, while practical limitations of the controller's knowledge and dexterity further challenge the attainable thermodynamic efficiency of computation. Here we develop a few of the fundamental thermodynamic consequences of transforming metastable memories and identify several practical opportunities for greater energetic efficiency.

The chapter contains several new results, including a new decomposition of the nonequilibrium free energy that shows under what circumstances a coarse-grained description is sufficient to understand the thermodynamics of metastable memory transformations; implications for composite memory systems and the role of knowledge in work extraction; the thermodynamic cost of modular computation, which generalizes a recent result by Boyd, Mandal, and Crutchfield (2018); and the minimal work expected of any two-input–one-output logic function and the dissipation incurred when these circuits are not designed for the statistics of the memories they transform. We close with a short tutorial that explicitly calculates the fundamental thermodynamic limits of the universal NAND gate.

Metastable Memory Systems

We start by considering a memory system, which is simply a physical system meant to store information. During computations, the dynamics of the memory system is driven by an external work reservoir to transform the memory from its initial state to its final state. For the memory system to be of much practical utility, it should be able to store memories robustly between computations. One way to achieve this is with nonvolatile memory elements that—through metastability—retain their memories over long time scales without active power consumption, even when the computer is turned off.[1]

---

[1] Two types of memory are common in practical computers. The first uses a nonvolatile metastable memory that does not require energetic upkeep and retains its memory even when the computer is powered off. The second type requires active power to retain the memory, as in CMOS transistor architectures, where the inevitable leakage of currents implies constant power consumption. Without power, the volatile memory is lost. For reasons of both anticipated supremacy in energetic efficiency and clarity of our exposition, we choose to describe transformations of the former nonvolatile type of memory in the following. However, we expect our results to maintain at least some relevance in the energetic limits of transformations of active memories, where the work and dissipation discussed here should be roughly

Chapter 13: Transforming Metastable Memories

At each moment, the work reservoir can exert influences according to the vector quantities $x \in \chi$. For example, x may represent the configuration of the applied electromagnetic field, a collection of piston positions, or any other controllable factors that influence the Hamiltonian of the memory system. The instantaneous Hamiltonian $\mathcal{H}_x$ of the memory system determines the instantaneous energies $\{E_x(s)\}_{s \in \mathcal{S}}$ of the system's microstates $\mathcal{S}$. The control parameter x is held fixed while the memory is to be retained. Changes to the memory system are implemented by a trajectory of time-varying control $x_{0:\tau}$ (often called a *protocol* in the literature) over a duration τ that drives the system to a new state.

Computations utilizing metastable memories imply a strong separation of time scales in the nondriven dynamics of the memory system such that the dynamics of the various metastable regions are nearly autonomous with respect to each other and can quickly establish local equilibria. The autonomy within certain regions of state space suggests that we partition the set of microstates $\mathcal{S}$ into a set of metastable *memory states* $\mathcal{M}$.

The system is also in contact with an effectively memoryless heat bath at temperature T with which it exchanges energy, which enables the system's relaxation to both local and global equilibria.

Over very long time frames (time frames much longer than any computation performed by the system, and much longer even than the waiting times between computations), the memory system—if left undriven, experiencing only the static control setting x—would asymptotically relax to the *global-equilibrium distribution* π_x, which is exactly stationary

---

additive to the background "housekeeping" power consumption by active circuits.

under the combined influence of the Hamiltonian $\mathcal{H}_x$ and the interaction with the heat bath. We assume that the global-equilibrium distribution is the canonical one: $\pi_x(s) = \frac{e^{-\beta E_x(s)}}{Z_x}$, where $\beta = (k_\mathrm{B}T)^{-1}$ and Z_x is the standard partition function $Z_x = \sum_{s \in \mathcal{S}} e^{-\beta E_x(s)}$ that normalizes the distribution and yields the equilibrium free energy $F_x^\mathrm{eq} = -k_\mathrm{B}T \ln Z_x$.

However, metastability implies that this time scale of global relaxation is much longer than the time scale of computation. On the time scale *between* computations, all probability density within each memory state $m \in \mathcal{M}$ is assumed to relax approximately to its *local-equilibrium distribution* $\pi_x^{(m)}$, as Esposito (2012) discusses for the case of strong separation of time scales, with

$$\begin{aligned}
\pi_x^{(m)}(s) &= \delta_{s \in m} \frac{e^{-\beta E_x(s)}}{\sum_{s' \in m} e^{-\beta E_x(s')}} \\
&= \delta_{s \in m}\, \pi_x(s) \frac{Z_x}{Z_x^{(m)}} \\
&= \delta_{s \in m}\, e^{-\beta\left(E_x(s) - F_x^{(m)}\right)},
\end{aligned} \qquad (13.1)$$

where $Z_x^{(m)}$ is the memory's *local partition function*: $Z_x^{(m)} \equiv \sum_{s \in m} e^{-\beta E_x(s)}$. This quantity strongly suggests defining the *local-equilibrium free energy*,

$$F_x^{(m)} \equiv -k_\mathrm{B}T \ln Z_x^{(m)}, \qquad (13.2)$$

which turns out to provide useful intuition for the thermodynamics of transformations between metastable states, as we shall soon see.

Between computations, the distribution relaxes quickly to a classical superposition of local equilibria determined by the net probability in each memory state at the end of the last computation. Given a postcomputation distribution over memory states $\Pr(\mathcal{M}_\tau)$, the distribution over microstates quickly approaches the metastable superposition:

$$\Pr(\mathcal{S}_{\tau+\delta t}) \approx \sum_{m \in \mathcal{M}} \Pr(\mathcal{M}_\tau = m)\, \pi^{(m)}_{x_\tau}, \qquad (13.3)$$

where $\mathcal{S}_t$ is the random variable for the microstate at time t and $\mathcal{M}_t$ is the random variable for the memory state at time t.

However, during a computation, the dynamic control protocol $x_{0:\tau}$ induces a net state-to-state stochastic transition dynamic $\mathsf{T}_{x_{0:\tau}}$ over $\mathcal{S}$ that can strongly couple and transform memory states, as required of a computation.

Driven Dynamics and Computations

A deterministic *computation* $\mathcal{C} : \mathcal{M} \to \mathcal{M}$ is an operation mapping the set of memory states $\mathcal{M}$ to itself. In practice, it is implemented by a driving protocol $x_{0:\tau}$ that controls the evolution of the system for a duration τ. Figure 13.1 (left) shows schematically how the protocol can change the energy landscape to induce (right) a net transition among microstates that corresponds to a (generically stochastic) computation on the coarse-grained memory states (depicted as the different shaded regions). The system should start and end with the same resting influence $x_0 = x_\tau$, as can be seen in the top-left bubble of figure 13.1, if a consistent metastable memory landscape is desired between computations. The set of all protocols that *reliably* implement a computation $\mathcal{C}$ is

$$\chi_\mathcal{C} \equiv \left\{ x_{0:\tau} \in \chi^{[0,\tau]} : \Pr_{x_{0:\tau}}\left(\mathcal{S}_\tau \notin \mathcal{C}(m) \,\middle|\, \mathcal{S}_0 \sim \pi^{(m)}_{x_0}\right) < \epsilon \text{ for all } m \in \mathcal{M} \right\} \qquad (13.4)$$

for some error tolerance ϵ. The assumed separation of time scales allows us to employ the local-equilibrium distribution $\pi^{(m)}_{x_0}$ as the initial distribution in the test for reliable memory evolution.

THE ENERGETICS OF COMPUTING IN LIFE & MACHINES

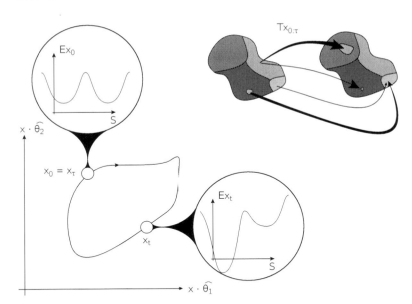

Figure 13.1. A generic driving protocol $x_{0:\tau}$ (the time series of the control parameters over a duration τ) induces a net stochastic transition dynamic $\mathsf{T}_{x_{0:\tau}}$ over the set of microstates $\mathcal{S}$. A thermodynamically efficient computation on the coarse-grained set of memory states $\mathcal{M}$ is implemented via a control protocol chosen from χ_C that reliably implements the logical transformation with minimal dissipation.

Different protocols implementing the same computation typically dissipate different amounts of work. The grand challenge for energy-efficient computation is to identify the control protocols, given realistic control restrictions, that reliably implement a computation in finite time with the minimal resultant work dissipation.

Work, Nonequilibrium Free Energy, and Dissipation

The work dissipated in a computation is the work W that is irretrievably lost to the environment:

$$W_{\text{diss}} = W - \Delta E_{x_t}(s_t) - k_\text{B} T \Delta \ln\bigl(\Pr_{\overleftarrow{x}_t}(s_t)\bigr), \qquad (13.5)$$

where the right-most term is recognized as the change in the nonaveraged precursor to nonequilibrium entropy.

Equivalently, W_{diss} is the amount of heat that is *not* offset by a corresponding change in this trial-specific internal entropy. Notably, equation (13.5) is valid over *any* time interval $t \in [0, \tau]$ and therefore (by adjusting the considered duration τ) also tracks the dynamics of the dissipation associated with the computation as the system relaxes during and after the work performed to implement the computation. As the memory system relaxes to a new metastable distribution on a relatively short time scale relevant to computation, the dissipation should appear to saturate to the total dissipation associated with the computation. In the preceding, $\Pr_{\overleftarrow{x}_t}(s_t)$ is the *driving-induced* probability of the microstate s_t at time t. Similarly, $\Pr_{\overleftarrow{x}_t}(\mathcal{S}_t)$ is the driving-induced probability *distribution* over microstates—it can be thought of as the distribution over microstates *conditioned* on the full driving history $\overleftarrow{x}_t$, including both the controlled preparation of the system prior to the computation and the protocol implementing the computation, up to time t.

The *expected* dissipation, given some initial preparation of the memory system and a particular driving protocol, is thus

$$\langle W_{\text{diss}} \rangle = \langle W \rangle - \Delta \mathcal{F}, \qquad (13.6)$$

where the expected nonequilibrium free energy at time t is

$$\mathcal{F} = U - k_{\text{B}} T\, H\bigl(\Pr_{\overleftarrow{x}_t}(\mathcal{S}_t)\bigr) = F^{\text{eq}}_{x_t} + k_{\text{B}} T\, D_{\text{KL}}\bigl(\Pr_{\overleftarrow{x}_t}(\mathcal{S}_t) \big\| \pi_{x_t}\bigr). \qquad (13.7)$$

In the preceding equation, $U = \langle E_{x_t}(s_t)\rangle_{\Pr_{\overleftarrow{x}_t}(\mathcal{S}_t)}$ is the expected internal energy of the system at time t. $H(\cdot)$ is the Shannon entropy (in nats) of its argument, and we will use $H_{\overleftarrow{x}_t}(\mathcal{S}_t)$ to denote $H\bigl(\Pr_{\overleftarrow{x}_t}(\mathcal{S}_t)\bigr)$. Finally, $D_{\text{KL}}(\cdot)$ is the Kullback–Leibler divergence, which is always nonnegative. $D_{\text{KL}}\bigl(\Pr_{\overleftarrow{x}_t}(\mathcal{S}_t)\big\|\pi_{x_t}\bigr)$ is the *nonequilibrium addition* to free energy—the thermodynamic resource that scales with the distribution's distance from equilibrium. It should be noted that $\Delta F_{x_t}^{\text{eq}} = 0$ over the full course of a computation, because a computation starts and ends with the same resting influence $x_0 = x_\tau$.

Recent finite-time fluctuation theorems (most directly, equations (38) and (42) of Riechers and Crutchfield (2017)) guarantee that the average dissipated work, when starting in any potentially nonequilibrium and non-steady-state distribution, is always nonnegative:

$$\langle W_{\text{diss}} \rangle \geq 0 . \qquad (13.8)$$

This implies that the average amount of work that must be performed is bounded by

$$\langle W \rangle \geq \Delta \mathcal{F} . \qquad (13.9)$$

Equation (13.9) can also be derived by other means, as in the work of Takara, Hasegawa, and Driebe (2010) and Esposito and Van den Broeck (2011). The work performed in surplus to $\Delta \mathcal{F}$ is eventually dissipated and contributes to the entropy production by the computation.

So, how much work is actually dissipated? Surely the average work dissipated in transforming a distribution depends on the particulars of the protocol, with plenty of room for wastefulness if the protocol is not carefully designed. However, the *minimal* dissipated work is characterized by the allowed duration τ to implement the transformation and also by the degree of control one has over the system's Hamiltonian. Let us briefly consider the case of perfect control, in which the controller can apply any Hamiltonian to the system. By instantaneously changing the initial Hamiltonian—to make any initial distribution the canonical distribution of the new

Hamiltonian—before subsequent finite-speed driving of the system, we can immediately apply the recent results of finite-time thermodynamics (Salamon and Berry 1983; Sivak and Crooks 2012; Zulkowski and DeWeese 2014; Bonança and Deffner 2014; Mandal and Jarzynski 2016) to conclude that the work dissipated by an optimal protocol, meant to transform between two distributions in a finite time τ with minimal dissipation, is generically (to first order of approximation) inversely proportional to the allowed duration τ; that is, $\langle W_{\text{diss}} \rangle_{\text{min}} \sim \tau^{-1}$. In part, the next section shows that this same τ^{-1} scaling of the dissipation can be achieved at intermediate time scales as long as the distribution stays close to a local-equilibrium distribution, even if it is never close to a global-equilibrium distribution. More generally, the next section contains the fundamental thermodynamics of computation through the transformation of metastable memories.

Nonequilibrium Thermodynamics at the Level of Memory States

The above is all now-standard nonequilibrium thermodynamics. However, we seek thermodynamic implications for transformation of *memory states* rather than microstates. Fortunately, a rigorous hierarchical description can be achieved through a series of decompositions of familiar thermodynamic quantities.

To start, we note that since $\mathcal{M}$ is a coarse graining of $\mathcal{S}$, we have

$$H_{\overleftarrow{x}_t}(\mathcal{S}_t) = H_{\overleftarrow{x}_t}(\mathcal{M}_t, \mathcal{S}_t) = H_{\overleftarrow{x}_t}(\mathcal{M}_t) + H_{\overleftarrow{x}_t}(\mathcal{S}_t | \mathcal{M}_t). \tag{13.10}$$

We will use equation 13.10 together with a novel decomposition of the expected internal energy, which is valid at any time given any coarse graining of microstates $\mathcal{S}$ into the coarse-grained set

$\mathcal{M}$. In particular, the expected internal energy of the system can be decomposed as

$$U = \langle E_{x_t}(s_t)\rangle_{\Pr_{\overleftarrow{x}_t}(\mathcal{S}_t)} = \big\langle \langle E_{x_t}(s_t)\rangle_{\Pr_{\overleftarrow{x}_t}(\mathcal{S}_t|\mathcal{M}_t=m)}\big\rangle_{\Pr_{\overleftarrow{x}_t}(\mathcal{M}_t)}. \tag{13.11}$$

Crucially, utilizing the identity $E_x(s) = -k_B T \ln(\pi_x^{(m)}(s)) + F_x^{(m)}$, we find that the expected internal energy of a system that has been driven into memory state m is

$$\langle E_{x_t}(s_t)\rangle_{\Pr_{\overleftarrow{x}_t}(\mathcal{S}_t|\mathcal{M}_t=m)}$$
$$= k_B T\, H_{\overleftarrow{x}_t}(\mathcal{S}_t|\mathcal{M}_t=m) + F_{x_t}^{(m)} + F_{\overleftarrow{x}_t,\text{add}}^{(m)}. \tag{13.12}$$

$F_{x_t}^{(m)}$ is the local-equilibrium free energy, and

$$F_{\overleftarrow{x}_t,\text{add}}^{(m)} \equiv k_B T D_{\text{KL}}\big(\Pr_{\overleftarrow{x}_t}(\mathcal{S}_t|\mathcal{M}_t=m)\,\big\|\,\pi_{x_t}^{(m)}\big) \tag{13.13}$$

is the *local nonequilibrium addition to free energy* in region m. The expected internal energy thus always has the decomposition $U = k_B T H_{\overleftarrow{x}_t}(\mathcal{S}_t|\mathcal{M}_t) + \langle F_{x_t}^{(m)}\rangle_{\Pr_{\overleftarrow{x}_t}(\mathcal{M}_t)} + \langle F_{\overleftarrow{x}_t,\text{add}}^{(m)}\rangle_{\Pr_{\overleftarrow{x}_t}(\mathcal{M}_t)}$. At the same time, we always have that $U = \mathcal{F} + k_B T H_{\overleftarrow{x}_t}(\mathcal{S}_t)$. Putting these together, we find that the nonequilibrium free energy can always be decomposed according to the contributions commensurate with the coarse-grained description:

$$\mathcal{F} = \langle F_{x_t}^{(m)}\rangle_{\Pr_{\overleftarrow{x}_t}(\mathcal{M}_t)} + \langle F_{\overleftarrow{x}_t,\text{add}}^{(m)}\rangle_{\Pr_{\overleftarrow{x}_t}(\mathcal{M}_t)} - k_B T H_{\overleftarrow{x}_t}(\mathcal{M}_t). \tag{13.14}$$

Moreover, when the coarse graining accords with well-designed metastable memory states, the separation of time scales implies that $F_{\overleftarrow{x}_t,\text{add}}^{(m)} \to 0$ quickly after any driving.[2] Hence,

---

[2] It is important to note that this is an *assumption* about the dynamics that is well suited to the memory systems typically used in practical computations. The results of the following are only as reliable as this assumption is valid.

before and shortly after a computation, we can decompose the nonequilibrium entropy into two very manageable parts:

$$\mathcal{F} \approx \langle F_{x_t}^{(m)} \rangle_{\mathrm{Pr}_{\overleftarrow{x}_t}(\mathcal{M}_t)} - k_\mathrm{B} T H_{\overleftarrow{x}_t}(\mathcal{M}_t), \qquad (13.15)$$

that is, the expected local-equilibrium free energy, less the coarse-grained entropy of the memory states. Any difference from equality is due to work already performed that is expected to soon be dissipated in the relaxation to local equilibria. Equation (13.15) was previously highlighted by Parrondo, Horowitz, and Sagawa (2015). However, the local nonequilibrium addition to the free energy, as in equation (13.14), is a new finding that offers further insight.

The local nonequilibrium addition to free energy is a thermodynamic resource that, in principle, can be traded to perform useful work. However, if either the time scale of relaxation within each memory state is faster than the relevant speed of the driving protocol, or the control parameters are too coarse or otherwise incapable of influencing the fine degrees of freedom within the memory state, then the local nonequilibrium addition to free energy will inevitably be lost to dissipation as the local distributions relax to their local equilibria. Conversely, the coarse-grained memory probabilities are assumed to be metastable and controllable, and so nonequilibrium changes at the coarse-grained level can be implemented thermodynamically reversibly.

With these considerations in mind, equation (13.14) suggests that a driving protocol that keeps the distribution close to a metastable one (i.e., a weighted average of local-equilibrium distributions) at all times, such that $F^{(m)}_{\overleftarrow{x}_t, \mathrm{add}}$ always stays close to zero in each metastable region, can be used to implement thermodynamically efficient computations with $\langle W_\mathrm{diss} \rangle \sim \tau^{-1} \to 0$ as $\tau \to \infty$. In this nearly quasistatic limit, such processes will be thermodynamically reversible.

Finally, coming back to equation (13.6), this implies that the minimum average work necessary to implement a computation on metastable memory states is

$$\langle W \rangle_{\min} = \Delta \langle F_{x_t}^{(m)} \rangle_{\Pr_{\overleftarrow{x}_t}(\mathcal{M}_t)} - k_\mathrm{B} T \Delta H_{\overleftarrow{x}_t}(\mathcal{M}_t). \quad (13.16)$$

In the computational setting, equation (13.16) can be interpreted as a generalization of Landauer's principle for the minimum work necessary to implement computations that transform metastable memories of *different* local-equilibrium free energies. In a more general setting, when $\langle W \rangle_{\min}$ is negative, $\langle W_\text{extracted} \rangle_{\max} = -\langle W \rangle_{\min}$ can also be interpreted as the maximal work that can be *extracted* from a metastable system. The heterogeneity of local free energy then offers an easy explanation of how a single bit of macroscopic information (e.g., "Is the apple in the left or right box?") can carry a macroscopically huge (much larger than $k_\mathrm{B} T \ln 2$) energetic gain, as in the work of Gokler *et al.* (2017), when interacting with far-from-equilibrium systems. Figure 13.2 gives further insight into the meaning of the local-equilibrium free energies in equation (13.16): larger local-equilibrium free energy can result from either larger average energy or more certainty in the local-equilibrium microstate distribution.

Clearly, if all local-equilibrium free energies are equal (either through identically constructed potentials or otherwise through some delicate balance of local energies and entropies) such that $F_{x_0}^{(m)} = F_{x_0}^{(m')}$ for all $m, m' \in \mathcal{M}$, then $\Delta \langle F_{x_t}^{(m)} \rangle_{\Pr_{\overleftarrow{x}_t}(\mathcal{M}_t)} = 0$, and the minimal work necessary to implement a computation reduces to

$$\begin{aligned} \langle W \rangle_{\min} &= -k_\mathrm{B} T \Delta H_{\overleftarrow{x}_t}(\mathcal{M}_t) \\ &= k_\mathrm{B} T H_{\overleftarrow{x}_0}(\mathcal{M}_0) - k_\mathrm{B} T H_{\overleftarrow{x}_\tau}(\mathcal{M}_\tau). \end{aligned} \quad (13.17)$$

The minimum work thus depends solely on the change in entropy over memory states. Equation (13.17) constitutes

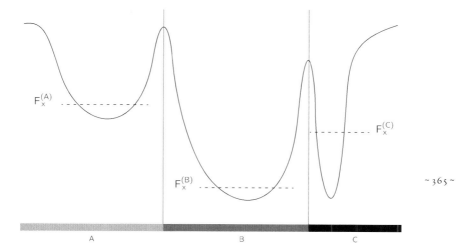

Figure 13.2. Schematic of a simple one-dimensional potential with three metastable wells of different character. Metastability requires barriers of at least several times the thermal energy $k_B T$. Larger local-equilibrium free energy $F_x^{(m)}$ can result from higher energy, greater certainty, or both.

the modern understanding of Landauer's principle for the minimum work necessary to implement computations that transform metastable memories of equal free energy.

If the computation reduces the internal entropy of the memory system, then it will require work, although this work can later be reclaimed and recycled if the computation was performed without dissipation. Conversely, when $\langle W \rangle_{\min}$ is negative, work can be *extracted* as a result of the transformation, for example, to lift a weight or to fuel other computations. In this latter case, the memory system can perform as an *information engine*, trading certainty for useful work. In this regime, $\langle W_{\text{extracted}} \rangle_{\max} = -\langle W \rangle_{\min} = k_B T \Delta H_{\overleftarrow{x}_t}(\mathcal{M}_t)$ is the *maximal* average work that can be extracted from transformations resulting in this memory–entropy change.

Implications for Composite Memory Systems

If the system is composed of N multistable memory elements, then it is natural to treat the memory state as a composite state of N random variables: $\mathcal{M}_t = \mathcal{M}_t^{(1:N+1)} = (\mathcal{M}_t^{(1)}, \mathcal{M}_t^{(2)}, \ldots \mathcal{M}_t^{(N)})$. In general, the memory elements are correlated such that $\Pr_{\overleftarrow{x}_t}(\mathcal{M}_t^{(1)}, \mathcal{M}_t^{(2)}, \ldots \mathcal{M}_t^{(N)}) \neq \prod_{n=1}^{N} \Pr_{\overleftarrow{x}_t}(\mathcal{M}_t^{(n)})$. Moreover, this correlation has important thermodynamic consequences.

Consider a nonequilibrium system of N identical memory elements, each having K identical metastable regions (such that $F_{x_0}^{(m)} = F_{x_0}^{(m')}$ for all $m, m' \in \mathcal{M}$), and let $h_{\overleftarrow{x}_t} \equiv \frac{H_{\overleftarrow{x}_t}(\mathcal{M}_t^{(1:N+1)})}{N}$ denote the coarse-grained entropy *density* of the memory system. If the system is transformed by some driving protocol $x_{0:\tau}$ that increases the system's entropy, then, from equation (13.17), the maximal extractable work per memory element is given by

$$\tfrac{\beta}{N} \langle W_{\text{extracted}} \rangle_{\max} = h_{\overleftarrow{x}_\tau} - h_{\overleftarrow{x}_0}. \qquad (13.18)$$

In the case where the memory elements are arranged in a one-dimensional topology, the entropy density has been taken to mean the Shannon entropy *rate* of the sequence as it is scanned spatially (Boyd, Mandal, and Crutchfield 2017). While this is technically correct, it is sufficiently nuanced to require careful interpretation. In particular, the entropy density $h \equiv \lim_{L \to \infty} \frac{H(\boldsymbol{p}_L)}{L}$ that can be inferred from the frequentist statistics gathered along the sequence of the instantaneous configuration will, in general, converge to a value that is *not* the same as $h_{\overleftarrow{x}_t}$.[3] Rather,

$$\ln(K) \geq h \geq h_{\overleftarrow{x}} \geq 0. \qquad (13.19)$$

Crucially, it is the entropy *conditioned on the driving history* that matters in the theoretical limit of what orderliness can be thermodynamically leveraged—and this is *a priori* distinct from anything that could be inferred from the instantaneous configuration, even in the limit of $N \to \infty$. Moreover, this thermodynamic entropy density is independent of spatial dimension or even of any spatial topology of the memory elements. Indeed, the topology of the memory elements— being arranged in a one-dimensional string, for example—is *a priori independent* of the topology of the correlations among random variables, and it is only the latter that fundamentally matters for the thermodynamics of information processing. However, spatial locality occasionally does correspond to the correlational structure, especially when correlations develop as a consequence of local physical interactions.

Turing machines and related models of computation require not only a bit string but also a memoryful read–write head that can operate on the tape. Treating these two components inclusively as part of the memory system makes the system self-contained and provides important lessons

---

[3] For example, the string of the digits of pi (3.14159...) has a Shannon entropy rate of $h = \ln(K)$ nats per symbol in any K-ary expansion for $K \in \{2, 3, \ldots\}$. (For example, $h = \ln(10)$ in the given decimal expansion of pi, whereas $h = \ln(2)$ in its binary expansion.) It would naïvely seem to provide no thermodynamic fuel, as it appears to be a completely "random" sequence. To the contrary, if stored into memory, the sequence contains *full* thermodynamic fuel (i.e., it can be fully leveraged to do work), because the memory system will be *driven* by $\overleftarrow{x}_t$ to hold this sequence uniquely so that $H_{\overleftarrow{x}_t}(\mathcal{M}_t) = 0$, because $\Pr(\mathcal{M}_t) = \delta_{\mathcal{M}^{(1)},3}\delta_{\mathcal{M}^{(2)},1}\delta_{\mathcal{M}^{(3)},4}\cdots$. That is, in the space *of sequences*, the state of the memory system is delta distributed. This particular example points to the proper way of thinking about entropy and about what *kind* of information can be thermodynamically leveraged in computer memory in general.

about the thermodynamics of such functionally segregated systems (Boyd *et al.* 2017; Boyd, Mandal, and Crutchfield 2017).

This sort of inclusiveness[4] also sheds light on "Maxwell's demon"–type scenarios, where the net system is decomposed into a subsystem and a "demon." If the subsystem is initially out of equilibrium, then the demon can simply extract work by extracting the subsystem's nonequilibrium addition to free energy—no mystery there. The more complicated story arises when the subsystem is initially *in* equilibrium, as in Maxwell's original *Gedankenexperiment*, where the subsystem is a two-compartment box of gas starting in equilibrium. The demon can nevertheless decrease the entropy of the subsystem by increasing its knowledge of the subsystem (Lloyd 1989). Work can then be extracted as the subsystem is subsequently brought back to equilibrium. But if this process is to reset to form a full cycle—if the demon's memory is to be erased, for example—then no net work can be extracted on average. Interestingly, it is not *necessarily* erasure where cost is incurred (Boyd and Crutchfield 2016), but whether the cost is incurred in measurement or in erasure can be explained via heterogeneous local-equilibrium free energies of the memory states as an application of equation (13.16).

---

[4] Including two subsystems (e.g., "subsystem" and "demon") explicitly as components of the same memory system means that the coarse-grained memory entropy decomposes as $H(\mathcal{M}^{(\text{sub})}, \mathcal{M}^{(\text{demon})}) = H(\mathcal{M}^{(\text{sub})}) + H(\mathcal{M}^{(\text{demon})}) - I(\mathcal{M}^{(\text{sub})}; \mathcal{M}^{(\text{demon})})$, where I denotes mutual information. Interpreting the latter quantity as the knowledge the demon has of the system, we can see from substitution into equation (13.16) that knowledge is another thermodynamic resource that can be exchanged for entropy reduction, free-energy gain, or work extraction. However, we also see that its origination carries either an energetic or entropic cost of at least what the knowledge was later worth; knowledge is a *medium* for thermodynamic transactions rather than a free source of energy.

To close this topic, we note that work *can* be extracted at a constant rate when an active environment continuously drives the subsystem out of equilibrium, at no cost to the extractor (formerly known as the "demon"). The extractor can then siphon off the power that it needs to sustain its luxurious nonequilibrium lifestyle—perhaps even to appease its greed for massive, speedy computations.

The Thermodynamic Cost of Ignorance and Neglect

If one has sufficient control over the energy levels of a system, and if quasistatically slow transformations are tolerable, then a transformation between any two distributions is always possible *without dissipation*. Example methods to implement such dissipationless computations are given, for example, in the work of Åberg (2013), Garner *et al.* (2017), Parrondo, Horowitz, and Sagawa (2015), and Boyd, Mandal, and Crutchfield (2018). We have further argued that the zero-dissipation limit is also approached with the slightly weaker requirement that metastable distributions (rather than strictly the global-equilibrium distribution) be maintained throughout the transformation. However, even in this nearly quasistatic case, if correlations are ignored, or if other features of the distribution are neglected or misrepresented for whatever reason in the manipulation of the distribution, then there is necessarily extra work incurred and dissipated when the driving protocol is run.

Suppose a driving protocol $x^*_{0:\tau}$ is chosen to minimize dissipation while implementing a computation $\mathcal{C}$ and starting in the distribution $\boldsymbol{q}_0$; that is, $x^*_{0:\tau} \equiv \mathrm{argmin}_{x_{0:\tau} \in \chi_\mathcal{C}} \langle W_\mathrm{diss} \rangle_{\mathrm{Pr}_{x_{0:\tau}}(\mathcal{S}_{0:\tau} | \mathcal{S}_0 \sim \boldsymbol{q}_0)}$. For simplicity, let's further assume that we take the $\tau \to \infty$ limit so that this computation is achieved with no dissipation at all when the system is initiated in the distribution q_0.

Moreover, $x_{0:\tau}^*$ and q_0 are a pair, each minimizing dissipation for the other. Indeed, with $\langle W_{\text{diss}} \rangle = 0$ and the generic requirement that $\langle W_{\text{diss}} \rangle \geq 0$, it follows that, among possible ways to initiate the distribution over system states, q_0 minimizes the dissipation incurred when running the drive protocol $x_{0:\tau}^*$. This latter fact allows us to utilize the recent theorem by Kolchinsky and Wolpert (2017) that, if q_0 minimizes the dissipation for some drive protocol $x_{0:\tau}$, then there is necessarily an extra dissipation incurred by starting in some other distribution μ_0, given by

$$\beta \langle W_{\text{diss}}(\mu_0) \rangle - \beta \langle W_{\text{diss}}(q_0) \rangle \\ = D_{\text{KL}}(\mu_0 \| q_0) - D_{\text{KL}}(\mu_\tau \| q_\tau), \quad (13.20)$$

where $\langle W_{\text{diss}}(\mu_0) \rangle \equiv \langle W_{\text{diss}} \rangle_{\Pr_{x_{0:\tau}^*}(\mathcal{S}_{0:\tau} | \mathcal{S}_0 \sim \mu_0)}$ and $\langle W_{\text{diss}}(q_0) \rangle \equiv \langle W_{\text{diss}} \rangle_{\Pr_{x_{0:\tau}^*}(\mathcal{S}_{0:\tau} | \mathcal{S}_0 \sim q_0)}$, which is equal to 0 in this case. In the foregoing calculations, $\mu_\tau = \Pr_{x_{0:\tau}^*}(\mathcal{S}_\tau | \mathcal{S}_0 \sim \mu_0)$ and $q_\tau = \Pr_{x_{0:\tau}^*}(\mathcal{S}_\tau | \mathcal{S}_0 \sim q_0)$ are the time-evolved versions of μ_0 and q_0, respectively, under the influence of the driving $x_{0:\tau}^*$.

Several immediate novel consequences of equation (13.20) when it is applied to our framework are worth teasing out because they yield important general lessons about dissipation incurred during computation.

DISSIPATION THROUGH MODULARITY AND NEGLECTED CORRELATION

Let us say that the system is *actually* in distribution μ_0 but the controller *thinks*—or otherwise acts *as if*—the distribution is q_0.

One case in which this happens in practice is when correlations exist among parts of a memory system but computations are implemented only modularly. By implicitly marginalizing some of the memory elements, modular computing necessarily ignores the correlations among modular units.

Suppose, for example, that we partition the memory system into two composite pieces $\mathcal{S}_t = \left(\mathcal{S}_t^{(1)}, \mathcal{S}_t^{(2)}\right)$ and that the two memory subsystems are correlated: $\mu_t = \Pr(\mathcal{S}_t^{(1)}, \mathcal{S}_t^{(2)}) \neq \Pr(\mathcal{S}_t^{(1)})\Pr(\mathcal{S}_t^{(2)})$; however, the two memory subsystems are operated on independently (i.e., modularly), which means $q_t = \Pr(\mathcal{S}_t^{(1)})\Pr(\mathcal{S}_t^{(2)})$. That is, the distribution, although correlated, is operated on *as if* the two components were statistically independent. The implications are immediately accessible:

$$\beta \langle W_{\text{diss}}^{(\text{mod})} \rangle = D_{\text{KL}}(\mu_0 \| q_0) - D_{\text{KL}}(\mu_\tau \| q_\tau)$$
$$= -\Delta D_{\text{KL}}\bigl(\Pr(\mathcal{S}_t^{(1)}, \mathcal{S}_t^{(2)}) \,\|\, \Pr(\mathcal{S}_t^{(1)})\Pr(\mathcal{S}_t^{(2)})\bigr)$$
$$= \mathbf{I}\bigl(\mathcal{S}_0^{(1)}; \mathcal{S}_0^{(2)}\bigr) - \mathbf{I}\bigl(\mathcal{S}_\tau^{(1)}; \mathcal{S}_\tau^{(2)}\bigr),$$
(13.21)

where $\mathbf{I}(\cdot\,;\cdot)$ is the mutual information between its arguments. This means that work is necessarily dissipated whenever a modular computation discards information between two subsystems.[5]

When the modular computations are being performed on metastable memory states, then, assuming the memory starts and ends in a metastable distribution with microstate probabilities $\mu_t(s) = \sum_{m \in \mathcal{M}} \mu'_t(m) \pi_{xt}^{(m)}(s)$ for $t = 0$ and $t = \tau$, we find that we can formulate the result in terms of the memory states of the two subsystems:

$$\beta \langle W_{\text{diss}}^{(\text{mod})} \rangle = \mathbf{I}\bigl(\mathcal{M}_0^{(1)}; \mathcal{M}_0^{(2)}\bigr) - \mathbf{I}\bigl(\mathcal{M}_\tau^{(1)}; \mathcal{M}_\tau^{(2)}\bigr). \quad (13.22)$$

---

[5] The opposite situation (i.e., mutual information *increasing* between the two subsystems) does not happen under the current assumption of modularity, and so we are not in danger of deriving $\langle W_{\text{diss}}(\mu_0)\rangle < 0$ here, which would be counter to the second law of thermodynamics. If a computation creates correlation between two subsystems, then q_τ would not be separable, and the analysis would have proceeded differently.

Although we have arrived at this result by rather different means, equation (13.22) is essentially the same as the main result of (Boyd, Mandal, and Crutchfield 2018) (although there it was assumed that $\mathcal{M}^{(2)}$ is unchanged by the computation, which led to $\mathcal{M}_\tau^{(2)} = \mathcal{M}_0^{(2)}$, and it was also assumed there that the local-equilibrium free energies were all the same). It is notable that our result does *not* require any assumption about the local free energies of the memory subsystems—they can be arbitrarily heterogeneous.

With modular computations happening on N different subsystems, the result generalizes easily. With $\boldsymbol{\mu}_t = \Pr(\mathcal{S}_t^{(1)}, \mathcal{S}_t^{(2)}, \ldots \mathcal{S}_t^{(N)})$ and $\boldsymbol{q}_t = \prod_{n=1}^{N} \Pr(\mathcal{S}_t^{(n)})$, we find that

$$\beta \langle W_{\text{diss}}^{(\text{mod})} \rangle = -\Delta D_{\text{KL}}\left(\Pr(\mathcal{S}_t^{(1)}, \mathcal{S}_t^{(2)}, \ldots \mathcal{S}_t^{(N)}) \,\Big\|\, \prod_{n=1}^{N} \Pr(\mathcal{S}_t^{(n)})\right)$$
$$= C_{\text{tot}}(\mathcal{S}_0^{(1)}, \mathcal{S}_0^{(2)}, \ldots \mathcal{S}_0^{(N)}) - C_{\text{tot}}(\mathcal{S}_\tau^{(1)}, \mathcal{S}_\tau^{(2)}, \ldots \mathcal{S}_\tau^{(N)}), \quad (13.23)$$

where $C_{\text{tot}}(\mathcal{S}_t^{(1)}, \mathcal{S}_t^{(2)}, \ldots \mathcal{S}_t^{(N)}) = \left(\sum_{n=1}^{N} H(\mathcal{S}_t^{(n)})\right) - H(\mathcal{S}_t^{(1)}, \mathcal{S}_t^{(2)}, \ldots \mathcal{S}_t^{(N)})$ is the so-called *total correlation* among its arguments. This generalization is necessary for predicting the dissipation that will be incurred when many modular computations are performed in parallel.

Suppose instead that we want to consider the problem at the level of metastable memory states, with the joint distribution over the memory states of the subsystems $\boldsymbol{\mu}'_t = \Pr(\mathcal{M}_t^{(1)}, \mathcal{M}_t^{(2)}, \ldots \mathcal{M}_t^{(N)})$. If the memory system is assumed to start and end the computation in a classically superposed metastable distribution such that $\boldsymbol{\mu}_t = \sum_{m \in \mathcal{M}} \boldsymbol{\mu}'_t(m) \boldsymbol{\pi}_{xt}^{(m)}$ at $t = 0, \tau$, as in equation (13.3), then using $\frac{\mu_0(s)}{q_0(s)} = \frac{\mu'_0(m(s))}{q'_0(m(s))}$ and making use of $\boldsymbol{\pi}_{xt}^{(m)}(s) = \delta_{s \in m} \boldsymbol{\pi}_{xt}^{(m)}(s)$ in our calculation, regardless of any heterogeneity among the local-equilibrium free energies, we again find that

$$\begin{aligned}\beta \langle W_{\text{diss}}^{(\text{mod})}\rangle = {} & C_{\text{tot}}(\mathcal{M}_0^{(1)}, \mathcal{M}_0^{(2)}, \ldots \mathcal{M}_0^{(N)}) \\ & - C_{\text{tot}}(\mathcal{M}_\tau^{(1)}, \mathcal{M}_\tau^{(2)}, \ldots \mathcal{M}_\tau^{(N)}).\end{aligned} \quad (13.24)$$

Whether framed in terms of microstates or memory states, our general result means that *the minimal extra dissipation incurred by modular computation is exactly $k_B T$ times the reduction in total correlation among all memory subsystems.*

DISSIPATION THROUGH FAILING TO MODEL STATISTICS OF MANIPULATED MEMORY

Let us consider the implications for the common logic gates that serve as the building blocks for practical computers. Recall that the simple NAND gate is sufficient for universal computation. It is therefore worthwhile to consider what dissipation is commonly incurred in these important logic gates—and to show how this dissipation can be avoided.

It is important to note that *even without correlation*, it is critical to model the input statistics of a computation correctly to avoid dissipating work. Modeling correlation is, then, a requirement on top of this. Because we have already briefly discussed the role of correlations, let us focus here on the more basic point of modeling any input statistics whatsoever.

To address this, we explicitly consider a physical instantiation of the memory components of a NAND gate, which include two memory elements whose memory states are to be used as the input for the NAND gate and a third memory element that will store the value of the output. We assume that only the output is overwritten during the computation—the input memory states may be kept around for later use.

Note that this is already a particular physical model of the NAND computation; indeed, alternatives exist, such as storing the output in the location of one of the former inputs by overwriting one of the inputs. However, we analyze the proposed two-input–one-output model here because it is arguably the most relevant to the typical desired use of a NAND gate. Other ancillary memory elements may be used in the computation, as in the work of Owen, Kolchinsky, and Wolpert (2017), but, because they will return to their original states by the end of the computation, these ancillary memories do not need to result in any additional dissipation and so can be left implicit in the self-consistency of the current analysis.

Each of the three explicitly considered memory elements is assumed to be bistable (i.e., each of the three memory elements is assumed to have two metastable regions of state space).[6] Let the microstate of each memory element be specified by its position in the interval $(-\pi, \pi]$. Between computations, including at $t = 0$ and $t = \tau$, the metastable regions for each memory element are $\mathbf{0} \equiv (-\pi, 0]$ and $\mathbf{1} \equiv (0, \pi]$, giving a natural partition for the memory states.

The microstate of the memory system at any time t can be treated as a composite random variable $\mathcal{S}_t = (\mathcal{S}_t^{(\text{in}_1)}, \mathcal{S}_t^{(\text{in}_2)}, \mathcal{S}_t^{(\text{out})})$, with $\mathcal{S}_t^{(\cdot)} \in (-\pi, \pi]$. Similarly, the memory state is the composite random variable $\mathcal{M}_t = (\mathcal{M}_t^{(\text{in}_1)}, \mathcal{M}_t^{(\text{in}_2)}, \mathcal{M}_t^{(\text{out})})$, with $\mathcal{M}_t^{(\cdot)} \in \{\mathbf{0}, \mathbf{1}\}$ corresponding to the two metastable regions for each memory element. Thus the joint state space $\mathcal{S} = \mathbb{R}^3_{(-\pi, \pi]}$ has eight metastable regions,

---

[6] For example, each memory element may be realized physically by the bistable magnetic moment of a superparamagnetic nanocrystal in the so-called blocked regime, where the Néel relaxation time between metastable regions is much larger than the time scale of computation in the system. We can assume that there is sufficient uniaxial anisotropy (or sufficiently low temperature) to create a potential barrier many times the thermal energy between the potential wells of the two metastable regions. Although it is nice to have several realistic physical systems in mind, ultimately, the physical details of the bistable memory element will be largely irrelevant, and the analysis transcends these specifics.

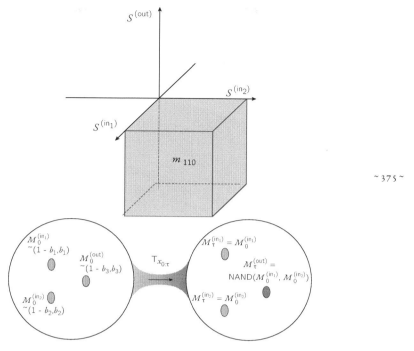

Figure 13.3. Composite state space, memory states, and physical transformation associated with the logical NAND operation.

which we identify as the joint memory system's eight memory states: $\mathcal{M} = \{m_{000}, m_{001}, m_{010}, \ldots m_{111}\}$, where each memory state is labeled according to its corresponding region of state space, $m_{jk\ell} = \{s \in \mathcal{S} : s^{(\text{in}_1)} \in (j\pi - \pi, j\pi], s^{(\text{in}_2)} \in (k\pi - \pi, k\pi], s^{(\text{out})} \in (\ell\pi - \pi, \ell\pi]\}$. That is, each of the memory states is one of the octants of the state space, as shown in figure 13.3 (top).

As a first analysis of this system, let us suppose that all memory elements are initially uncorrelated: $\Pr(\mathcal{M}_0) = \Pr(\mathcal{M}_0^{(\text{in}_1)}) \Pr(\mathcal{M}_0^{(\text{in}_2)}) \Pr(\mathcal{M}_0^{(\text{out})})$. Suppose, though, that each memory element has an initial bias such that $\Pr(\mathcal{M}_0^{(\text{in}_1)} = \mathbf{1}) = b_1$, $\Pr(\mathcal{M}_0^{(\text{in}_2)} = \mathbf{1}) = b_2$, and $\Pr(\mathcal{M}_0^{(\text{out})} = \mathbf{1}) = b_3$. Let us call this distribution over memory states $\boldsymbol{\mu}'_0 = \Pr(\mathcal{M}_0)$. The corresponding initial distribution over microstates is $\Pr(\mathcal{S}_0) = \sum_{m \in \mathcal{M}} \Pr(\mathcal{M}_0 = m)\, \delta_{\mathcal{S}_0 \in m}\, \pi_{x_0}^{(m)}$, which we will call $\boldsymbol{\mu}_0$. This physical setup, and the logical transformation of the output bit, is diagrammed in figure 13.3 (bottom).

Suppose that a NAND gate is constructed such that it does not dissipate any work when it transforms the memory system in its equilibrium state: $\boldsymbol{q}_0 = \sum_{m \in \mathcal{M}} \delta_{\mathcal{S}_0 \in m}\, \pi_{x_0}^{(m)}/8$. That is, the transformation is designed to dissipate no work when the memory states are all initialized in the uniform distribution $\boldsymbol{q}_0' = \frac{1}{8} \begin{bmatrix} 1 & 1 & \ldots & 1 \end{bmatrix}$. The minimal extra dissipation incurred by using this NAND transformation optimized for minimal dissipation in the case of uniform distribution over memory states, given that the initial statistics of the memory elements are actually biased by the b_i, is

$$\begin{aligned}
\beta \left\langle W_{\text{diss}}^{(\text{mismatch})} \right\rangle &= D_{\text{KL}}(\boldsymbol{\mu}_0 \| \boldsymbol{q}_0) - D_{\text{KL}}(\boldsymbol{\mu}_\tau \| \boldsymbol{q}_\tau) \\
&= D_{\text{KL}}(\boldsymbol{\mu}_0' \| \boldsymbol{q}_0') - D_{\text{KL}}(\boldsymbol{\mu}_\tau' \| \boldsymbol{q}_\tau') \\
&= \sum_{m \in \mathcal{M}} \boldsymbol{\mu}_0'(m) \ln\left(\frac{\boldsymbol{\mu}_0'(m)}{1/8}\right) - \boldsymbol{\mu}_\tau'(m) \ln\left(\frac{\boldsymbol{\mu}_\tau'(m)}{1/4}\right) \quad (13.25) \\
&= \ln 8 - H(\boldsymbol{\mu}_0') - \ln 4 + H(\boldsymbol{\mu}_\tau') \\
&= \ln 2 - \boldsymbol{H}_2(b_3),
\end{aligned}$$

where $\boldsymbol{H}_2(b) \equiv -b \ln b - (1-b) \ln(1-b)$.

We reflect that the full dissipation of operating the NAND gate (when not optimized for the correct memory biases) is essentially the entropy production from ignoring the single bias of b_3 associated with the output. The protocol *could* have been designed to be dissipation free, but the current NAND implementation does not erase $\mathcal{M}_0^{(\text{out})}$ in a way that salvages its original nonequilibrium addition to free energy.

More generally, if we allow any kind of initial correlation within the initial configuration, such that the input and output bits *are* initially correlated according to $\boldsymbol{\mu}_0' = \Pr(\mathcal{M}_0) \neq$

$\Pr(\mathcal{M}_0^{(\text{in}_1)}) \Pr(\mathcal{M}_0^{(\text{in}_2)}) \Pr(\mathcal{M}_0^{(\text{out})})$, then the resulting dissipation generalizes to

$$\left\langle W_{\text{diss}}^{(\text{mismatch})} \right\rangle = k_B T \ln 2 - k_B T H_{\mu_0'}\left(\mathcal{M}_0^{(\text{out})} \mid \mathcal{M}_0^{(\text{in}_1)}, \mathcal{M}_0^{(\text{in}_2)}\right), \tag{13.26}$$

where the right-most term is the conditional Shannon entropy of the *initialized* value of the output bit (i.e., before the NAND operation is implemented), given the initial values of the input.

Again, this $\beta \left\langle W_{\text{diss}}^{(\text{mismatch})} \right\rangle$ turns out to be the irreversible entropy production of ignoring the nonequilibrium distribution of the original output bit. A smarter protocol could have instead leveraged this ordered nonequilibrium addition to free energy to perform the NAND computation with less work, but, because the protocol was not altered to take advantage of this initial nonequilibrium situation, the thermodynamic resource is forever lost through dissipation by the end of the computation.

DISSIPATION AND MINIMAL WORK FOR ANY TWO-INPUT–ONE-OUTPUT LOGIC GATE

Having gone through this analysis, it should be clear that the NAND function played no essential role in determining the minimal dissipation from neglected initial biases, other than the fact that the NAND operation is a deterministic two-input–one-output function. Hence equation (13.26) describes the minimal dissipation of *all* two-input–one-output logic gates—NAND, AND, XOR, and so on—when the output memory element is overwritten by a computation that does not leverage the initial biases of the memory elements it is manipulating.

We have focused herein on the work *dissipated*, because this is the undesirable waste of which designers of future hyperefficient computers should be hyperaware. It is noteworthy, though, that, even when no work is dissipated, the minimal work to implement the computation will also depend on the initial biases of the memory elements that are to be manipulated. From equation (13.17), we see that the minimal work necessary to implement the NAND gate—indeed, to implement *any* two-input–one-output gate where the output memory element is to be overwritten—is

$$\langle W \rangle_{\min} = k_{\rm B} T H_{\mu'_0}(\mathcal{M}_0^{(\text{in}_1)}, \mathcal{M}_0^{(\text{in}_2)}, \mathcal{M}_0^{(\text{out})})$$
$$\qquad - k_{\rm B} T H_{\mu'_\tau}(\mathcal{M}_0^{(\text{in}_1)}, \mathcal{M}_0^{(\text{in}_2)}, \mathcal{M}_\tau^{(\text{out})}) \quad (13.27)$$
$$= k_{\rm B} T H_{\mu'_0}(\mathcal{M}_0^{(\text{out})} | \mathcal{M}_0^{(\text{in}_1)}, \mathcal{M}_0^{(\text{in}_2)}).$$

In the case of biased but uncorrelated initial inputs and outputs, this reduces to $\langle W \rangle_{\min} = k_{\rm B} T \boldsymbol{H}_2(b_3)$. However, as long as this work is not dissipated, it can continue to be salvaged and recycled in future computations.

Looking back at $\langle W_{\text{diss}} \rangle$, we see that our result for the minimum work puts the dissipated work in a new context. In particular,

$$\left\langle W_{\text{diss}}^{(\text{mismatch})} \right\rangle = k_{\rm B} T \ln 2 - \langle W \rangle_{\min}. \quad (13.28)$$

We interpret this as the minimum work that would need to be performed given the uniform distribution of initial memory states for which the system was designed minus the minimum work given the actual biases. In the case of biased but uncorrelated initial memory states, this can be framed as

$$\left\langle W_{\text{diss}}^{(\text{mismatch})} \right\rangle = k_{\rm B} T \boldsymbol{H}_2(\tfrac{1}{2}) - k_{\rm B} T \boldsymbol{H}_2(b_3). \quad (13.29)$$

When all memory states have the same local-equilibrium free energy, biases in the input should be treated as a resource—a nonequilibrium addition to free energy. To ignore these biases is to waste this resource, resulting in unnecessary dissipation.

Onward

The results of our analysis highlight the fundamental thermodynamic limits of conventional computation—a limit we are steadily approaching but from which we are still quite far. Constructively, our analysis also points to ways *around* these limits for future hyperefficient computers. First, these hypothetical future computers could have implementations that adapt to input biases and thus eliminate needless dissipation within each modular computation. Second, we propose that future hyperefficient computers compute common composite routines in a single global transformation to reduce modular dissipation.

REFERENCES

Åberg, J. 2013. "Truly Work-like Work Extraction via a Single-Shot Analysis." *Nature Communications* 4:1925.

Bonança, M. V. S., and S. Deffner. 2014. "Optimal Driving of Isothermal Processes Close to Equilibrium." *Journal of Chemical Physics* 140 (24): 244119.

Boyd, A. B., and J. P. Crutchfield. 2016. "Maxwell Demon Dynamics: Deterministic Chaos, the Szilard Map, and the Intelligence of Thermodynamic Systems." *Physical Review Letters* 116 (19): 190601.

Boyd, A. B., D. Mandal, and J. P. Crutchfield. 2017. "Leveraging Environmental Correlations: The Thermodynamics of Requisite Variety." *Journal of Statistical Physics* 167, no. 6 (June): 1555–1585.

———. 2018. "Thermodynamics of Modularity: Structural Costs Beyond the Landauer Bound." *Phys. Rev. X* 8 (3): 031036.

Boyd, A. B., D. Mandal, P. M. Riechers, and J. P. Crutchfield. 2017. "Transient Dissipation and Structural Costs of Physical Information Transduction." *Physical Review Letters* 118 (22): 220602.

Esposito, M. 2012. "Stochastic Thermodynamics under Coarse Graining." *Physical Review E* 85 (4): 041125.

Esposito, M., and C. Van den Broeck. 2011. "Second Law and Landauer Principle Far from Equilibrium." *Europhysics Letters* 95 (4): 40004.

Garner, A. J. P., J. Thompson, V. Vedral, and M. Gu. 2017. "Thermodynamics of Complexity and Pattern Manipulation." *Physical Review E* 95 (4): 042140.

Gokler, C., A. Kolchinsky, Z. Liu, I. Marvian, P. Shor, O. Shtanko, K. Thompson, D. Wolpert, and S. Lloyd. 2017. "When Is a Bit Worth Much More than kT ln2?" arXiv:1705.09598.

Kolchinsky, A., and D. H. Wolpert. 2017. "Dependence of Dissipation on the Initial Distribution over States." *Journal of Statistical Mechanics: Theory and Experiment* 2017 (8): 083202.

Lloyd, S. 1989. "Use of Mutual Information to Decrease Entropy: Implications for the Second Law of Thermodynamics." *Physical Review A* 39 (10): 5378.

Mandal, D., and C. Jarzynski. 2016. "Analysis of Slow Transitions between Nonequilibrium Steady States." *Journal of Statistical Mechanics: Theory and Experiment* 2016 (6): 063204.

Owen, J. A., A. Kolchinsky, and D. H. Wolpert. 2017. "Number of Hidden States Needed to Physically Implement a Given Conditional Distribution." arXiv:1709.00765.

Parrondo, J. M. R., J. M. Horowitz, and T. Sagawa. 2015. "Thermodynamics of Information." *Nature Physics* 11 (2): 131.

Riechers, P. M., and J. P. Crutchfield. 2017. "Fluctuations When Driving between Nonequilibrium Steady States." *Journal of Statistical Physics* 168, no. 4 (August): 873–918.

Salamon, P., and R. S. Berry. 1983. "Thermodynamic Length and Dissipated Availability." *Physical Review Letters* 51 (13): 1127–1130.

Sivak, D. A., and G. E. Crooks. 2012. "Thermodynamic Metrics and Optimal Paths." *Physical Review Letters* 108 (19): 190602.

Takara, K., H.-H. Hasegawa, and D. J. Driebe. 2010. "Generalization of the Second Law for a Transition between Nonequilibrium States." *Physics Letters, Series A* 375 (2): 88–92.

Zulkowski, P. R., and M. R. DeWeese. 2014. "Optimal Finite-Time Erasure of a Classical Bit." *Physical Review E* 89 (5): 052140.

PHYSICAL LIMITATIONS OF WORK EXTRACTION FROM TEMPORAL CORRELATIONS

Elan Stopnitzky, University of Hawaiʻi
Susanne Still, University of Hawaiʻi
Thomas E. Ouldridge, Imperial College London
Lee Altenberg, University of Hawaiʻi

Introduction

Leo Szilard proposed a simple *Gedankenexperiment* almost ninety years ago to resolve the paradox of Maxwell's demon, arguing that information about a system could be converted to work by an automated mechanism in place of a sentient being (Szilard 1929). Szilard's proposed information engine cyclically repeats two distinct phases: first, acquiring information and recording it into a stable memory, and then using this information to extract work with a given mechanism. This allowed him to compute a bound on the costs associated with acquiring and recording information, necessary to prevent a violation of the Second Law (Szilard 1929). Many extensions to Szilard's engine have been explored in the literature (e.g., Zurek 1986; Jeon and Kim 2016; Marathe and Parrondo 2010; Kim *et al*. 2011; Vaikuntanathan and Jarzynski 2011), and modern formulations of nonequilibrium thermodynamics naturally incorporate correlations as a potential source of work (see, e.g., Esposito and Van den Broeck 2011).

Much recent interest has nonetheless focused on the information utilization side by building on the idea of a device that exploits a data-carrying tape to extract work from a single heat bath (Mandal and Jarzynski 2012). Such a device advances along a sequence of 0s and 1s that contains an overall bias toward either 0 or 1. The device couples to one input bit at a time and, while in contact with the bit, undergoes free-running dynamics that can alter the bit. This interaction increases the entropy of the tape upon output of the changed bit, and it is this entropy increase that is used to compensate for the entropy decrease of the heat bath. Extensions of these machines exploit statistical information within the tape in the form of temporal correlations (Boyd, Mandal, and Crutchfield 2017), or spatial correlations between tapes (McGrath *et al.* 2017), rather than an overall bias in the input bits. The resulting simple dynamical models of all these proposals help develop a concrete physical understanding of the role information plays in thermodynamics. To serve this purpose, it is important that these devices be physically realizable.

Real physical systems have underlying time-continuous dynamics. Moreover, whenever the work extraction device is designed to operate without a time-dependent, externally applied driving protocol during the periods of interaction with an individual bit, the time-continuous dynamics must also be time homogeneous and obey detailed balance to be physical.

We explore here how this fact constrains possible designs of the class of *temporal* correlation–powered devices proposed by Boyd, Mandal, and Crutchfield (2017, and some references therein). We find that demanding underlying time-continuous, time-homogeneous dynamics drastically limits the set of allowable transition matrices, thereby dramatically reducing the resulting efficiency (see the section "Work Extraction by Time-Continuous, Free-Running Devices"). This fact is demonstrated first through the relative performance of randomly generated transition matrices, and second via an evolutionary algorithm; in both cases, we compare the situation with and without the constraint. We explain the difference in performance by the mathematical properties of the relevant matrices and their associated physical implications. Finally, we show that these limitations disappear when the restriction to time-homogeneous dynamics is lifted and transition rates are modulated by external manipulation (see the section "Time-Inhomogeneous Protocols").

Model of a Temporal Correlation Powered Work Extraction Device

This model largely follows the work of Boyd, Mandal, and Crutchfield (2017). Imagine a work extraction system with two internal states, $s \in \{A, B\}$, which can be coupled to and decoupled from a work reservoir (such as a weight), an input tape with bits $b^{in} \in \{0, 1\}$, an output tape with $b^{out} \in \{0, 1\}$, and a heat bath. The joint state-input value of the coupled system is then $(s, b) \in \{A, B\} \times \{0, 1\} = \{A0, A1, B0, B1\}$, where b denotes a coupled bit. Each of these four joint states possesses a potential energy, E_i, $i \in \{A0, A1, B0, B1\}$. The dynamics are described by a time-dependent vector containing the probabilities that the system is in one of the joint states at a given time: $\mathbf{p}_{sb} \equiv [p_{A0}, p_{A1}, p_{B0}, p_{B1}]^\top$ ($^\top$ denotes the transpose).

The engine alternates between an "interaction step," during which the internal state interacts with a bit, and a "switching step," during which the bit is changed. Any changes in energy during an interaction step are due to the exchange of heat with the heat bath, and changes in energy during a switching step are due to exchange of work with the work reservoir. We therefore talk about "heat" steps and "work" steps and will use these words to label transformations, as a reminder.

An interaction step is represented by the transformation $\mathbf{p}_{sb} \xrightarrow{\text{heat}} \mathbf{M}\mathbf{p}_{sb}$, where the joint state evolves under the action of a reversible, column stochastic matrix, $\mathbf{M}$, for a duration of time, τ. During interaction steps, the system undergoes a free-running relaxation toward equilibrium. Thus $\mathbf{M}$ must be reversible, and the stationary distribution, $\boldsymbol{\rho}^{(\mathbf{M})} = \mathbf{M}\boldsymbol{\rho}^{(\mathbf{M})}$, satisfies detailed balance

$$M_{ij}\rho_j^{(\mathbf{M})} = M_{ji}\rho_i^{(\mathbf{M})} \quad \text{for all } i, j \in \{A0, A1, B0, B1\}. \quad (14.1)$$

The coupled device and bit relax toward thermodynamic equilibrium, described by the Boltzmann distribution $\rho_i^{(\mathbf{M})} = e^{-E_i/kT}/Z$. For simplicity, we choose units such that $kT = 1$ and set the energy scale so that $Z = 1$. We can then write the energy of each joint state as $E_i = -\ln \rho_i^{(\mathbf{M})}$, with $i \in \{A0, A1, B0, B1\}$.

In a switching step, the internal state is held fixed, and a new bit is coupled to the device. Whichever bit composed the joint state prior to the switching step is printed to the output tape; that is, switching from the machine's nth cycle to cycle $n+1$ changes the state of the bit that is interacting with the machine from $b = b_n^{\text{out}}$ to $b = b_{n+1}^{\text{in}}$. Depending on whether the input is 0 or 1, switching corresponds to the transformation of the joint state probability vector

$$\begin{aligned}\mathbf{p}_{sb} &= [p_{A0}, p_{A1}, p_{B0}, p_{B1}]^\top \xrightarrow{\text{input 0}} \\ \bar{\mathbf{p}}_{sb} &= [p_{A0}+p_{A1}, 0, p_{B0}+p_{B1}, 0]^\top \equiv \mathbf{F}_0 \mathbf{p}_{sb}\end{aligned} \quad (14.2)$$

$$\begin{aligned}\mathbf{p}_{sb} &= [p_{A0}, p_{A1}, p_{B0}, p_{B1}]^\top \xrightarrow{\text{input 1}} \\ \bar{\mathbf{p}}_{sb} &= [0, p_{A0}+p_{A1}, 0, p_{B0}+p_{B1}]^\top \equiv \mathbf{F}_1 \mathbf{p}_{sb},\end{aligned} \quad (14.3)$$

where the matrices $\mathbf{F}_0$ and $\mathbf{F}_1$ represent the switching:

$$\mathbf{F}_0 = \begin{bmatrix} 1 & 1 & 0 & 0 \\ 0 & 0 & 0 & 0 \\ 0 & 0 & 1 & 1 \\ 0 & 0 & 0 & 0 \end{bmatrix} \quad \text{and} \quad \mathbf{F}_1 = \begin{bmatrix} 0 & 0 & 0 & 0 \\ 1 & 1 & 0 & 0 \\ 0 & 0 & 0 & 0 \\ 0 & 0 & 1 & 1 \end{bmatrix}.$$

Chapter 14: Physical Limitations of Work Extraction

To extract work, the machine must on average raise the energy of the joint (s, b)-state during interaction steps (i.e., absorb heat) and lower the energy during switching steps (i.e., deposit energy into the work reservoir).

In the following, for simplicity, we limit our analysis to an input tape consisting of alternating 1s and 0s. This is an interesting special case, because the per-symbol entropy of the input tape is maximal, as $\text{Prob}(b^{\text{in}} = 0) = \text{Prob}(b^{\text{in}} = 1) = \frac{1}{2}$, and hence cannot be leveraged for work extraction. Any net gain is thus due to exploiting temporal correlations.

A single complete cycle of operation is defined by the product of transition matrices $\mathbf{C} = \mathbf{MF_0MF_1}$ (reflecting the alternating switching and interaction steps), taking the probability distribution of the four states from $\mathbf{p}_{sb}^{(n)}$ to $\mathbf{p}_{sb}^{(n+2)} = \mathbf{Cp}_{sb}^{(n)}$. We require that repeated application of the matrix $\mathbf{C}$ to any starting distribution $\mathbf{p}_{sb}^{(0)}$ converges to a steady-state distribution $\pi_0(s, b)$, defined by $\mathbf{C}\pi_0(s, b) = \pi_0(s, b)$. Thus $\mathbf{C}$ must be a primitive matrix (irreducible and aperiodic), which is assured if $M_{ij} > 0$ for all $i, j \in \{A0, A1, B0, B1\}$. We can define a steady-state distribution if we census the system at each step of the cycle (π_1 to π_3 in the following equation array). Starting with feeding in a 1, a cycle is then characterized by the following changes (we use b' to denote a bit that is about to be transferred to the output tape):

$$\pi_0(s_{n-1}, b'_{n-1}) \xrightarrow{\text{work}} \pi_1(s_{n-1}, b_n) = \mathbf{F}_1\pi_0(s_{n-1}, b'_{n-1}), \quad (14.4)$$

$$\pi_1(s_{n-1}, b_n) \xrightarrow{\text{heat}} \pi_2(s_n, b'_n) = \mathbf{M}\pi_1(s_{n-1}, b_n), \quad (14.5)$$

$$\pi_2(s_n, b'_n) \xrightarrow{\text{work}} \pi_3(s_n, b_{n+1}) = \mathbf{F}_0\pi_2(s_n, b'_n), \quad (14.6)$$

$$\pi_3(s_n, b_{n+1}) \xrightarrow{\text{heat}} \pi_0(s_{n+1}, b'_{n+1}) = \mathbf{M}\pi_3(s_n, b_{n+1}). \quad (14.7)$$

The average work supplied to the work reservoir per input symbol is given by the sum of the average energy changes in the two switching steps: $\langle W \rangle = -\frac{1}{2}\big[\langle E \rangle_{\pi_0(s_{n-1}, b'_{n-1})} - \langle E \rangle_{\pi_1(s_{n-1}, b_n)} +$

$\langle E \rangle_{\pi_2(s_n,b'_n)} - \langle E \rangle_{\pi_3(s_n,b_{n+1})}$]. The factor of $\frac{1}{2}$ is due to two bits being encountered per cycle.[1]

Work Extraction by Time-Continuous, Free-Running Devices

We now depart from the approach of Boyd, Mandal, and Crutchfield (2017). To be physically realizable in the absence of externally applied driving during the interaction period, the matrix **M** should correspond with a continuous-time equilibration process for some time τ. In other words, we require there be a generator, **G**, such that $\mathbf{M} = e^{\tau \mathbf{G}}$, where **G** is a reversible rate matrix with nonnegative off-diagonal elements, in which every column sums to 0. If **M** can be constructed in this way, then **M** is said to be "reversibly embeddable" (Jia 2016). The following results from Jia (2016) are crucial: if **M** is reversible, then **M** is diagonalizable and the eigenvalues of **M** are real; if **M** is also embeddable, then the eigenvalues of **M** are all positive, and the generator, **G**, of **M** is unique.

To improve upon random sampling, we constructed an evolutionary algorithm to explore the search space. The algorithm applied mutations to individual machine dynamics and fixed those mutations whenever they led to improved performance. Owing to the high dimension of the space of transition matrices, this algorithm performed better than a grid search. When ignoring the embeddability constraint, the evolutionary algorithm readily returned the best design of Boyd, Mandal, and Crutchfield (2017) and never found one better. However, enforcing embeddability lowered the efficiency drastically. The best embeddable design that the evolutionary algorithm found is shown in figure 14.3. It

---

[1] Note that this is work *extracted* from the joint (s, b) system. By convention, work done on the system is positive, as is heat flowing into the system.

Chapter 14: Physical Limitations of Work Extraction

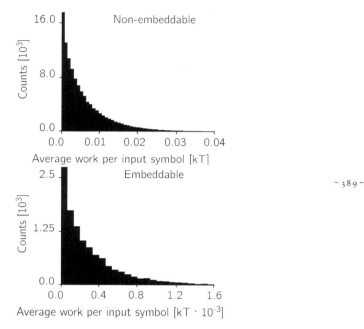

Figure 14.1. Histogram of work, W_{out}, extracted by randomly generated reversible transition matrices. Only positive work values are shown. With 10^6 randomly generated matrices of each type, 11% of nonembeddable matrices and 0.3% of embeddable matrices achieved positive work production.

achieved only $\approx 0.0146\ kT$ per input symbol. That is less than 4% of the yield of the nonembeddable best design of Boyd, Mandal, and Crutchfield (2017).

Owing to these properties, processes governed by reversible and embeddable transition matrices generally extract much less work than those governed by reversible but not embeddable ones, as we will see shortly. In the rest of this chapter, we will only be considering matrices that are reversible. For brevity, we henceforth use the terms *embeddable* and *nonembeddable* to refer to the two different classes of matrices. In figure 14.1, we display histograms showing, for the two categories, the number of randomly generated matrices that achieve various values of positive work. The procedure used to make these

THE ENERGETICS OF COMPUTING IN LIFE & MACHINES

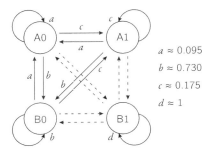

Figure 14.2. Design of the best embeddable transition matrix found by the evolutionary algorithm, with $W_{\text{out}} \approx 0.0146\ kT$. Dotted arrows denote transition probabilities close to zero. The transition matrix for this graph has a rank 1 submatrix, a second eigenvalue very close to 1, and the two smallest eigenvalues very close to 0.

matrices is detailed in the appendix. The best randomly found embeddable matrices extract roughly a factor twenty less work than the best randomly found nonembeddable ones. In comparison, the construction given by Boyd, Mandal, and Crutchfield (2017, fig. 6), which is nonembeddable, extracts $\frac{kT}{\epsilon} \approx 0.368\ kT$ of work per input symbol. We see from the histogram that finding a machine with a comparable efficiency by chance is rather unlikely.

We now discuss why the performance of embeddable designs is so poor. Optimal performance requires that the internal state of the device, s_n, contain predictive information,[2] $I(s_n; b_{n+1})$, about the next incoming bit, b_{n+1}. To see why, note that the nonequilibrium free energy of system coupled to bit, $F[\pi] = \langle E \rangle_\pi - kTH[\pi]$, cannot increase spontaneously during an interaction step. Thereby, the heat absorbed in one interaction step is upper bounded by the entropy change, which can be written as $\langle Q_n \rangle \leq kT\left[I(s_n; b_{n+1}) - I(s_{n+1}; b'_{n+1}) + H(s_{n+1}) - H(s_n) + H(b'_{n+1}) - H(b_{n+1})\right]$. Adding two of these heat contributions to account

---

[2] We detail in the appendix how to calculate information.

Chapter 14: Physical Limitations of Work Extraction

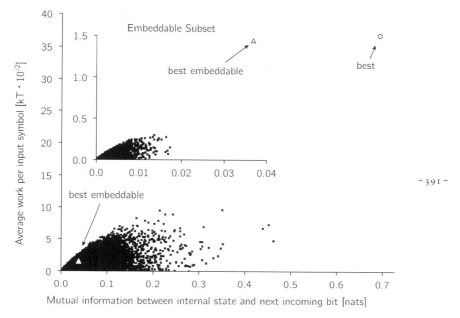

Figure 14.3. Average work per input symbol, W_{out}, versus average predictive information per symbol. The triangle denotes the best embeddable design found by the evolutionary algorithm. The circle denotes the design of Boyd, Mandal, and Crutchfield (2017). The inset shows only the subset of embeddable designs. See the appendix for further details.

for the full cycle of equations (14.4)–(14.7), we get a cancellation, because $H(s_{n+1}) - H(s_{n-1}) = 0$, owing to the fact that we are in the same stationary distribution π_0 at the beginning and end of the cycle.

Taking the average then sets an upper bound on the extractable work per input symbol, W_{out}. The bound depends on how the average predictive information about the input, $I_{\text{pred}} = (I(s_{n-1}; b_n) + I(s_n; b_{n+1}))/2$, compares to the average memory about the output, $I_{\text{mem}} = (I(s_n; b'_n) + I(s_{n+1}; b'_{n+1}))/2$, and on how the average output entropy, $H_B^{\text{out}} = (H(b'_n) + H(b'_{n+1}))/2$, compares to the average input entropy, $H_B^{\text{in}} = (H(b_n) + H(b_{n+1}))/2$:

$$W_{\text{out}} \leq kT\left[I_{\text{pred}} - I_{\text{mem}} + \Delta H_B\right], \qquad (14.8)$$

where $\Delta H_B \equiv H_B^{\text{out}} - H_B^{\text{in}}$. We display W_{out} as a function of I_{pred} in figure 14.3 for each of the two classes of transition matrices. It is clear from the plot that predictive information between device and next incoming bit is severely limited for embeddable systems and that there is a consequent reduction in extractable work.

Perfect prediction of the next incoming bit requires synchronization with the input. For a period 2 input, the internal state must change in each interaction step from A to B, or vice versa, necessitating a bipartite graph structure whose associated transition matrix has negative eigenvalues (Gallager 2012) and therefore is nonembeddable. Tracking a periodic signal with period greater than 2 would require complex eigenvalues (Gallager 2012) and is therefore impossible for any reversible matrix, embeddable or otherwise.

Synchronization is hampered by the tendency of embeddable systems to undergo "self-transitions," in which the system starts and ends in the *same* state during an interaction interval. These self-transitions are also undesirable because they are associated with no net exchange of energy with the bath, thus wasting the input. Self-transitions arise from the diagonal entries in **M**, which can be set to zero for nonembeddable M (Boyd, Mandal, and Crutchfield 2017), but not if M is embeddable. To see why, note that **M** being a stochastic matrix implies that it has an eigenvalue of 1, and M being embeddable implies that all other eigenvalues are positive. Thus the trace of M must be greater than 1, and self-transitions cannot be neglected. The average fraction of such self-transitions in interaction steps can be written as

$$\frac{1}{2}\mathbf{d}(\mathbf{M})^\top [\boldsymbol{\pi}(s_n, b_n) + \boldsymbol{\pi}(s_{n+1}, b_{n+1})], \qquad (14.9)$$

where $\mathbf{d}(\mathbf{M})$ is a vector of the diagonal elements of M. We have found numerically that self-transitions occur on

Chapter 14: Physical Limitations of Work Extraction

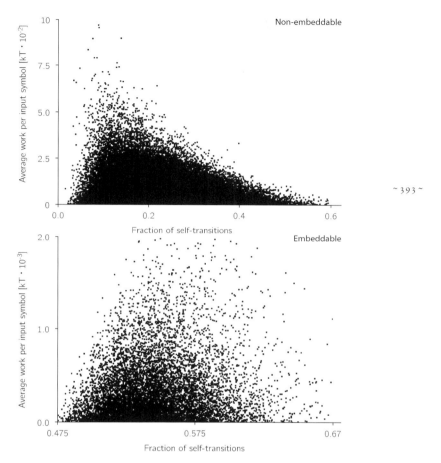

Figure 14.4. The relationship between W_{out} and the fraction of self-transitions.

average at a minimum of one-fourth of interaction steps. We provide in the appendix an example of a matrix satisfying this property. Note that this feature implies that no embeddable matrices exist "close" to the nonembeddable design of Boyd, Mandal, and Crutchfield (2017). The relationship between self-transition rate and average work extracted for randomly chosen embeddable and nonembeddable designs is shown in figure 14.4.

We display only those designs that lead to positive work extraction. There are fewer points in the right panel because a smaller fraction of embeddable matrices lead to positive average work. Because the nonembeddable matrices can be made to have a trace of 0, their rate of self-transitions can also be set to 0, while, for the embeddable matrices, a minimum trace of 1 forces the rate of self-transitions to be at least one-fourth, leading to inefficiency. Moreover, approaching the minimal trace of 1 and the corresponding minimal self-transition rate of one-fourth is not a viable strategy for maximizing efficiency, because it would mean that all the smaller eigenvalues would have to approach 0. This requirement is in tension with the fact that the modulus of the second largest eigenvalue, $\lambda_2 < 1$, bounds the distance from equilibrium at the end of an interaction step through $\mathbf{Mp} = \boldsymbol{\rho}^{(M)} + \lambda_2 c_2 \mathbf{v}_2 + \lambda_3 c_3 \mathbf{v}_3 + \lambda_4 c_4 \mathbf{v}_4$, where $\mathbf{p} = \boldsymbol{\rho}^{(M)} + c_2 \mathbf{v}_2 + c_3 \mathbf{v}_3 + c_4 \mathbf{v}_4$ is an arbitrary starting distribution expanded in the basis of $\mathbf{M}$'s eigenvectors, $\boldsymbol{\rho}^{(M)}, \mathbf{v}_2, \mathbf{v}_3, \mathbf{v}_4$. Thus taking the limit as the trace approaches 1 would cause instantaneous relaxation to equilibrium in every interaction step, which prevents work from being extracted.

Intuitively, the device cannot extract work if it relaxes fully to equilibrium in each interaction step, because this would prevent it from retaining memory. More formally, let E_{eq} denote the energy of the equilibrium distribution. With complete relaxation to equilibrium in each interaction step, the sum of the energy changes over the two switching steps is then $\frac{1}{Z}[E_{A1}e^{-E_{A0}} + E_{A0}e^{-E_{A1}} + E_{B0}e^{-E_{B1}} + E_{B1}e^{-E_{B0}}] - E_{eq} \geq 0$. This quantity is nonnegative because the equilibrium distribution pairs the highest energies with the smallest Boltzmann factors, so any reordering of the factors cannot decrease the average energy.

Altogether, there exists a trade-off between the inefficiency resulting from excessively frequent stasis and the inefficiency of relaxing too close to equilibrium. This trade-off would be less severe if more internal states were included, because, with a transition matrix of higher dimension, the trace could be kept relatively small, even with a large second eigenvalue. However, adding additional internal states would not necessarily guarantee a substantial improvement on the overall performance, because the prohibition on bipartite graphs prevents synchronization with the input—an issue that persists.

Much of the inefficiency of embeddable designs arises from an inability to track the input reliably by switching the device state at each step, but embeddable designs also suffer from a second drawback that would limit work extraction even if the input were a pure string of 1s. To extract work, it is vital that the energy tend to increase during the interaction window. The work extracted is equal to the average number of transitions during these windows multiplied by the increase in energy per transition. However, if we increase the energy of the high-energy states to which we hope the system will transition, then we decrease the net number of transitions, because we decrease their occupancy in equilibrium—and embeddable designs can only relax toward equilibrium during the interaction window. Thus there is an unavoidable trade-off for embeddable designs between facilitating many transitions that each contribute only a little to the work extracted and allowing only a few that contribute a large amount. The overall power is maximized at intermediate values (see, e.g., Mandal and Jarzynski 2012).

Time-Inhomogeneous Protocols

We have shown that devices that are free running during the interaction period, which are restricted to reversibly embeddable, time-homogeneous dynamics, can extract only a small fraction of the work available from an alternating input of 0s and 1s. One might ask whether devices connected to a time-dependent, externally applied protocol during the interaction period, resulting in time-inhomogeneous dynamics that need not satisfy detailed balance, could perform better. In this setting, for example, it can be ensured that the device's state *must* change during a cycle, allowing a better match to the input's periodicity.

Here we show that it is relatively straightforward to construct a device of this type that extracts, in the quasistatic limit, all the work stored in an input tape of alternating 1s and 0s. External manipulations correspond to changing the energy levels of the system over time (Van den Broeck *et al.* 2013) and allow the input and/or extraction of work during the interaction period of duration τ. For our purposes, it is sufficient to consider only devices in which the energies of the joint states (s, b) vary during τ but are restricted to all being *identical* at the beginning and end of each window. In this case, the work of switching to the next input bit on the tape is zero, and only the window τ need be considered to compute the extracted work.

To demonstrate an optimal device, let us compose it of two standard operations: erasure and relaxation. Let a given pair of states within a larger state space each have an occupation probability of $p/2$. Erasure shifts all of this probability to just one state, leaving it occupied with probability p, and the other with a probability 0. Famously, erasure can in principle be performed at a work cost of $pk_B T \ln 2$ (Landauer 1961; Wolpert, Kolchinsky, and Owen 2017). Relaxation is the inverse of erasure, and therefore work of $pk_B T \ln 2$ can be extracted. In both cases, these optimal limits on the work are reached by *thermodynamically* reversible processes, in which manipulations must be applied quasistatically and the reversal of the protocol would restore the initial probability distribution. We also consider switching, or the transfer of probability from one state to another that has zero initial probability. Switching can be decomposed as a relaxation followed by an erasure and therefore has no total work requirement if performed in a thermodynamically reversible manner.

Let us consider the following transition matrix for the interaction step:

$$\mathbf{T} = \begin{bmatrix} 0 & 1/2 & 0 & 1/2 \\ 0 & 1/2 & 0 & 1/2 \\ 1/2 & 0 & 1/2 & 0 \\ 1/2 & 0 & 1/2 & 0 \end{bmatrix}, \qquad (14.10)$$

where $\mathbf{T}\mathbf{p}_{sb}$ gives the evolution of $\mathbf{p}_{sb}$ during a single interaction window. This transition matrix ensures that the machine transitions to the A state if the input bit is in state 1 and to the B state if the input bit is in state 0. This oscillation is the central switching motif that allows the device to track the input. $\mathbf{T}$ is produced by composition of the following sequential operations: a switch from $(A, 0)$ to $(B, 0)$; a switch from $(B, 1)$ to $(A, 1)$; a relaxation from $(B, 0)$ to $(B, 0)$ or $(B, 1)$; and, finally, a relaxation from $(A, 1)$ to $(A, 0)$ or $(A, 1)$. The states $(A, 1)$ and $(B, 0)$ are effectively used as ancillary states (Wolpert, Kolchinsky, and Owen 2017) to facilitate the necessary reversing of the machine's state at each step, prior to relaxation.

Regardless of the initial condition, a single application of $\mathbf{F}_1 \mathbf{T} \mathbf{F}_0 \mathbf{T}$ will bring the system to

$$\hat{\mathbf{p}}_{sb} = \begin{bmatrix} p_{A0}, p_{A1}, p_{B0}, p_{B1} \end{bmatrix}^\top = \begin{bmatrix} 0, 0, 0, 1 \end{bmatrix}^\top, \qquad (14.11)$$

in which the states of the device and tape are perfectly coordinated. Because $\hat{\mathbf{p}}_{sb}$ is an eigenvector of $\mathbf{F}_1\mathbf{T}\mathbf{F}_0\mathbf{T}$ with eigenvalue 1, subsequent applications of $\mathbf{F}_1\mathbf{T}\mathbf{F}_0\mathbf{T}$ will also return $\hat{\mathbf{p}}_{sb}$. Given the initial condition of $\hat{\mathbf{p}}_{sb}$, the switch and relaxation procedures underlying T can be implemented in a thermodynamically reversible manner, yielding $k_B T \ln 2$ of work per T operation (or $2k_B T \ln 2$ per full cycle) owing to the relaxation steps. Thus the device extracts all of the available work after an initial alignment cycle.

Discussion and Conclusion

We have learned that building physically realistic devices that exploit temporal correlations with a well-defined period to extract work from a heat bath at high efficiency can be challenging. Specifically, devices with time-continuous dynamics cannot extract much work from an alternating input of 0s and 1s if they operate in a free-running fashion during interaction with the input bit. External manipulation by a time-dependent protocol alleviates this issue. We leave exploration of work extraction from inputs with a different temporal correlational structure for future work. However, as per the discussion in the section "Time-Inhomogeneous Protocols," external manipulation is likely to be generally key for optimal work extraction, as it can guarantee that necessary transitions occur.

For highly efficient work-extraction systems to emerge (e.g., due to an evolutionary process), they would then have to develop the ability to operate in an actively rather than passively driven fashion to reach maximum efficiency. An interesting implication that arises from this is the need for a higher-order control structure for active driving. Hierarchical organizations are ubiquitous in biology, and it would be interesting to modify our evolutionary algorithm to explore the emergence of hierarchical structures for greater work extraction. For such systems, there will be a trade-off between the speed at which they operate and the amount of energy they can extract. In our simple example, we can extract all the work we put in plus net gain from the heat bath, because external manipulations are applied reversibly. However, if we put constraints on the execution time, then we should see a trade-off between power and efficiency, similar to effects discussed, for example, in Proesmans, Cleuren, and Van den Broeck (2016, and references therein). We leave a systematic study of this effect to future work.

Acknowledgments

We would like to thank Jim Crutchfield, Alexander Boyd, Christopher Jarzynski, David Sivak, Rob Shaw, David Wolpert, Artemy Kolchinsky, Karoline Wiesner, and Michael Lachmann for relevant discussions. This work was partially supported by the Foundational Questions Institute (grant FQXi-RFP3-1345). L. A. received support from the University of Hawai'i Office of the Vice Chancellor for Research; the John Templeton Foundation; the Stanford Center for Computational, Evolutionary, and Human Genomics; and the Morrison Institute for Population and Resources Studies, Stanford University. T. E. O. is supported by a Royal Society University Research Fellowship.

Appendix

PREDICTIVE INFORMATION IN STEADY STATE

In steady state, the distribution of internal states prior to receiving a 1 is $\boldsymbol{\pi}_{0,s} = [\pi_{0,A}, \pi_{0,B}] = [\pi_{0,A0} + \pi_{0,A1}, \pi_{0,B0} + \pi_{0,B1}]$ (see equation (14.4)). The distribution prior to receiving a 0 is

$\pi_{2,s} = [\pi_{2,A}, \pi_{2,B}] = [\pi_{2,A0} + \pi_{2,A1}, \pi_{2,B0} + \pi_{2,B1}]$ (see equation (14.6)). The overall probability of being in state A at the end of an interaction step is then $\pi(A) = \frac{1}{2}[\pi_{0,A0} + \pi_{0,A1} + \pi_{2,A0} + \pi_{2,A1}]$, and $\pi(B) = 1 - \pi(A)$. With these expressions, and noting that the overall probability of receiving each bit is 1/2, the mutual information between the internal state at the end of an interaction interval and the next incoming bit simplifies to $I_{\text{pred}} = \frac{1}{2}[\pi_{0,A} \ln \frac{\pi_{0,A}}{\pi(A)} + \pi_{0,B} \ln \frac{\pi_{0,B}}{\pi(B)} + \pi_{2,A} \ln \frac{\pi_{2,A}}{\pi(A)} + \pi_{2,B} \ln \frac{\pi_{2,B}}{\pi(B)}]$.

MAKING REVERSIBLE RANDOM MATRICES

The following procedure for making reversible matrices at random is taken from Bordenave, Caputo, and Chafai (2011). Let $\{K_{ij} | j \leq i \leq 4\}$ be a set of 10 real, random variables created by sampling uniformly from an interval $(0, N_{\max}]$. This set forms the lower triangle of a symmetric 4×4 matrix with $K_{ij} = K_{ji}$. Define $\pi_j \equiv \sum_i K_{ij}$. Then the matrix $\mathbf{M}$ given by $M_{ij} = \frac{K_{ij}}{\pi_j}$ is a reversible stochastic matrix with stationary distribution $\boldsymbol{\pi}$. We can also make generators $\mathbf{G}$ via $\mathbf{G} = \mathbf{M} - \mathbf{I}$, where $\mathbf{I}$ is the identity matrix. We generated 10^6 $\mathbf{M}$ by this procedure with $N_{\max} = 100$, as well as the corresponding generators given by $\mathbf{G} = \mathbf{M} - \mathbf{I}$. Note that this procedure gives transition rates in the range $(0, 1]$. From this set of generators, we made 3×10^6 embeddable transition matrices, with 10^6 each for $\tau = 1$, $\tau = 0.01$, and $\tau = 100$. The interaction intervals with τ other than 1 led to poor performance, because the other values produced very high self-transition rates ($\tau = .01$) and full equilibration during the interaction interval ($\tau = 100$).

LOWER BOUND ON SELF-TRANSITIONS

The overall chance of making a self-transition under the action of $\mathbf{M}$ on distribution $\mathbf{p}$ is $\mathbf{d}(\mathbf{M})^\top \mathbf{p}$, where $\mathbf{d}(\mathbf{M})$ is the vector of the diagonal entries of $\mathbf{M}$. We can construct a matrix with the smallest numerically found self-transition rate by assuming that this is achieved when the trace is minimal. For embeddable stochastic matrices, the lower bound on the trace is 1 (see the argument in the section "Work Extraction by Time-Continuous, Free-Running Devices"). In this case, **M** has a single eigenvalue of 1 and all other eigenvalues 0. Thus **M** is rank 1 with four repeats of the same column $\mathbf{m} = [m_0, m_1, m_2, m_3]^\top$. The matrix $\mathbf{C} = \mathbf{MF}_0\mathbf{MF}_1$ is then equal to **M**, and the steady state of the complete cycle is nothing more than the repeated column, that is, $\pi_0(s_n, b'_{n-1}) = \mathbf{m}$. Equation (14.9) says that the average number of self-transitions over the cycle is $L = \frac{1}{2}\left[(m_0 + m_1)^2 + (m_2 + m_3)^2\right]$. Minimizing this number is a simple optimization problem with the constraint $m_0 + m_1 + m_2 + m_3 = 1$. The solution is $(m_0 + m_1) = (m_2 + m_3) = \frac{1}{2}$. Substituting these in gives $L = \frac{1}{4}$. This condition can be satisfied for approximately embeddable **M**, which can be constructed, for example, by perturbing the matrix with all entries equal to $\frac{1}{4}$ so that the smaller eigenvalues are just slightly positive and not exactly zero.

REFERENCES

Bordenave, C., P. Caputo, and D. Chafai. 2011. "Spectrum of Large Random Reversible Markov Chains: Heavy-Tailed Weights on the Complete Graph." *Annals of Probability* 39, no. 4 (July): 1544–1590.

Boyd, A. B., D. Mandal, and J. P. Crutchfield. 2017. "Correlation-Powered Information Engines and the Thermo-dynamics of Self-Correction." *Physical Review E* 95 (1): 012152.

Esposito, M., and C. Van den Broeck. 2011. "Second Law and Landauer Principle Far from Equilibrium." *Europhysics Letters* 95 (4): 40004.

Gallager, R. G. 2012. *Discrete Stochastic Processes.* Vol. 321. New York: Springer Science & Business Media.

Jeon, H. J., and S. W. Kim. 2016. "Optimal Work of the Quantum Szilard Engine under Isothermal Processes with Inevitable Irreversibility." *New Journal of Physics* 18 (4): 043002.

Jia, C. 2016. "A Solution to the Reversible Embedding Problem for Finite Markov Chains." *Statistics and Probability Letters* 116 (Supplement C): 122–130.

Kim, S. W., T. Sagawa, S. De Liberato, and M. Ueda. 2011. "Quantum Szilard Engine." *Physical Review Letters* 106 (7): 070401.

Landauer, R. 1961. "Irreversibility and Heat Generation in the Computing Process." *IBM Journal of Research and Development* 5 (3): 183–191.

Mandal, D., and C. Jarzynski. 2012. "Work and Information Processing in a Solvable Model of Maxwell's Demon." *Proceedings of the National Academy of Sciences of the United States of America* 109 (29): 11641–11645.

Marathe, R., and J. M. R. Parrondo. 2010. "Cooling Classical Particles with a Microcanonical Szilard Engine." *Physical Review Letters* 104 (24): 245704.

McGrath, T., N. S. Jones, P. R. ten Wolde, and T. E. Ouldridge. 2017. "Biochemical Machines for the Interconversion of Mutual Information and Work." *Physical Review Letters* 118 (2): 028101.

Proesmans, K., B. Cleuren, and C. Van den Broeck. 2016. "Power–Efficiency–Dissipation Relations in Linear Thermodynamics." *Physical Review Letters* 116 (22): 220601.

Szilard, L. 1929. "Über die Entropieverminderung in einem thermodynamischen System bei Eingriffen intelligenter Wesen." *Zeitschrift für Physik* 53 (11–12): 840–856.

Vaikuntanathan, S., and C. Jarzynski. 2011. "Modeling Maxwell's Demon with a Microcanonical Szilard Engine." *Physical Review E* 83 (6): 061120.

Van den Broeck, C., *et al.* 2013. "Stochastic Thermodynamics: A Brief Introduction." *Phys. Complex Colloids* 184:155–193.

Wolpert, D. H., A. Kolchinsky, and J. A. Owen. 2017. "The Minimal Hidden Computer Needed to Implement a Visible Computation." arXiv 1708:08494.

Zurek, W. H. 1986. "Maxwell's Demon, Szilard's Engine and Quantum Measurements." In *Frontiers of Nonequilibrium Statistical Physics*, 151–161. New York: Springer.

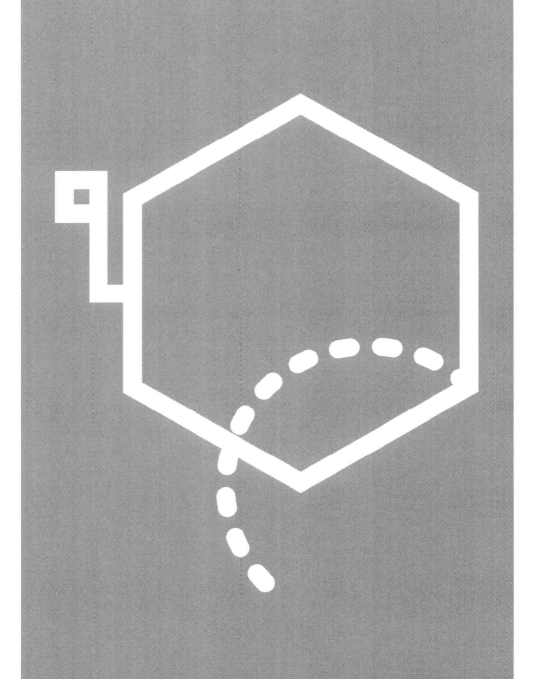

DETAILED FLUCTUATION THEOREMS: A UNIFYING PERSPECTIVE

Riccardo Rao, University of Luxembourg
Massimiliano Esposito, University of Luxembourg

Introduction

The discovery of different fluctuation theorems (FTs) over the last two decades constitutes major progress in nonequilibrium physics (Harris and Schütz 2007; Esposito, Harbola, and Mukamel 2009; Jarzynski 2011; Campisi, Hänggi, and Talkner 2011; Seifert 2012; Van den Broeck and Esposito 2015). These relations are exact constraints that some fluctuating quantities satisfy arbitrarily far from equilibrium. They have been verified experimentally in many different contexts, ranging from biophysics to electronic circuits (Ciliberto 2017). However, they come in different forms—detailed fluctuation theorems (DFTs) or integral fluctuation theorems (IFTs)—and concern various types of quantities. Understanding how they are related and to what extent they involve mathematical quantities or interesting physical observables can be challenging.

The aim of this chapter is to provide a simple, yet elegant, method to identify a class of finite-time DFTs for time-inhomogeneous Markov jump processes. The method is based on splitting the entropy production (EP) in three contributions by introducing a reference probability mass function (PMF). The latter is parameterized by the time-dependent driving protocol, which renders the dynamics time-inhomogeneous. The first contribution quantifies the EP as if the system were in the reference PMF, the second by the extent to which the reference PMF changes with the driving protocol, and the last by the mismatch between the actual and the reference PMFs. We show that when the system is initially prepared in the reference PMF, the joint probability distribution for the first two terms always satisfies a DFT. We then show that various known DFTs can be immediately recovered as special cases. We emphasize at which level our results make contact with physics and also clarify the nontrivial connection between DFTs and EP fluctuations. Our EP splitting is also shown to be connected to information theory. Indeed, it can be used to derive a generalized Landauer principle identifying the minimal cost needed to move the actual PMF away from the reference PMF. While unifying, we emphasize that our approach by no means encompasses all previously derived FTs and that other FT generalizations have been made (see, e.g., Chetrite and Gupta 2011; Seifert 2012; Pérez-Espigares, Kolton, and Kurchan 2012; Verley, Chétrite, and Lacoste 2012; Baiesi and Falasco 2015).

The plan of the chapter is as follows. Time-inhomogeneous Markov jump processes are introduced in the following section. Our main results are presented in the third section: we first introduce the EP as a quantifier of detailed balance breaking, and we then show that, by choosing a reference PMF, EP splitting ensues. This enables us to identify the fluctuating quantities satisfying a DFT and an IFT when the system is initially prepared in the reference PMF. Whereas IFTs hold for arbitrary reference PMFs, DFTs require reference PMFs to be determined solely by the driving protocol encoding the time dependence of the rates. The EP decomposition is also shown to lead to a generalized Landauer principle. The remaining sections are devoted to specific reference PMFs and show that they give rise to interesting mathematics or physics: first, the steady-state PMF of the Markov jump process is chosen, giving rise to the adiabatic–nonadiabatic split of the EP (Esposito, Harbola, and Mukamel 2007); then, the equilibrium PMF of a spanning tree of the graph defined by the Markov jump process is chosen and gives rise to a cycle–cocycle decomposition of the EP (Polettini 2014). Physics is introduced next, and the properties that the Markov jump process must satisfy to describe the thermodynamics of an open system are described. In the next section, the microcanonical distribution is chosen as the reference PMF leading to the splitting of the EP into system and reservoir entropy change. Finally, the generalized Gibbs equilibrium PMF is chosen as a reference and leads to a conservative–nonconservative splitting of the EP (Rao and Esposito 2018b). Conclusions are drawn in the final section, and some technical proofs are discussed in appendices.

Markov Jump Process

We introduce time-inhomogeneous Markov jump processes and set the notation.

We consider an externally driven open system described by a finite number of states, which we label n. Transitions allowed between pairs of states are identified by directed edges,

$$e \equiv (nm, \nu), \quad \text{for } n \xleftarrow{\nu} m, \tag{15.1}$$

where the label ν indexes different transitions between the same pair of states (e.g., transitions due to different reservoirs). The evolution in time of the probability of finding the system in the state n, $p_n \equiv p_n(t)$, is ruled by the *master equation* (ME)

$$d_t p_n = \sum_m W_{nm} p_m, \tag{15.2}$$

where the elements of the *rate matrix* are represented as

$$W_{nm} = \sum_e w_e \{\delta_{n,\mathfrak{t}(e)}\delta_{m,\mathfrak{o}(e)} - \delta_{n,m}\delta_{m,\mathfrak{o}(e)}\}. \quad (15.3)$$

The latter is written in terms of stochastic transition rates, $\{w_e\}$, and the functions

$$\mathfrak{o}(e) := m \quad \text{and} \quad \mathfrak{t}(e) := n, \quad \text{for } e = (nm, \nu), \quad (15.4)$$

which map each transition to the state from which it originates (origin) and to which it leads (target), respectively. The off-diagonal entries of the rate matrix (first term in brackets in equation 15.3) give the probability per unit time to transition from m to n. The diagonal entries (second term in brackets in equation 15.3) are the escape rates denoting the probability per unit time of leaving the state m. For thermodynamic consistency, we assume that each transition $e \equiv (nm, \nu)$ is reversible, namely, if w_e is finite, the corresponding backward transition $-e \equiv (mn, \nu)$ is allowed and has a finite rate w_{-e}, too. For simplicity, we also assume that the rate matrix is irreducible at all times, so that the stochastic dynamics is ensured to be ergodic. The Markov jump process is said to be *time-inhomogeneous* when the transition rates depend on time. The driving *protocol value* π_t determines the values of all rates at time t, $\{w_e \equiv w_e(\pi_t)\}$.

The ME (15.2) can be rewritten as a continuity equation,

$$d_t p_n = \Sigma_e D_e^n \langle j^e \rangle, \quad (15.5)$$

where we introduce the averaged transition *probability fluxes*,

$$\langle j^e \rangle = w_e p_{\mathfrak{o}(e)}, \quad (15.6)$$

and the *incidence matrix* D,

$$D_e^n := \delta_{n,\mathfrak{t}(e)} - \delta_{n,\mathfrak{o}(e)} = \begin{cases} +1 & \text{if } \xrightarrow{e} n, \\ -1 & \text{if } \xleftarrow{e} n, \\ 0 & \text{otherwise}, \end{cases} \quad (15.7)$$

Chapter 15: Detailed Fluctuation Theorems

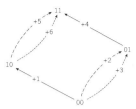

Figure 15.1. A network of transitions.

which couples each transition to the pair of states that it connects and hence encodes the *network topology*. On the graph identified by the vertices $\{n\}$ and the edges $\{e\}$, the incidence matrix D defined in equation (15.7) can be viewed as a (negative) divergence operator when acting on edge-space vectors—as in the ME (15.5)—or as a gradient operator when acting on vertex-space vectors. It satisfies the symmetry $D^n_{-e} = -D^n_e$.

Example. Let us consider the Markov jump process in the network in figure 15.1, in which only the six forward transitions are depicted. It is characterized by four states, $\{00, 01, 10, 11\}$, connected by transitions as described by the incidence matrix

$$D = \begin{matrix} & \begin{matrix} +1 & +2 & +3 & +4 & +5 & +6 \end{matrix} \\ \begin{matrix} 00 \\ 10 \\ 01 \\ 11 \end{matrix} & \begin{pmatrix} -1 & -1 & -1 & 0 & 0 & 0 \\ 1 & 0 & 0 & 0 & -1 & -1 \\ 0 & 1 & 1 & -1 & 0 & 0 \\ 0 & 0 & 0 & 1 & 1 & 1 \end{pmatrix} \end{matrix}.$$

(15.8)

Backward transitions are obtained from $D^n_{-e} = -D^n_e$.

Notation. From now on, upper–lower indices and Einstein summation notation will be used: repeated upper–lower indices imply the summation over all the allowed values for those indices. Time derivatives are denoted by d_t or δ_t, whereas the overdot [˙] is reserved for rates of change of quantities that are not exact time derivatives of state functions. We also take the Boltzmann constant k_B to be equal to 1.

General Results

This section constitutes the core of the chapter. The main results are presented in their most general form.

EP DECOMPOSITION AT THE ENSEMBLE AVERAGE LEVEL

After defining the ensemble-averaged EP, we will show how to decompose it generically in terms of a reference PMF.

A PMF p_n satisfies the *detailed-balance* property if and only if

$$w_e p_{\mathfrak{o}(e)} = w_{-e} p_{\mathfrak{o}(-e)}, \qquad (15.9)$$

for all transitions e. This implies that all net transition probability currents vanish: $\langle j^e \rangle - \langle j^{-e} \rangle = 0$. The central quantity that we will consider is the *EP rate*:

$$\langle \dot{\Sigma} \rangle = \tfrac{1}{2} A_e \left[\langle j^e \rangle - \langle j^{-e} \rangle \right] = A_e \langle j^e \rangle \geq 0, \qquad (15.10)$$

where the *affinities* are given by

$$A_e = \ln \frac{w_e p_{\mathfrak{o}(e)}}{w_{-e} p_{\mathfrak{o}(-e)}}. \qquad (15.11)$$

The EP rate is a measure of the amount by which the system breaks detailed balance or, equivalently, time-reversal symmetry. Indeed, its form ensures that it is always nonnegative and vanishes if and only if equation (15.9) holds. Notice that $A_{-e} = -A_e$. As we will see in the section "System–Reservoir Decomposition," in physical systems, the EP quantifies the total entropy change in the system plus the environment (Schnakenberg 1976).

We now decompose the EP rate into two contributions using a generic PMF $p_n^{\text{ref}} \equiv p_n^{\text{ref}}(t)$ as a *reference*. We make no assumption about the properties of p_n^{ref} at this stage, and we define the reference potential and the reference affinities as

$$\psi_n^{\text{ref}} := -\ln p_n^{\text{ref}} \qquad (15.12)$$

and

$$A_e^{\text{ref}} := \ln \frac{w_e p_{\mathfrak{o}(e)}^{\text{ref}}}{w_{-e} p_{\mathfrak{o}(-e)}^{\text{ref}}} = \ln \frac{w_e}{w_{-e}} + \psi_n^{\text{ref}} D_e^n, \qquad (15.13)$$

respectively. The former can be thought of as the entropy associated with p_n^{ref}, that is, as its *self-information*, whereas the latter measures the extent to which p_n^{ref} breaks detailed balance. Merely by adding and subtracting $\psi_n^{\text{ref}} D_e^n$ from the EP rate, equation (15.10) can be formally decomposed as

$$\langle \dot{\Sigma} \rangle = \langle \dot{\Sigma}_{\text{nc}} \rangle + \langle \dot{\Sigma}_{\text{c}} \rangle \geq 0, \qquad (15.14)$$

where the *reference nonconservative contribution* is an EP with affinities replaced by reference affinities

$$\langle \dot{\Sigma}_{\text{nc}} \rangle := A_e^{\text{ref}} \langle j^e \rangle \qquad (15.15)$$

and the *reference conservative contribution* is

$$\langle \dot{\Sigma}_{\text{c}} \rangle := -\Sigma_n d_t p_n \ln \left\{ p_n / p_n^{\text{ref}} \right\}. \qquad (15.16)$$

Using the ME (15.5), equation (15.16) can be further decomposed as

$$\langle \dot{\Sigma}_{\text{c}} \rangle = -d_t \mathcal{D}(p \,||\, p^{\text{ref}}) + \langle \dot{\Sigma}_d \rangle, \qquad (15.17)$$

where the first term quantifies the change in time of the *dissimilarity* between p_n and p_n^{ref}, because

$$\mathcal{D}(p \,||\, p^{\text{ref}}) := \Sigma_n p_n \ln \left\{ p_n / p_n^{\text{ref}} \right\} \qquad (15.18)$$

is a *relative entropy*, whereas the second term,

$$\langle \dot{\Sigma}_d \rangle := -\Sigma_n p_n d_t \ln p_n^{\text{ref}} = \Sigma_n p_n d_t \psi_n^{\text{ref}}, \quad (15.19)$$

accounts for possible time-dependent changes of the reference state; we name it the *driving contribution*. The reason for assigning this particular name will become clear later, as we will request p_n^{ref} to depend parametrically on time only via the driving protocol, that is, $p_n^{\text{ref}}(t) = p_n^{\text{ref}}(\pi_t)$.

Using these equations, one can easily rearrange equation (15.14) into

$$\langle \dot{\Sigma}_d \rangle + \langle \dot{\Sigma}_{nc} \rangle \geq d_t \mathcal{D}(p \,||\, p^{\text{ref}}). \quad (15.20)$$

When $p_n^{\text{ref}}(t) = p_n^{\text{ref}}(\pi_t)$, one can interpret this equation as follows. The left-hand side describes the EP contribution due to the time-dependent protocol, $\langle \dot{\Sigma}_d \rangle$, and to the break of detailed balance required to sustain the reference PMF, $\langle \dot{\Sigma}_{nc} \rangle$. The right-hand side, when positive, thus represents the minimal cost (ideally achieved at vanishing EP) to move the PMF further away from the reference PMF. When the right-hand side is negative, its absolute value becomes the maximal amount by which the two EP contributions can decrease as the PMF approaches the reference PMF. This result can be seen as a *mathematical generalization of the Landauer principle*, as it provides a connection between an information-theoretical measure of the dissimilarity between two PMFs and the driving and break of detailed balance needed to achieve it. Its precise physical formulation, discussed in detail in Rao and Esposito (2018a), is obtained when expressing equation (15.20) in terms of the reference PMF used in the section "Conservative–Nonconservative Decomposition."

EP DECOMPOSITION AT THE TRAJECTORY LEVEL

We now perform the analogue of the EP decomposition (15.14) at the level of single stochastic trajectories.

A stochastic *trajectory* of duration t, $\boldsymbol{n}_t$, is defined as a set of transitions $\{e_i\}$ sequentially occurring at times $\{t_i\}$ starting from n_0 at time 0. If not stated otherwise, the transitions index i runs from $i = 1$ to the last transition prior to time t, N_t, whereas the state at time $\tau \in [0, t]$ is denoted by n_τ. The whole trajectory is encoded in the *instantaneous fluxes*,

$$j^e(\tau) := \Sigma_i \delta_{e,e_i} \delta(\tau - t_i), \qquad (15.21)$$

as they encode the transitions that occur and their timing. Its corresponding trajectory probability measure is given by

$$\mathfrak{P}[\boldsymbol{n}_t; \pi_t] = \prod_{i=1}^{N_t} w_{e_i}(\pi_{t_i}) \prod_{i=0}^{N_t} \exp\left\{-\int_{t_i}^{t_{i+1}} d\tau \, \Sigma_e w_e(\pi_\tau) \delta_{n_\tau, o(e)}\right\}, \qquad (15.22)$$

where the first term accounts for the probability of transitioning along the edges, and the second term accounts for the probability of the system spending $\{t_{i+1} - t_i\}$ time in the state $\{n_{t_i}\}$. When averaging equation (15.21) over all stochastic trajectories, we obtain the averaged fluxes (eq. 15.6)

$$\langle j^e(\tau) \rangle = \int \mathfrak{D}\boldsymbol{n}_t \, \mathfrak{P}[\boldsymbol{n}_t; \pi_t] \, p_{n_0}(0) \, j^e(\tau), \qquad (15.23)$$

where $\int \mathfrak{D}\boldsymbol{n}_t$ denotes the integration over all stochastic trajectories.

The change along $\boldsymbol{n}_t$ of a state function like ψ_n^{ref} can be expressed as

$$\Delta \psi^{\text{ref}}[\boldsymbol{n}_t] = \psi_{n_t}^{\text{ref}}(t) - \psi_{n_0}^{\text{ref}}(0)$$
$$= \int_0^t d\tau \left\{ \left[d_\tau \psi_n^{\text{ref}}(\tau)\right]\bigg|_{n=n_\tau} + \psi_n^{\text{ref}}(\tau) \, D_e^n \, j^e(\tau) \right\}. \qquad (15.24)$$

The first term on the right-hand side accounts for the instantaneous changes of p_n^{ref}, whereas the second accounts for its finite changes due to stochastic transitions. Analogously, the trajectory EP—which is not a state function—can be written as

$$\Sigma[\boldsymbol{n}_t; \pi_t] = \int_0^t d\tau\, j^e(\tau) \ln \frac{w_e(\pi_\tau)}{w_{-e}(\pi_\tau)} - \ln \frac{p_{n_t}(t)}{p_{n_0}(0)}. \quad (15.25)$$

Adding and subtracting the terms of equation (15.24) from the EP, we readily obtain the fluctuating expressions of the nonconservative and conservative contributions of the EP,

$$\Sigma[\boldsymbol{n}_t; \pi_t] = \Sigma_{\mathrm{nc}}[\boldsymbol{n}_t; \pi_t] + \Sigma_{\mathrm{c}}[\boldsymbol{n}_t]. \quad (15.26)$$

The former reads

$$\Sigma_{\mathrm{nc}}[\boldsymbol{n}_t; \pi_t] = \int_0^t d\tau\, A^{\mathrm{ref}}(\tau)\, j^e(\tau), \quad (15.27)$$

while, for the latter,

$$\Sigma_{\mathrm{c}}[\boldsymbol{n}_t] = -\Delta\mathcal{D}[\boldsymbol{n}_t] + \Sigma_{\mathrm{d}}[\boldsymbol{n}_t], \quad (15.28)$$

where

$$\Delta\mathcal{D}[\boldsymbol{n}_t] := \ln \frac{p_{n_t}(t)}{p_{n_t}^{\mathrm{ref}}(t)} - \ln \frac{p_{n_0}(0)}{p_{n_0}^{\mathrm{ref}}(0)} \quad (15.29)$$

and

$$\Sigma_{\mathrm{d}}[\boldsymbol{n}_t] := \int_0^t d\tau \left[d_\tau \psi_n^{\mathrm{ref}}(\tau) \right]_{n=n_\tau}. \quad (15.30)$$

We emphasize that equation (15.26) holds for any reference PMF p_n^{ref} exactly as it was for its ensemble-averaged rate counterpart, equation (15.14).

FLUCTUATION THEOREMS

We proceed to show that a class of FTs ensues from the decomposition equations (15.14–15.26). To do so, we now need to assume that the reference PMF depends instantaneously *solely* on the protocol value $p_n^{\mathrm{ref}}(\tau) = p_n^{\mathrm{ref}}(\pi_\tau)$. In other words, p_n^{ref} at time τ is completely determined by $\{w_e(\pi_\tau)\}$. This justifies *a posteriori* the name "driving contribution" for equation (15.19).

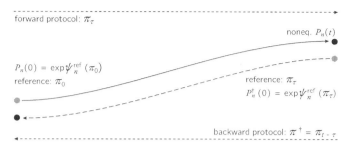

Figure 15.2. Schematic of the forward and backward processes related by our DFT.

Various instances of such PMFs will be provided in the following sections. We define a *forward process* where the system is initially prepared in $p_n(0) = p_n^{\text{ref}}(\pi_0)$ at a value of the protocol π_0 and then evolves under the Markov jump process driven by a protocol π_τ, for $\tau \in [0, t]$. The corresponding *backward process*, denoted with a dagger, is defined as follows. The system is initially prepared in the reference PMF corresponding to the final value of the forward process, $p_n^\dagger(0) = p_n^{\text{ref}}(\pi_t)$, and then evolves under the Markov jump process driven by the forward protocol reversed in time:

$$\pi_\tau^\dagger := \pi_{t-\tau}, \quad \text{for } \tau \in [0, t] \qquad (15.31)$$

(see fig. 15.2).

Our main result is that the forward and backward processes are related by the following *finite-time DFT*:

$$\frac{P_t(\Sigma_{\text{d}}, \Sigma_{\text{nc}})}{P_t^\dagger(-\Sigma_{\text{d}}, -\Sigma_{\text{nc}})} = \exp\left\{\Sigma_{\text{d}} + \Sigma_{\text{nc}}\right\}. \qquad (15.32)$$

Here $P_t(\Sigma_{\text{d}}, \Sigma_{\text{nc}})$ is the probability of observing a driving contribution to the EP Σ_{d} and a nonconservative contribution Σ_{nc} along the forward process. Instead, $P_t^\dagger(-\Sigma_{\text{d}}, -\Sigma_{\text{nc}})$ is the probability of observing a driving contribution equal to $-\Sigma_{\text{d}}$ and a nonconservative contribution $-\Sigma_{\text{nc}}$ along the backward process.

We now mention two direct implications of our DFT. First, by marginalizing the joint probability, one easily verifies that the sum of nonconservative and driving EP contributions also satisfies a DFT:

$$\frac{P_t(\Sigma_d + \Sigma_{nc})}{P_t^\dagger(-\Sigma_d - \Sigma_{nc})} = \exp\{\Sigma_d + \Sigma_{nc}\}. \quad (15.33)$$

Second, when averaging equation (15.32) over all possible values of Σ_d and Σ_{nc}, an IFT ensues:

$$\langle \exp\{-\Sigma_d - \Sigma_{nc}\}\rangle = 1. \quad (15.34)$$

The proofs of equations (15.32–15.34) are given in the appendix and use the generating function techniques developed in Rao and Esposito (2018b) and Esposito, Harbola, and Mukamel (2007).

We note that the IFT holds for any reference PMF regardless of the requirement that $p_n^{\text{ref}}(\tau) = p_n^{\text{ref}}(\pi_\tau)$ (see appendix). In contrast, this requirement must hold for the DFT, else the probability $P_t^\dagger(\Sigma_d, \Sigma_{nc})$ would no longer describe a physical backward process in which solely the protocol function is time reversed. Indeed, if one considers an arbitrary p_n^{ref}, the backward process corresponds not only to reversing the protocol but also to the stochastic dynamics itself (see eq. 15.117).

Another noteworthy observation is that the fluctuating quantity $\Sigma_d + \Sigma_{nc}$ can be seen as the ratio between the probability of observing a trajectory $\boldsymbol{n}_t$ along the forward process (eq. 15.22) and the probability of observing the time-reversed trajectory along the backward process:

$$\Sigma_{nc}[\boldsymbol{n}_t; \pi_t] + \Sigma_d[\boldsymbol{n}_t; \pi_t] = \ln \frac{\mathfrak{P}[\boldsymbol{n}_t; \pi_t]\, p_{n_0}^{\text{ref}}(\pi_0)}{\mathfrak{P}[\boldsymbol{n}_t^\dagger; \pi_t^\dagger]\, p_{n_t}^{\text{ref}}(\pi_t)}. \quad (15.35)$$

The latter trajectory is denoted by $\boldsymbol{n}_t^\dagger$; it starts from n_t and is defined by

$$j^{\dagger e}(\tau) := \sum_i \delta_{e,-e_i} \delta(t - \tau - t_i). \quad (15.36)$$

This result follows using equation (15.22) and the observation that the contribution due to the waiting times vanishes in the ratio on the right-hand side. It can also be used to prove the DFT in two alternative ways, the first inspired by García-García et al. (2010) and the second using trajectory probabilities (see appendix). These proofs rely on the fact that both the driving and the nonconservative EP contributions satisfy the *involution* property

$$\Sigma_{\text{nc}}[\bm{n}_t^\dagger; \pi_t^\dagger] = -\Sigma_{\text{nc}}[\bm{n}_t; \pi_t] \quad \text{and} \quad \Sigma_{\text{d}}[\bm{n}_t^\dagger; \pi_t^\dagger] = -\Sigma_{\text{d}}[\bm{n}_t; \pi_t], \tag{15.37}$$

namely, the change of Σ_{d} and Σ_{nc} for the backward trajectory along the backward process is minus the change along the forward trajectory of the forward process. This result follows from direct calculation of equations (15.27) and (15.30) (see appendix).

Let us finally get back to the generalized Landauer principle for systems initially prepared in the reference state, as we did in this subsection for the FTs to hold. Using equation (15.20), we see that the arguments of the FTs 15.33 and 15.34 (i.e., the driving and the nonconservative contributions to the EP) can be interpreted, on average, as the cost to generate a dissimilarity (or a lag) between the actual and reference PMFs at the end of the forward protocol. A special case of this result is discussed in Vaikuntanathan and Jarzynski (2009).

EP FLUCTUATIONS

We now discuss the properties of the fluctuating EP and its relation to the previously derived FTs.

An IFT for the EP always holds,

$$\langle \exp\{-\Sigma\} \rangle = 1, \tag{15.38}$$

regardless of the initial condition (Seifert 2005). In our framework, this can be seen as the result of choosing the actual $p_n(\tau)$ as the reference for the IFT 15.34.

In contrast, a general DFT for the EP does not hold. This can be easily understood at the level of trajectory probabilities. Indeed, the fluctuating EP can be written as the ratio of forward and backward probabilities, as in equation (15.35), but the initial condition of the forward process is arbitrary and that of the backward process is the final PMF of the forward process:

$$\Sigma[\boldsymbol{n}_t; \pi_t] = \ln \frac{\mathfrak{P}[\boldsymbol{n}_t; \pi_t] p_{n_0}(0)}{\mathfrak{P}[\boldsymbol{n}_t^\dagger; \pi_t^\dagger] p_{n_t}(t)}. \qquad (15.39)$$

As a result, the involution property is generally lost, $\Sigma[\boldsymbol{n}_t^\dagger; \pi_t^\dagger] \neq -\Sigma[\boldsymbol{n}_t; \pi_t]$, because $p_{n_0}^\dagger(t) \neq p_{n_0}(0)$, and hence the DFT is lost, too (Seifert 2005).

However, in special cases, the fluctuating quantity $\Sigma_d + \Sigma_{nc}$ that satisfies a DFT can be interpreted as an EP. This happens if, at the end of the forward (resp. backward) process, the protocol stops changing in time in such a way that the system relaxes from p_{n_t} to an equilibrium $p_{n_t}^{\text{ref}}$ (resp. from $p_n^\dagger(t)$ to an equilibrium $p_n^{\text{ref}}(\pi_0)$) and thus without contributing either to Σ_d or to Σ_{nc} (even at the trajectory level). In such cases, $\Sigma_d + \Sigma_{nc}$ can be seen as the EP of the *extended* process including the relaxation. On average, it is greater than or equal to the EP of the same process without the relaxation, because the nonnegative EP during the relaxation is given by $\mathcal{D}(p(t) \,||\, p^{\text{ref}}(\pi_t)) \geq 0$.

A GAUGE THEORY PERSPECTIVE

We now show that the decomposition in equation (15.14) can be interpreted as the consequence of the gauge freedom discussed by Polettini (2012). Indeed, Polettini shows that the following gauge transformation leaves the stochastic dynamics (eq. 15.5) and the EP rate (eq. 15.10) unchanged:

$$\begin{aligned} p_n &\to p_n \exp \psi_n, & w_e &\to w_e \exp -\psi_{o(e)}, \\ D_e^n &\to D_e^n \exp \psi_n, \text{ and} & \Sigma_n &\to \Sigma_n \exp -\psi_n. \end{aligned} \qquad (15.40)$$

When considering a gauge term ψ_n changing in time, one needs also to shift the time derivative as

$$d_t \to d_t - \partial_t, \qquad (15.41)$$

where ∂_t behaves as a normal time derivative but acts only on ψ_n. Let us now consider the EP rate rewritten as

$$\langle \dot{\Sigma} \rangle = \langle j^e \rangle \ln \frac{w_e}{w_{-e}} + d_t \sum_n p_n \left[-\ln p_n \right]. \qquad (15.42)$$

One readily sees that the transformation in equations (15.40–15.41) changes the first term into the nonconservative term, equation (15.15), whereas the second is changed into the conservative term, equation (15.16). Finally, we note that connections between gauge transformations and FTs were also discussed in Chetrite and Gupta (2011) and Garrahan (2016).

This concludes the presentation of our main results. In the following pages, we consider various specific choices for p_n^{ref} that depend solely on the driving protocol and thus give rise to DFTs. Each choice provides a specific meaning to Σ_{nc} and Σ_{c}. Table 15.1 summarizes the reference potentials, affinities, and conservative contributions for these different choices.

Adiabatic–Nonadiabatic Decomposition

We now provide a first instance of reference PMF based on the fixed point of the Markov jump process.

The Perron–Frobenius theorem ensures that the ME 15.5 has, at all times, a unique instantaneous *steady-state PMF*,

$$\sum_m W_{nm}(\pi_t) p_m^{\text{ss}}(\pi_t) = 0, \quad \text{for all } n \text{ and } t. \qquad (15.43)$$

When using this PMF as the reference, $p_n^{\text{ref}} = p_n^{\text{ss}}$, we recover the *adiabatic–nonadiabatic EP rate* decomposition (Esposito, Harbola, and Mukamel 2007; Esposito and Van den Broeck 2010a, 2010b; Ge and Qian 2010; García-García

Table 15.1. Summary of the reference potentials, affinities, and conservative EP contributions for the specific references discussed in the text. The nonconservative EP contribution follows from $\langle \dot{\Sigma}_{nc} \rangle = A_e^{ref} \langle j^e \rangle$, whereas the driving contribution follows from $\langle \dot{\Sigma}_d \rangle = \sum_n p_n d_t \psi_n^{ref}$. Overall, $\langle \dot{\Sigma} \rangle = \langle \dot{\Sigma}_{nc} \rangle + \langle \dot{\Sigma}_c \rangle = \langle \dot{\Sigma}_{nc} \rangle + \langle \dot{\Sigma}_d \rangle - d_t \mathcal{D}(p \,\|\, p^{ref})$, where $\mathcal{D}$ is the relative entropy.

Decomposition	ψ_n^{ref}	A_e^{ref}	$\langle \dot{\Sigma}_c \rangle$
Adiabatic–nonadiabatic	$-\ln p_n^{\text{ss}}$	$\ln \dfrac{w_e p_{o(e)}^{\text{ss}}}{w_{-e} p_{o(-e)}^{\text{ss}}}$	$-\langle j^e \rangle D_e^n \ln\{p_n/p_n^{\text{ss}}\}$
Cycle–co-cycle	$-\ln\{\prod_{e \in \mathcal{T}_n} w_e - Z\}$	$\begin{cases} 0, & \text{if } e \in \mathcal{T}, \\ \mathcal{A}_e, & \text{if } e \in \mathcal{T}^* \end{cases}$	$\sum_{e \in \mathcal{T}} \langle \mathcal{J}_e \rangle A_e$
System–reservoir	$\mathcal{S}_{\text{mc}} - S_n$	$\delta S_e^{\text{r}} = -f_y \delta X_e^y$	$[S_n - \ln p_n] D_e^n \langle j^e \rangle$
Conservative–nonconservative	$\Phi_{\text{gg}} - [S_n - F_\lambda L_n^\lambda]$	$\mathcal{F}_{y_f} \delta X_e^{y_f}$	$[S_n - F_\lambda L_n^\lambda - \ln p_n] D_e^n \langle j^e \rangle$

et al. 2010; García-García *et al.* 2012). More specifically, the nonconservative term gives the *adiabatic* contribution, which is zero only if the steady state satisfies detailed balance, and the conservative term gives the *nonadiabatic* contribution, which characterizes transient and driving effects. A specific feature of this decomposition is that both terms are nonnegative, as proved in the appendix: $\langle \dot{\Sigma}_{nc} \rangle \geq 0$ and $\langle \dot{\Sigma}_{c} \rangle \geq 0$. In turn, the nonadiabatic contribution decomposes into a relative entropy term and a driving entropy term.

Provided that the forward and backward processes start in the steady state corresponding to the initial value of the respective protocol, the general DFT and IFT derived in equations (15.32) and (15.34) hold for the adiabatic and driving contributions of the adiabatic–nonadiabatic EP decomposition (Esposito, Harbola, and Mukamel 2007; Esposito and Van den Broeck 2010a).

In detailed-balanced systems, the adiabatic contribution is vanishing, $\langle \dot{\Sigma}_a \rangle = 0$, and we obtain a FT for the sole driving contribution:

$$\frac{P_t(\Sigma_d)}{P_t^\dagger(-\Sigma_d)} = \exp \Sigma_d. \qquad (15.44)$$

The celebrated Crooks's DFT (Crooks 1998, 1999, 2000) and Jarzynski's IFT (Jarzynski 1997) are of this type.

ADDITIONAL FTS

Due to the particular mathematical properties of the steady-state PMF, additional FTs for the adiabatic and driving terms ensue. These are not covered by our main DFT, equation (15.32), and their proofs are discussed in the appendix.

For the former, the forward process is produced by the original dynamics initially prepared in an arbitrary PMF. The backward process instead has the same initial PMF and the same driving protocol as the forward process, but the dynamics is governed by the rates

$$\hat{w}_e := w_{-e} p^{ss}_{o(-e)} / p^{ss}_{o(e)}. \qquad (15.45)$$

At any time, the following DFT relates the two processes:

$$\frac{P_t(\Sigma_a)}{\hat{P}_t(-\Sigma_a)} = \exp\Sigma_a, \qquad (15.46)$$

where $\hat{P}(-\Sigma_a)$ is the probability of observing $-\Sigma_a$ adiabatic EP during the backward process. The Speck–Seifert IFT for the housekeeping heat is the IFT version of this DFT (Speck and Seifert 2005).

For the driving term, the forward process is again produced by the original dynamics but is now initially prepared in a steady state. The backward process is instead produced by the rates in equation (15.45) with time-reversed driving protocol, and the system must initially be prepared in a steady state. Under these conditions, one has

$$\frac{P_t(\Sigma_d)}{\hat{P}^{\dagger}_t(-\Sigma_d)} = \exp\Sigma_d, \qquad (15.47)$$

where $\hat{P}^{\dagger}(-\Sigma_d)$ is the probability of observing $-\Sigma_d$ driving EP during the backward process. The Hatano–Sasa IFT (Hatano and Sasa 2001) is the IFT version of this DFT.

Cycle–Cocycle Decomposition

We proceed by providing a second instance of reference PMF based on the equilibrium PMF for a spanning tree of the graph defined by the incidence matrix of the Markov jump process.

We partition the edges of the graph into two disjoint subsets: $\mathcal{T}$ and $\mathcal{T}^*$. The former identifies a *spanning tree*, namely, a minimal subset of paired edges, $(e, -e)$, that connects all states. These edges are called *co-chords*. All the other edges form what we term *chords*. Equivalently, $\mathcal{T}$ is a maximal subset of edges that does not enclose any cycle—the trivial loops composed by forward and backward transitions, $(e, -e)$, are not regarded as cycles. The graph obtained by combining $\mathcal{T}$ and $e \in \mathcal{T}^*$ identifies one and only one cycle, denoted by $\mathcal{C}_e$, for $e \in \mathcal{T}^*$. Algebraically, cycles are characterized as

$$\sum_{e' \in \mathcal{C}_e} D_{e'}^n = \sum_{e'} D_{e'}^n \mathcal{C}_e^{e'} = 0, \quad \text{for all } n, \qquad (15.48)$$

where $\{\mathcal{C}_e^{e'}\}$, for $e \in \mathcal{T}^*$, represent the vectors in the edge space whose entries are all 0, except for those corresponding to the edges of the cycle, which are equal to 1.

We now note that if $\mathcal{T}$ were the sole allowed transitions, the PMF, defined as follows, would be an equilibrium steady state (Schnakenberg 1976):

$$p_n^{\text{st}}(\pi_t) := \frac{1}{Z} \prod_{e \in \mathcal{T}_n} w_e(\pi_t), \qquad (15.49)$$

where $Z = \sum_m \prod_{e \in \mathcal{T}_m} w_e$ is a normalization factor and $\mathcal{T}_n$ denotes the spanning tree *rooted* in n, namely, the set of edges of $\mathcal{T}$ that are oriented toward the state n. Indeed, p_n^{st} would satisfy the property of detailed balance (eq. 15.9):

$$\begin{aligned}
w_e p_{\mathfrak{o}(e)}^{\text{st}} &= \frac{w_e}{Z} \prod_{e' \in \mathcal{T}_{\mathfrak{o}(e)}}, \\
w_{e'} &= \frac{w_{-e}}{Z} \prod_{e' \in \mathcal{T}_{\mathfrak{o}(-e)}}, \qquad (15.50) \\
w_{e'} &= w_{-e} p_{\mathfrak{o}(-e)}^{\text{st}}, \quad \text{for all } e \in \mathcal{T}.
\end{aligned}$$

We now pick this equilibrium PMF as a reference for our EP decomposition, $p_n^{\text{ref}} = p_n^{\text{st}}$. However, to derive the specific expressions for $\langle \dot{\Sigma}_{nc} \rangle$ and $\langle \dot{\Sigma}_c \rangle$, the following result is necessary: the edge probability fluxes can be decomposed as

$$\langle j^e \rangle = \sum_{e' \in \mathcal{T}} \langle \mathcal{J}_{e'} \rangle \mathcal{E}_{e'}^e + \sum_{e' \in \mathcal{T}^*} \langle \mathcal{J}_{e'} \rangle \mathcal{C}_{e'}^e, \quad (15.51)$$

where $\{\mathcal{E}_e\}$ denotes the canonical basis of the edge vector space, $\mathcal{E}_e^{e'} = \delta_e^{e'}$ (Knauer 2011). Algebraically, this decomposition hinges on the fact that the set $\{\mathcal{C}_e\}_{e \in \mathcal{T}^*} \cup \{\mathcal{E}_e\}_{e \in \mathcal{T}}$ is a basis of the edge vector space (Polettini 2014). Note that for $e \in \mathcal{T}^*$, the only nonvanishing contribution in equation (15.51) comes from the cycle identified by e, and hence $\langle j^e \rangle = \langle \mathcal{J}_e \rangle$. The coefficients $\{\langle \mathcal{J}_e \rangle\}$ are called *cocycle fluxes* for the co-chords, $e \in \mathcal{T}$, and *cycle fluxes* for the chords, $e \in \mathcal{T}^*$. They can be understood as follows Polettini (2014). Removing a pair of edges, e and $-e$, from the spanning tree $(e, -e \in \mathcal{T})$ disconnects two blocks of states. The cocycle flux $\{\langle \mathcal{J}_e \rangle\}$ of that edge is the probability flowing from the block identified by the origin of e, $\mathfrak{o}(e)$ to that identified by the target of e, $\mathfrak{t}(e)$. Instead, the cycle flux $\{\langle \mathcal{J}_e \rangle\}$ of an edge, $e \in \mathcal{T}^*$, quantifies the probability flowing along the cycle formed by adding that edge to the spanning tree. Graphical illustrations of cocycle and cycle *currents*, $\langle \mathcal{J}^e \rangle - \langle \mathcal{J}^{-e} \rangle$, can be found in Polettini (2014).

We can now proceed with our main task. Using equations (15.48) and (15.49), we verify that

$$\psi_n^{\text{ref}} D_e^n = \begin{cases} -\ln\{w_e/w_{-e}\}, & \text{if } e \in \mathcal{T}, \\ -\ln\{w_e/w_{-e}\} + \mathcal{A}_e, & \text{if } e \in \mathcal{T}^*, \end{cases} \quad (15.52)$$

where

$$\mathcal{A}_e = \sum_{e'} \mathcal{C}_e^{e'} \ln\{w_{e'}/w_{-e'}\}, \quad \text{for } e \in \mathcal{T}^*, \quad (15.53)$$

is the cycle affinity related to $\mathcal{C}_e$. It follows that

$$A_e^{\text{ref}} = \ln \frac{w_e}{w_{-e}} + \psi_n^{\text{ref}} D_e^n = \begin{cases} 0, & \text{if } e \in \mathcal{T}, \\ \mathcal{A}_e, & \text{if } e \in \mathcal{T}^*, \end{cases} \quad (15.54)$$

Chapter 15: Detailed Fluctuation Theorems

from which the nonconservative contribution readily follows:

$$\langle \dot{\Sigma}_{nc} \rangle = \sum_{e \in \mathcal{T}^*} \mathcal{A}_e \langle j_e \rangle = \sum_{e \in \mathcal{T}^*} \mathcal{A}_e \langle \mathcal{J}_e \rangle. \quad (15.55)$$

In the last equality, we used the property of cycle fluxes discussed after equation (15.51). Hence the nonconservative contribution accounts for the dissipation along network cycles. In turn, combining equation (15.16) with equations (15.51) and (15.52), one obtains the conservative contribution

$$\langle \dot{\Sigma}_c \rangle = \sum_{e \in \mathcal{T}} \mathcal{A}_e \langle \mathcal{J}_e \rangle, \quad (15.56)$$

which accounts for the dissipation along cocycles. Using these last two results, the E P decomposition in equation (15.14) becomes the *cycle-cocycle* decomposition found in Polettini (2014):

$$\langle \dot{\Sigma} \rangle = \sum_{e \in \mathcal{T}^*} \mathcal{A}_e \langle j_e \rangle + \sum_{e \in \mathcal{T}} \mathcal{A}_e \langle \mathcal{J}_e \rangle. \quad (15.57)$$

As for all decompositions, the conservative contribution—here the cocycle one—vanishes at steady state in the absence of driving. The cycle contribution instead disappears in detailed-balanced systems, when all the cycle affinities vanish. This statement is indeed the *Kolmogorov criterion* for detailed balance (Kolmogoroff 1936; Kelly 1979).

The flux decomposition equation (15.51) is also valid at the trajectory level, where the cycle and cocycle fluxes become fluctuating instantaneous fluxes, $\{\mathcal{J}_e\}$. Obviously, the same holds true for the cycle–cocycle EP decomposition. Therefore, if the system is in an equilibrium PMF of type in equation (15.49) at the beginning of the forward and the backward processes, a DFT and an IFT hold by applying equations (15.32) and (15.34). Note that the fluctuating quantity appearing in the DFT, $\Sigma_d + \Sigma_{nc}$, can be interpreted as

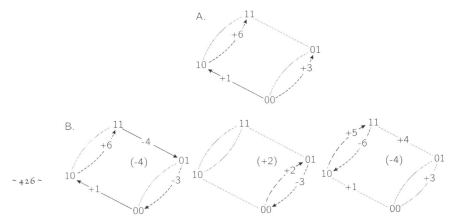

Figure 15.3. (A) Spanning tree and (B) corresponding cycles for the network in figure 15.1.

the EP of the extended process in which, at time t, the driving is stopped, all transitions in $\mathcal{T}^*$ are shut down, and the system is left to relax to equilibrium—which is the initial PMF of the backward process.

It is worth mentioning that one can easily extend the formulation of our DFT by considering the joint probability distribution for each subcontribution of Σ_d and Σ_{na} antisymmetrical under time reversal. This can be shown using either of the proofs in the appendix. In the case of the cycle–cocycle decomposition, it would lead to

$$\frac{P_t(\Sigma_d, \{\mathcal{A}_e(j_e - j_{-e})\}_{e \in \mathcal{T}^*})}{P_t^\dagger(-\Sigma_d, \{-\mathcal{A}_e(j_e - j_{-e})\}_{e \in \mathcal{T}^*})} = \exp\left\{\Sigma_d + \sum_{e \in \mathcal{T}^*} \mathcal{A}_e j_e\right\}, \quad (15.58)$$

which is a generalization of the DFT derived in Polettini and Esposito (2014) to time-inhomogeneous systems. In turn, the latter is a generalization of the steady-state DFT derived by Andrieux and Gaspard (2007) to finite times.

Example. A spanning tree for the network in figure 15.1 is depicted in figure 15.3A. The cycles defined by the corresponding chords are depicted in figure 15.3B. Algebraically, these cycles are represented as follows:

$$\mathcal{C} = \begin{matrix} & \begin{matrix} -4 & +2 & +5 \end{matrix} \\ \begin{matrix} +1 \\ +2 \\ +3 \\ +4 \\ +5 \\ +6 \end{matrix} & \begin{pmatrix} 1 & 0 & 0 \\ 0 & 1 & 0 \\ -1 & -1 & 0 \\ -1 & 0 & 0 \\ 0 & 0 & 1 \\ 1 & 0 & -1 \end{pmatrix} \end{matrix}, \quad (15.59)$$

where the negative entries must be regarded as transitions performed in the backward direction. The corresponding affinities, which determine the nonconservative contribution in equation (15.55), hence read

$$\begin{aligned} \mathcal{A}_{-4} &= \ln \frac{w_{+1}w_{+6}w_{-4}w_{-3}}{w_{-1}w_{-6}w_{+4}w_{+3}}, \\ \mathcal{A}_{+2} &= \ln \frac{w_{+2}w_{-3}}{w_{-2}w_{+3}}, \text{ and} \\ \mathcal{A}_{+5} &= \ln \frac{w_{+5}w_{-6}}{w_{-5}w_{+6}}. \end{aligned} \quad (15.60)$$

The affinities corresponding to the cycles taken in the backward direction follow from $\mathcal{A}_{-e} = -\mathcal{A}_e$. The expression of the cocycle fluxes can be verified as equal to

$$\begin{aligned} \langle \mathcal{J}_{+1} \rangle &= \langle j_{+1} \rangle - \langle j_{-4} \rangle, \\ \langle \mathcal{J}_{+3} \rangle &= \langle j_{+3} \rangle - \langle j_{-2} \rangle - \langle j_{+4} \rangle, \\ \langle \mathcal{J}_{+6} \rangle &= \langle j_{+6} \rangle - \langle j_{-5} \rangle - \langle j_{-4} \rangle, \\ \langle \mathcal{J}_{-1} \rangle &= \langle j_{-1} \rangle - \langle j_{+4} \rangle, \\ \langle \mathcal{J}_{-3} \rangle &= \langle j_{-3} \rangle - \langle j_{+2} \rangle - \langle j_{-4} \rangle, \\ \langle \mathcal{J}_{-6} \rangle &= \langle j_{-6} \rangle - \langle j_{+5} \rangle - \langle j_{+4} \rangle \end{aligned} \quad (15.61)$$

by expanding equation (15.57) into equation (15.10).

Table 15.2. Examples of system quantity–intensive field conjugated pairs in the entropy representation Note: $\beta_r := 1/T_r$ denotes the inverse temperature of the reservoir. Charges are carried by particles; the conjugated pair $(Q_n, -\beta_r V_r)$ is usually embedded in $(N_n, -\beta_r \mu_r)$.

System quantity X^κ	Intensive field $f_{(\kappa,r)}$
Energy, E_n	Inverse temperature, β_r
Particle number, N_n	Chemical potential, $-\beta_r \mu_r$
Charge, Q_n	Electric potential, $-\beta_r V_r$
Displacement, X_n	Generic force, $-\beta_r k_r$
Angle, θ_n	Torque, $-\beta_r \tau_r$

Stochastic Thermodynamics

The results obtained until are mathematical and have *a priori* no connection to physics. We now specify the conditions under which a Markov jump process describes the dynamics of an open physical system in contact with multiple reservoirs. This will enable us to introduce physically motivated decompositions and derive DFTs with a clear thermodynamic interpretation.

Each system state, n, is now characterized by given values of some *system quantities*, $\{X_n^\kappa\}$, for $\kappa = 1, \ldots, N_\kappa$, which include the internal energy, E_n, and possibly additional quantities (see table 15.2 for some examples). These must be regarded as globally *conserved quantities*, as their change in the system is always balanced by an opposite change in the reservoirs. When labeling the reservoirs with $\{r\}$, for $r = 1, \ldots, N_r$, the *balance equation* for X^κ along the transition e can be written as

$$X_{n'}^\kappa D_e^{n'} = \delta_i X_e^\kappa + \sum_r \delta X_e^{(\kappa,r)}. \quad (15.62)$$

The left-hand side is the overall change in the system, whereas $\delta_i X_e^\kappa$ denotes the changes due to internal transformations (e.g., chemical reactions, Schmiedl and Seifert 2007; Rao and Esposito 2018a), and $\delta X_e^{(\kappa,r)}$ quantifies the amount of X^κ supplied by the reservoir r to the system along the transition e. For the purpose of our discussion, we introduce the index $y = (\kappa, r)$, that is,

Chapter 15: Detailed Fluctuation Theorems

Table 15.3. Summary of the indices used throughout the chapter and the objects they label.

Index	Label for	Number
n	State	N_n
e	Transition	N_e
κ	System quantity	N_κ
r	Reservoir	N_r
$y \equiv (\kappa, r)$	Conserved quantity X^κ from reservoir r	N_y
λ	Conservation law and conserved quantity	N_λ
y_p	"Potential" y	N_λ
y_f	"Force" y	$N_y - N_\lambda$

the conserved quantity X^κ exchanged with the reservoir r, and we define the matrix δX whose entries are $\{\delta X_e^y \equiv \delta X_e^{(\kappa,r)}\}$. All indices used in the following discussion are summarized in table 15.3. Microscopic reversibility requires that $\delta X_e^y = -\delta X_{-e}^y$. We notice that more than one reservoir may be involved in each transition (see fig. 15.4).

In addition to the trivial set of conserved quantities $\{X^\kappa\}$, the system may be characterized by some additional quantities *specific* to each system. We now sketch the systematic procedure to

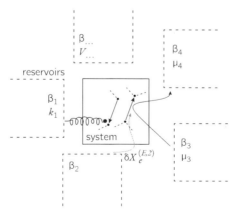

Figure 15.4. Pictorial representation of a system coupled to several reservoirs. Transitions may involve more than one reservoir and exchange between reservoirs. Work reservoirs are also taken into account.

identify these quantities and the corresponding conservation laws (Polettini, Bulnes Cuetara, and Esposito 2016; Rao and Esposito 2018b). Algebraically, conservation laws can be identified as a maximal set of independent vectors in the y-space, $\{\ell^\lambda\}$, for $\lambda = 1,\ldots,N_\lambda$ such that

$$\ell^\lambda_y \delta X^y_{e'} C^{e'}_e = 0 \quad \text{for all cycles, i.e., for all } e \in \mathcal{T}^*. \quad (15.63)$$

Indeed, the quantities $\{\ell^\lambda_y \delta X^y_e\}$, for $\lambda = 1,\ldots,N_\lambda$ are combinations of exchange contributions $\{\delta X^y_e\}$ for $y = 1,\ldots,N_\lambda$, which vanish along all cycles. They must therefore identify some state variables, $\{L^\lambda\}$, for $\lambda = 1,\ldots,N_y$ in the same way curl-free vector fields are conservative and identify scalar potentials:

$$L^\lambda_n D^n_e = \ell^\lambda_y \delta X^y_e \equiv \sum_r \left\{ \sum_\kappa \ell^\lambda_{(\kappa,r)} \delta X^{(\kappa,r)}_e \right\}. \quad (15.64)$$

This equation can be regarded as the balance equation for the conserved quantities. In the absence of internal transformations, $\delta_i X^\kappa_e$, trivial conservation laws correspond to $\ell^\kappa_y \equiv \ell^\kappa_{(\kappa',r)} = \delta^\kappa_{\kappa'}$ so that the balance equations (eq. 15.62) are recovered. Notice that each L^λ is defined up to a reference value.

Each reservoir r is characterized by a set of *entropic intensive fields* conjugated to the exchange of the system quantities $\{X^\kappa\}$, $\{f_{(\kappa,r)}\}$ for $\kappa = 1,\ldots,N_\kappa$ (see, e.g., Callen 1985, §2–3). A short list of X^κ–$f_{(\kappa,r)}$ conjugated pairs is reported in table 15.2. The thermodynamic consistency of the stochastic dynamics is ensured by the *local detailed balance*:

$$\ln \frac{w_e}{w_{-e}} = -f_y \delta X^y_e + S_n D^n_e. \quad (15.65)$$

This relates the log ratio of the forward and backward transition rates to the entropy change in the reservoirs resulting from the transfer of system quantities during that transition. The entropy change is evaluated using equilibrium thermodynamics (in the reservoirs) and reads $\{\delta S_e^r = -f_y \delta X_e^y\}$. The second term on the right-hand side is the internal entropy change occurring during the transition, as S_n quantifies the internal entropy of the state n. This term can be seen as the outcome of a coarse-graining procedure over a finer description in which multiple states with the same system quantities are collected in one single n (Esposito 2012). Using equation (15.65), the affinities of equation (15.11) can be rewritten as

$$A_e = \sum_r \left[-\sum_\kappa f_{(\kappa,r)} \delta X_e^{(\kappa,r)} \right] + [S_n - \ln p_n] D_e^n. \quad (15.66)$$

This relation shows that the affinity is the entropy change in all reservoirs plus the system entropy change. In other words, whereas equation (15.64) characterizes the balance of the conserved quantities along the transitions, equation (15.66) characterizes the corresponding lack of balance for entropy, namely, the second law.

As for the transition rates, the changes in time of the internal entropy S, the conserved quantities $\{X^\kappa\}$ (hence $\{\delta X_e^y\}$), and their conjugated fields $\{f_y\}$ are all encoded in the protocol function π_t. Physically, this modeling describes the two possible ways of controlling a system: either through $\{X^\kappa\}$ or S, which characterize the system states, or through $\{f_y\}$, which characterize the properties of the reservoirs.

Example. We illustrate the role of system-specific conservation laws by considering the double quantum dot (QD) depicted in figure 15.5A (Sánchez and Büttiker 2012; Strasberg *et al.* 2013; Thierschmann *et al.* 2015), whose networks of transition and energy landscapes are drawn in figures 15.1 and 15.5B, respectively. Electrons can enter empty dots from the reservoirs but cannot jump from one dot to the other. When the two dots are

THE ENERGETICS OF COMPUTING IN LIFE & MACHINES

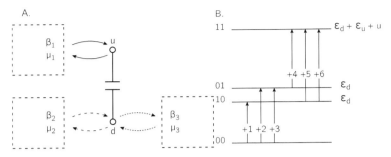

Figure 15.5. Double coupled QD in contact with three reservoirs. Transitions related to the first reservoir are depicted using solid lines, whereas those related to the second and third reservoirs use a dashed and a dotted line, respectively. The graphical rule was applied to the network of transitions in figure 15.1. (A) Pictorial representation of the system scheme. The upper dot u is in contact with the first reservoir, whereas the lower dot d is in contact with the second and third reservoirs. Energy and electrons are exchanged, but the dots cannot host more than one electron. (B) Energy landscape of the dot. When both dots are occupied, 11, a repulsive energy u adds to the occupied dots' energies, ϵ_u and ϵ_d.

occupied, an interaction energy, u, arises. Energy, E_n, and total number of electrons, N_n, characterize each state of the system:

$$E_{00} = 0, \quad E_{10} = \epsilon_u, \quad E_{01} = \epsilon_d, \quad E_{11} = \epsilon_u + \epsilon_d + u,$$
$$N_{00} = 0, \quad N_{10} = 1, \quad N_{01} = 1, \quad N_{11} = 2,$$

$$(15.67)$$

where the first entry in n refers to the occupancy of the upper dot, and the second to the lower. The entries of the matrix δX for the forward transitions are

$$\delta X = \begin{pmatrix} & +1 & +2 & +3 & +4 & +5 & +6 \\ (E,1) & \epsilon_u & 0 & 0 & \epsilon_u + u & 0 & 0 \\ (N,1) & 1 & 0 & 0 & 1 & 0 & 0 \\ (E,2) & 0 & \epsilon_d & 0 & 0 & \epsilon_d + u & 0 \\ (N,2) & 0 & 1 & 0 & 0 & 1 & 0 \\ (E,3) & 0 & 0 & \epsilon_d & 0 & 0 & \epsilon_d + u \\ (N,3) & 0 & 0 & 1 & 0 & 0 & 1 \end{pmatrix}$$

$$(15.68)$$

(see fig. 15.1), whereas the entries related to the backward transition follow from $\delta X^y_{-e} = -\delta X^y_e$. For instance, along the first transition, the system gains ϵ_u energy and one electron from the reservoir 1. The vector of entropic intensive fields is given by

$$f = \begin{pmatrix} \overset{(E,1)}{\beta_1} & \overset{(N,1)}{-\beta_1\mu_1} & \overset{(E,2)}{\beta_2} & \overset{(N,2)}{-\beta_2\mu_2} & \overset{(E,3)}{\beta_3} & \overset{(N,3)}{-\beta_3\mu_3} \end{pmatrix}. \quad (15.69)$$

Because the QDs and the electrons have no internal entropy, $S_n = 0$ for all n, the local detailed balance property, equation (15.65), can be easily recovered from the product $-f\delta X$. From a stochastic dynamics perspective, this property arises when considering fermionic transition rates, namely, $w_e = \Gamma_e(1 + \exp\{f_y\delta X_e^y\})^{-1}$ and $w_{-e} = \Gamma_e \exp\{f_y\delta X_e^y\}(1 + \exp\{f_y\delta X_e^y\})^{-1}$, for electrons entering and leaving the dot.

A maximal set of independent vectors in y-space satisfying equation (15.63) is composed of

$$\ell^E = \begin{pmatrix} \overset{(E,1)}{1} & \overset{(N,1)}{0} & \overset{(E,2)}{1} & \overset{(N,2)}{0} & \overset{(E,3)}{1} & \overset{(N,3)}{0} \end{pmatrix},$$

$$\ell^u = \begin{pmatrix} \overset{(E,1)}{0} & \overset{(N,1)}{1} & \overset{(E,2)}{0} & \overset{(N,2)}{0} & \overset{(E,3)}{0} & \overset{(N,3)}{0} \end{pmatrix},$$

$$\ell^d = \begin{pmatrix} \overset{(E,1)}{0} & \overset{(N,1)}{0} & \overset{(E,2)}{0} & \overset{(N,2)}{1} & \overset{(E,3)}{0} & \overset{(N,3)}{1} \end{pmatrix}. \quad (15.70)$$

The first vector identifies the energy state variable, E_n:

$$\ell^E \delta X = \begin{pmatrix} \overset{+1}{\epsilon_u} & \overset{+2}{\epsilon_d} & \overset{+3}{\epsilon_d} & \overset{+4}{\epsilon_u + u} & \overset{+5}{\epsilon_d + u} & \overset{+6}{\epsilon_d + u} \end{pmatrix} \equiv \{E_n D_e^n\}. \quad (15.71)$$

The other two, instead, give the occupancy of the upper and lower dots, N_n^u and N_n^d:

$$\ell^u \delta X = \begin{pmatrix} \overset{+1}{1} & \overset{+2}{0} & \overset{+3}{0} & \overset{+4}{1} & \overset{+5}{0} & \overset{+6}{0} \end{pmatrix} \equiv \{N_n^u D_e^n\}$$

$$\ell^d \delta X = \begin{pmatrix} \overset{+1}{0} & \overset{+2}{1} & \overset{+3}{1} & \overset{+4}{0} & \overset{+5}{1} & \overset{+6}{1} \end{pmatrix} \equiv \{N_n^d D_e^n\}. \quad (15.72)$$

A posteriori, we see that these conservation laws arise from the fact that no electron transfer from one dot to the other is allowed. The total occupancy of the system, N_n, is recovered from the sum of the last two vectors.

Now that a nonequilibrium thermodynamics has been built on top of the Markov jump process, we can proceed by considering two physically relevant p_n^{ref}.

System–Reservoir Decomposition

We start by considering a microcanonical PMF as reference:

$$p_n^{\text{ref}} = p_n^{\text{mc}} := \exp\{S_n - S_{\text{mc}}\}, \quad (15.73)$$

where

$$S_{\text{mc}} = \ln \sum_m \exp S_m \quad (15.74)$$

is the *Boltzmann's equilibrium entropy*. With this choice, the reference affinities become sums of entropy changes in the reservoirs:

$$A_e^{\text{ref}} = \delta S_e^{\text{r}} = -f_y \delta X_e^y; \quad (15.75)$$

hence the nonconservative contribution becomes the rate of entropy change in all reservoirs,

$$\langle \dot{\Sigma}_{\text{nc}} \rangle = \langle \dot{S}_{\text{r}} \rangle = -f_y \delta X_e^y \langle j^e \rangle. \quad (15.76)$$

For the conservative contribution, instead, one obtains

$$\langle \dot{\Sigma}_{\text{c}} \rangle = [S_n - \ln p_n] D_e^n \langle j^e \rangle. \quad (15.77)$$

Using equation (15.17), equation (15.77) can be rewritten in terms of the Gibbs–Shannon entropy,

$$\langle S \rangle = \sum_n p_n [S_n - \ln p_n], \quad (15.78)$$

and the Boltzmann entropy. Indeed,

$$\mathcal{D}(p \parallel p^{\text{mc}}) = S_{\text{mc}} - \langle S \rangle \quad (15.79)$$

and
$$\langle \dot{\Sigma}_d \rangle = d_t \mathcal{S}_{mc} - \sum_n p_n d_t S_n \qquad (15.80)$$
so that
$$\langle \dot{\Sigma}_c \rangle = d_t \langle \mathcal{S} \rangle - \sum_n p_n d_t S_n. \qquad (15.81)$$

The conservative contribution thus contains changes in the system entropy caused by the dynamics and the external drive.

The EP decomposition in equation (15.14) with equations (15.76) and (15.81) is thus the well-known *system–reservoir* decomposition, that is, the traditional *entropy balance*. Because the same decomposition holds at the trajectory level, if the initial PMF of the forward and backward processes are microcanonical, the DFT and IFT hold by applying equations (15.32) and (15.34). When the driving does not affect the internal entropy of the system states $\{S_n\}$, the DFT and IFT hold for the reservoir entropy alone. Finally, the fluctuating quantity appearing in the DFT, $\Sigma_d + \Sigma_{nc}$, can be interpreted as the EP of the extended process in which, at time t, the driving is stopped, all temperatures are raised to infinity, $\beta_r \to 0$, and the system is left to relax to equilibrium—the initial PMF of the backward process.

Conservative–Nonconservative Decomposition

We now turn to a reference PMF that accounts for conservation laws: the *generalized Gibbs PMF*.

To characterize this PMF, we observe that because $\{\ell^\lambda\}$ are linearly independent—otherwise, we would have linearly dependent conserved quantities—one can always identify a set of ys, denoted by $\{y_p\}$, such that the matrix whose rows are $\{\ell^\lambda_{y_p}\}$, for $\lambda = 1, \ldots, N_\lambda$, is nonsingular. We denote by $\{\overline{\ell}^{y_p}_\lambda\}$, for $\lambda = 1, \ldots, N_\lambda$, the columns of the inverse matrix. All other ys are denoted by $\{y_f\}$. Using the splitting $\{y_p\}$–$\{y_f\}$ and the properties of $\{\ell_{y_p}^\lambda\}$, in combination with the balance equation for conserved quantities (eq. 15.64), the local detailed-balance equation (15.65) can be decomposed as

$$\ln\frac{w_e}{w_{-e}} = \mathcal{F}_{y_f}\delta X_e^{y_f} + \left[S_n - F_\lambda L_n^\lambda\right] D_e^n, \qquad (15.82)$$

where

$$F_\lambda = f_{y_p}\overline{\ell}_\lambda^{y_p} \qquad (15.83)$$

are the system-specific intensive fields conjugated to the conserved quantities and

$$\mathcal{F}_{y_f} := F_\lambda \ell_{y_f}^\lambda - f_{y_f} \qquad (15.84)$$

are differences of intensive fields called nonconservative *fundamental forces*. Indeed, these nonconservative forces are responsible for breaking detailed balance. When they all vanish, $\mathcal{F}_{y_f} = 0$ for all y_f, the system is indeed detailed balanced, and the PMF

$$p_n^{gg} := \exp\left\{S_n - F_\lambda L_n^\lambda - \Phi_{gg}\right\}, \qquad (15.85)$$

with $\Phi_{gg} := \ln\sum_n \exp\{S_n - F_\lambda L_n^\lambda\}$, satisfies the detailed balance property in equation (15.9). The potential corresponding to equation (15.85), ψ_n^{gg}, is minus the *Massieu potential*, which is constructed by using all conservation laws (see, e.g., § 5-4 and 19-1 of Callen 1985; Peliti 2011, § 3.13). Choosing the PMF (eq. 15.85) as a reference, $p_n^{ref} = p_n^{gg}$, the reference affinity straightforwardly ensues from equation (15.82):

$$A_e^{ref} = A_e^{gg} = \mathcal{F}_{y_f}\delta X_e^{y_f}. \qquad (15.86)$$

Hence

$$\langle\dot{\Sigma}_{nc}\rangle = \mathcal{F}_{y_f}\langle I^{y_f}\rangle, \qquad (15.87)$$

where

$$\langle I^{y_f}\rangle = \delta X_e^{y_f}\langle j^e\rangle \qquad (15.88)$$

are the fundamental currents conjugated to the forces. For the conservative contribution, one obtains

$$\langle \dot{\Sigma}_c \rangle = \left[S_n - F_\lambda L_n^\lambda - \ln p_n \right] D_e^n \langle j^e \rangle. \tag{15.89}$$

When written as in equation (15.17), its two contributions are

$$\mathcal{D}(p \parallel p^{\text{gg}}) = \Phi_{\text{gg}} - \sum_n p_n \left[S_n - F_\lambda L_n^\lambda - \ln p_n \right], \tag{15.90}$$

which relates the equilibrium Massieu potential to its averaged nonequilibrium counterpart, and

$$\langle \dot{\Sigma}_d \rangle = d_t \Phi_{\text{gg}} - \sum_n p_n d_t \left[S_n - F_\lambda L_n^\lambda - \ln p_n \right], \tag{15.91}$$

which quantifies the dissipation due to external manipulations of $\{ S_n \}$, the fields $\{ F_\lambda \}$, and the conserved quantities $\{ L^\lambda \}$. We emphasize that because ψ_n^{gg} encompasses all conserved quantities, $\langle \dot{\Sigma}_c \rangle$ captures all dissipative contributions due to conservative forces. Hence $\langle \dot{\Sigma}_{nc} \rangle$ consists of a minimal number, $N_y - N_\lambda$, of purely nonconservative contributions. The EP decomposition equation (15.14) with equations (15.87) and (15.89) is the *conservative–nonconservative* decomposition of the EP obtained in Rao and Esposito (2018b).

The conservative–nonconservative splitting of the EP can be made at the trajectory level, too. Hence, if the initial condition of the forward and backward processes is of the form in equation (15.85), the DFT and IFT given by equations (15.32) and (15.34) hold.

Here, too, the fluctuating quantity appearing in the DFT, $\Sigma_d + \Sigma_{nc}$, can be interpreted as the EP of an extended process including relaxation, but for nonisothermal processes, the procedure can be significantly more involved. The details of this discussion can be found in Rao and Esposito (2018b).

Example. We now provide the expressions of ψ_n^{ref} and A_e^{ref} for the double QD discussed in the previous example (fig. 15.5). Therefore we split the set $\{y\}$ in $\{y_p\} = \{(E,1), (N,1), (N,2)\}$ and $\{y_f\} = \{(E,2), (E,3), (N,3)\}$, which is valid because the matrix whose entries are $\{\ell_{y_p}^\lambda\}$ is an identity matrix (see eq. 15.70). The fields conjugated with the complete set of conservation laws (eq. 15.83) are

$$F_E = \beta_1, \quad F_u = -\beta_1\mu_1, \text{ and } \quad F_d = -\beta_2\mu_2, \quad (15.92)$$

from which the reference potential of the state n (eq. 15.85) follows:

$$\psi_n^{gg} = \Phi^{gg} - \left[-\beta_1 E_n + \beta_1\mu_1 N_n^u + \beta_2\mu_2 N_n^d\right]. \quad (15.93)$$

Instead, the fundamental forces (eq. 15.84) are given by

$$\begin{aligned}\mathcal{F}_{(E,2)} &= \beta_1 - \beta_2, \\ \mathcal{F}_{(E,3)} &= \beta_1 - \beta_3, \text{ and} \\ \mathcal{F}_{(N,3)} &= \beta_3\mu_3 - \beta_2\mu_2,\end{aligned} \quad (15.94)$$

from which the reference affinities follow (eq. 15.86). The first two forces drive the energy flowing into the first reservoir from the second and third reservoirs, respectively, whereas the third force drives the electrons flowing from the third to the second reservoir.

Conclusions

In this chapter, we presented a general method to construct DFTs for Markov jump processes. The strategy to identify the fluctuating quantities that satisfy the DFT consists of splitting the EP in two by making use of a reference PMF. The choice of the reference PMF is arbitrary for IFTs but must depend solely on the driving protocol for DFTs. Out of the infinite number of FTs that may be considered, we tried to select those that have interesting mathematical properties or that can be expressed in terms of physical quantities when the Markov jump process is complemented with a thermodynamic structure. Table 15.1 summarizes the terms of the EP for each of our choices. We also emphasize that the EP always satisfies an IFT but generically not a DFT. Connections to information theory were also made by formulating a generalized Landauer principle.

We do not claim to have been exhaustive, and many other reference PMFs may be interesting to examine. We can mention at least two more interesting cases. By considering the steady-state PMF obtained when removing some edges from the graph—but not all chords, as in the section "Cycle–Cocycle Decomposition"—the marginal thermodynamic theory presented in Polettini and Esposito (2017, 2018) emerges. One can also consider a reference PMF in between the microcanonical PMF, which takes no conserved quantity into account, and the generalized Gibbs PMF, which takes them all into account. This happens, for instance, when only the obvious conserved quantities are accounted for, $\{X^\kappa\}$, as discussed in Bulnes Cuetara, Esposito, and Imparato (2014). In this case, one uses the fields of a given reservoir to define the reference equilibrium potential:

$$\psi_n^{\text{ref}} = \Phi - \left[S_n - \sum_\kappa f_{(\kappa,1)} \delta X_n^\kappa\right],$$

where Φ is determined by the normalization. The number of nonconservative forces appearing in $\langle \dot{\Sigma}_{\text{nc}} \rangle$ will be $N_y - N_\kappa$. But in the case that additional conservation laws are present ($N_\lambda > N_\kappa$), some of these forces are dependent on others, and their number will be larger than the minimum, $N_y - N_\lambda$. ☙

Acknowledgments

This work was funded by the *Luxembourg National Research Fund* (AFR PhD Grant 2014-2, No. 9114110), the *European Research*

Council (project NanoThermo, ERC-2015-CoG Agreement No. 681456), and the *National Science Foundation* (NSF Grant No. PHY-1748958). The authors thank A. Wachtel and A. Lazarescu for valuable feedback on the chapter.

Appendix

The following abbreviations are used in this manuscript:

DFT   detailed fluctuation theorem
IFT   integral fluctuation theorem
PMF   probability mass function
EP   entropy production
ME   master equation
MGF   moment generating function

MOMENT GENERATING FUNCTION DYNAMICS AND PROOFS OF THE FTS

We describe the moment generating function (MGF) technique that we use to prove the finite-time DFTs (eq. 15.32) (Rao and Esposito 2018b).

MGF DYNAMICS

Let $P_t(n, \delta O)$ be the joint probability of observing a trajectory ending in the state n along which the change of a generic observable, O, is δO. The changes of O along edges are denoted as $\{\delta O_e\}$, whereas the changes due to time-dependent driving while in the state n are denoted as $\dot{O}_n$. To write an evolution equation for this probability, let us expand it as

$$P_{t+dt}(n, \delta O)$$
$$\simeq \sum_e w_e \delta_{n,\mathfrak{t}(e)} P_t\left(\mathfrak{o}(e), \delta O - \delta O_e - \dot{O}_{\mathfrak{o}(e)} dt\right) dt \quad (15.95)$$
$$+ \left[1 - \sum_e w_e \delta_{n,\mathfrak{o}(e)} dt\right] P_t(n, \delta O - \dot{O}_n dt).$$

The first term accounts for transitions leading to the state n and completing the change of O, whereas the second describes the probability of completing the change of O while dwelling in the state n (and not leaving it). When keeping only the linear term in dt and performing the limit d$t \to 0$, we get

$$\mathrm{d}_t P_t(n, \delta O) = \sum_e w_e \delta_{n,\mathfrak{t}(e)} P_t(\mathfrak{o}(e), \delta O - \delta O_e) \\ - \sum_e w_e \delta_{n,\mathfrak{o}(e)} P_t(n, \delta O) - \dot{O}_n \partial_{\delta O} P_t(n, \delta O). \quad (15.96)$$

Rather than working with this differential equation, it is much more convenient to deal with the bilateral Laplace transform of $p_t(n, \delta O)$, that is, the MGF up to a sign,

$$\Lambda_{n,t}(q) := \int_{-\infty}^{\infty} \mathrm{d}\,\delta O \exp\{-q \delta O\} P_t(n, \delta O), \quad (15.97)$$

because its evolution equation is akin to a ME (eq. 15.2):

$$\mathrm{d}_t \Lambda_{n,t}(q) = \sum_m W_{nm,t}(q) \Lambda_{m,t}(q), \quad (15.98)$$

where the *biased rate matrix* reads $W_{nm,t}(q) =$

$$\sum_e w_e \left\{ \exp\{-q \delta O_e\} \delta_{n,\mathfrak{t}(e)} \delta_{m,\mathfrak{o}(e)} - \delta_{n,m} \delta_{m,\mathfrak{o}(e)} \right\} - q \dot{O}_n \delta_{n,m}. \quad (15.99)$$

The field q is usually referred to as a *counting field*. This equation is obtained by combining equations (15.96) and (15.97), and its initial condition must be $\Lambda_{n,0}(\delta O) = p_n(0)$. Note that equation (15.98) is not a ME because $\sum_n \Lambda_{n,t}(\delta O)$ is not conserved.

For later convenience, we recast equation (15.98) into a bracket notation,

$$\mathrm{d}_t |\Lambda_t(q)\rangle = \mathcal{W}_t(q) |\Lambda_t(q)\rangle, \quad (15.100)$$

and we proceed to prove a preliminary result. A formal solution of equation (15.98) is $|\Lambda_t(q)\rangle = \mathcal{U}_t(q) |P(0)\rangle$, where the time-evolution operator reads $\mathcal{U}_t(q) = \mathsf{T}_+ \exp \int_0^t \mathrm{d}\tau\, \mathcal{W}_\tau(q)$, T_+

being the time-ordering operator. We clearly have $d_t \mathcal{U}_t(q) = \mathcal{W}_t(q)\mathcal{U}_t(q)$. Let us now consider the following transformed evolution operator:

$$\tilde{\mathcal{U}}_t(q) := \mathcal{X}_t^{-1}\mathcal{U}_t(q)\mathcal{X}_0, \qquad (15.101)$$

where $\mathcal{X}_t$ is a generic time-dependent invertible operator. Its dynamics is ruled by the following biased stochastic dynamics:

$$d_t \tilde{\mathcal{U}}_t(q) = d_t \mathcal{X}_t^{-1} \mathcal{U}_t(q)\mathcal{X}_0 + \mathcal{X}_t^{-1} d_t \mathcal{U}_t(q)\mathcal{X}_0$$
$$= \{d_t \mathcal{X}_t^{-1} \mathcal{X}_t + \mathcal{X}_t^{-1}\mathcal{W}_t(q)\mathcal{X}_t\}\tilde{\mathcal{U}}_t(q) \equiv \tilde{\mathcal{W}}_t(q)\tilde{\mathcal{U}}_t(q), \qquad (15.102)$$

which allows us to conclude that the transformed time-evolution operator is given by

$$\tilde{\mathcal{U}}(q) = \mathsf{T}_+ \exp\int_0^t d\tau\, \tilde{\mathcal{W}}_\tau(q). \qquad (15.103)$$

From equations (15.101), (15.102), and (15.103), we deduce that

$$\mathcal{X}_t^{-1}\mathcal{U}_t(q)\mathcal{X}_0 = \mathsf{T}_+ \exp\int_0^t d\tau\, \left[d_\tau \mathcal{X}_\tau^{-1}\mathcal{X}_\tau + \mathcal{X}_\tau^{-1}\mathcal{W}_\tau(q)\mathcal{X}_\tau\right]. \qquad (15.104)$$

PROOF OF THE DFT

To prove the DFT (eq. 15.32), we briefly recall its two assumptions:

1. The reference PMF depends on time solely via the protocol function.

2. For both the forward and backward processes, the system is initially prepared in a reference PMF.

Let $P_t(n, \Sigma_\mathrm{d}, \Sigma_\mathrm{nc})$ be the joint probability of observing a trajectory ending in the state n along which the driving contribution is Σ_d, while the nonconservative contribution is Σ_nc.

Chapter 15: Detailed Fluctuation Theorems

The aforementioned probabilities, one for each n, are stacked in the ket $|P_t(\Sigma_d, \Sigma_{nc})\rangle$. The time evolution of the related MGF, $|\Lambda_t(q_d, q_{nc})\rangle$

$$:= \int_{-\infty}^{\infty} d\Sigma_d d\Sigma_{nc} \exp\{-q_d \Sigma_d - q_{nc}\Sigma_{nc}\} |P_t(\Sigma_d, \Sigma_{nc})\rangle, \tag{15.105}$$

is ruled by the biased stochastic dynamics of equation (15.98):

$$d_t |\Lambda_t(q_d, q_{nc})\rangle = \mathcal{W}_t(q_d, q_{nc}) |\Lambda_t(q_d, q_{nc})\rangle, \tag{15.106}$$

where the entries of the biased generator are given by

$$\mathcal{W}_{nm}(q_d, q_{nc}) = \sum_e w_e \{\exp\{-q_{nc} A_e^{\text{ref}}\} \delta_{n,t(e)}\delta_{m,o(e)} - \delta_{n,m}\delta_{m,o(e)}\} - q_d d_t \psi_m \delta_{n,m}. \tag{15.107}$$

Using the definition of reference affinity (eq. 15.13), one can see that the rate matrix satisfies the following symmetry:

$$\mathcal{W}_t^{\mathsf{T}}(q_d, q_{nc}) = \mathcal{P}_t^{-1} \mathcal{W}_t(q_d, 1 - q_{nc}) \mathcal{P}_t, \tag{15.108}$$

where the entries of $\mathcal{P}_t$ are given by

$$\mathcal{P}_{nm,t} := \exp\{-\psi_m^{\text{ref}}(\pi_t)\} \delta_{n,m} \tag{15.109}$$

and ^{T} denotes the transposition. Also, the initial condition is given by the reference PMF,

$$|\Lambda_0(q_d, q_{nc})\rangle = |p_0^{\text{ref}}\rangle = \mathcal{P}_0 |1\rangle, \tag{15.110}$$

where $|1\rangle$ denotes the vector in the state space whose entries are all equal to 1.

Using the formal solution of equation (15.106), the MGF of $P_t(\Sigma_d, \Sigma_{nc})$ can be written as

$$\Lambda_t(q_d, q_{nc}) \qquad (15.111)$$
$$= \langle 1 | \Lambda_t(q_d, q_{nc}) \rangle$$
$$= \langle 1 | \mathcal{U}_t(q_d, q_{nc}) \mathcal{P}_0 | 1 \rangle$$
$$= \langle 1 | \mathcal{P}_t \mathcal{P}_t^{-1} \mathcal{U}_t(q_d, q_{nc}) \mathcal{P}_0 | 1 \rangle,$$

where $\mathcal{U}_t(q_d, q_{nc})$ is the related time-evolution operator. Using the relation in equation (15.104), the last term can be recast into

$$\Lambda_t(q_d, q_{nc})$$
$$= \langle p_t^{\text{ref}} | \mathsf{T}_+ \exp \left\{ \int_0^t d\tau \left[d_\tau \mathcal{P}_\tau^{-1} \mathcal{P}_\tau + \mathcal{P}_\tau^{-1} \mathcal{W}_\tau(q_d, q_{nc}) \mathcal{P}_\tau \right] \right\} | 1 \rangle. \qquad (15.112)$$

Because $d_\tau \mathcal{P}_\tau^{-1} \mathcal{P}_\tau = \text{diag}\{d_\tau \psi_n^{\text{ref}}\}$, the first term in square brackets can be added to the diagonal entries of the second term, thus giving

$$\Lambda_t(q_d, q_{nc}) = \langle p_t^{\text{ref}} | \mathsf{T}_+ \exp \left\{ \int_0^t d\tau \left[\mathcal{P}_\tau^{-1} \mathcal{W}_\tau(q_d - 1, q_{nc}) \mathcal{P}_\tau \right] \right\} | 1 \rangle. \qquad (15.113)$$

The symmetry (eq. 15.108) allows us to recast the latter into

$$\Lambda_t(q_d, q_{nc}) = \langle p_t^{\text{ref}} | \mathsf{T}_+ \exp \left\{ \int_0^t d\tau \, \mathcal{W}_\tau^{\mathsf{T}}(q_d - 1, 1 - q_{nc}) \right\} | 1 \rangle. \qquad (15.114)$$

The crucial step comes as we time-reverse the integration variable: $\tau \to t - \tau$. Accordingly, the time-ordering operator, T_+, becomes an anti-time-ordering operator T_-, while the diagonal entries of the biased generator become

$$W_{mm,t-\tau}(q_d, q_{nc}) = -\sum_e w_e(\pi_{t-\tau}) \delta_{m,o(e)} - q_d \, d_{t-\tau} \psi_m^{\text{ref}}(\pi_{t-\tau})$$
$$= -\sum_e w_e(\pi_\tau^\dagger) \delta_{m,o(e)} + q_d \, d_\tau \psi_m^{\text{ref}}(\pi_\tau^\dagger), \qquad (15.115)$$

from which we conclude that

$$W_{nm,t-\tau}(q_d, q_{nc}) = W_{nm,\tau}^\dagger(-q_d, q_{nc}). \qquad (15.116)$$

Crucially, the assumption that ψ_n^{ref} depends on time via π_τ ensures that $\mathcal{W}_\tau^\dagger(q_\text{d}, q_\text{nc})$ can be regarded as the biased generator of the dynamics subject to the time-reversed protocol, that is, the dynamics of the backward process. If we were to consider an arbitrary p_n^{ref}—that is, the forward process would start from an arbitrary PMF—then $\mathcal{W}_\tau^\dagger(q_\text{d}, q_\text{nc})$ would be the rate matrix of the time-reversed stochastic dynamics,

$$\begin{aligned} 0 &= \sum_m \left[\delta_{nm} \text{d}_{t-\tau} - W_{nm}(\pi_{t-\tau}) \right] p_m \\ &= \sum_m \left[-\delta_{nm} \text{d}_\tau - W_{nm}(\pi_\tau^\dagger) \right] p_m, \end{aligned} \quad (15.117)$$

which is unphysical. Equation (15.114) thus becomes

$$\Lambda_t(q_\text{d}, q_\text{nc}) = \langle p_t^{\text{ref}} | \mathsf{T}_- \exp\left\{ \int_0^t \text{d}\tau \, \mathcal{W}_\tau^{\dagger\mathsf{T}} (1 - q_\text{d}, 1 - q_\text{nc}) \right\} | 1 \rangle. \quad (15.118)$$

Upon a global transposition, we can write

$$\Lambda_t(q_\text{d}, q_\text{nc}) = \langle 1 | \mathsf{T}_+ \exp\left\{ \int_0^t \text{d}\tau \, \mathcal{W}_\tau^\dagger (1 - q_\text{d}, 1 - q_\text{nc}) \right\} | p_t^{\text{ref}} \rangle, \quad (15.119)$$

where we also used the relationship between transposition and time ordering:

$$\mathsf{T}_+ \left(\prod_i A_{t_i}^\mathsf{T} \right) = \left(\mathsf{T}_- \prod_i A_{t_i} \right)^\mathsf{T}, \quad (15.120)$$

in which A_t is a generic operator. From the last expression, we readily obtain the symmetry that we are looking for:

$$\Lambda_t(q_\text{d}, q_\text{nc}) = \Lambda_t^\dagger (1 - q_\text{d}, 1 - q_\text{nc}), \quad (15.121)$$

where $\Lambda_t^\dagger(q_\text{d}, q_\text{nc})$ is the MGF of $P_t^\dagger(\Sigma_\text{d}, \Sigma_\text{nc})$. Indeed, its inverse Laplace transform gives the DFT in equation (15.32).

PROOF OF THE DFT FOR THE SUM OF DRIVING AND NONCONSERVATIVE EP CONTRIBUTIONS

Let us define $\Sigma_\text{s} := \Sigma_\text{d} + \Sigma_\text{nc}$ as the sum of the driving and the nonconservative EP contributions. A straightforward calculation leads from equation (15.32) to the DFT for Σ_s, equation (15.33):

$$\begin{aligned} P_t(\Sigma s) &= \int d\Sigma_d d\Sigma_{nc}\, P_t(\Sigma_d, \Sigma_{nc})\, \delta\left(\Sigma_s - \Sigma_d - \Sigma_{nc}\right) \\ &= \int d\Sigma_d\, P_t(\Sigma_d, \Sigma s - \Sigma_d) \\ &= \exp\Sigma_s \int d\Sigma_d\, P_t^\dagger(-\Sigma_d, \Sigma d - \Sigma_s) = P_t^\dagger(-\Sigma_s)\, \exp\Sigma_s. \end{aligned} \quad (15.122)$$

PROOF OF THE IFT

We now prove the IFT (eq. 15.34) using the MGF technique developed in Esposito, Harbola, and Mukamel (2007). We have already mentioned that the dynamics (eq. 15.106) does not describe a stochastic process because the normalization is not preserved. However, for $q_d = q_{nc} = 1$, the biased generator (eq. 15.107) can be written as $W_{nm}(1, 1)$

$$= \left[\sum_e w_e p_{o(e)}^{\text{ref}}\left\{\delta_{n,o(e)}\delta_{m,t(e)} - \delta_{n,m}\delta_{m,o(e)}\right\} + d_t p_n^{\text{ref}}\delta_{n,m}\right]\frac{1}{p_m^{\text{ref}}}, \quad (15.123)$$

from which it readily follows that

$$d_t |p_{\text{ref}}\rangle = \mathcal{W}(1, 1)|p^{\text{ref}}\rangle, \quad (15.124)$$

namely, p_n^{ref} is the solution of the biased dynamics (eq. 15.106) for $q_d = q_{nc} = 1$. The normalization condition thus demands that

$$\begin{aligned} 1 &= \langle 1|\Lambda_t(1, 1)\rangle \\ &= \int_{-\infty}^{\infty} d\Sigma_d d\Sigma_{nc} \exp\{-\Sigma_d - \Sigma_{nc}\}\langle 1|P_t(\Sigma_d, \Sigma_{nc})\rangle \quad (15.125) \\ &\equiv \langle \exp\{-\Sigma_d - \Sigma_{nc}\}\rangle, \end{aligned}$$

which is the IFT in equation (15.34). Note that we do not assume any specific property for p_n^{ref} in this context.

Alternative Proofs of the DFT

We here show two alternative proofs of the DFT (eq. 15.32), which relies on the involution property (eq. 15.37). For the nonadiabatic contribution, this property can be proved as follows. By time-reversing equation (15.27), $\tau \to t - \tau$, we obtain

$$\Sigma_{\text{nc}}[\boldsymbol{n}_t; \pi_t] = \int_0^t d\tau \, A_e^{\text{ref}}(\pi_\tau) \, j^e(\tau) = \int_0^t d\tau \, A_e^{\text{ref}}(\pi_{t-\tau}) \, j^e(t-\tau). \tag{15.126}$$

Because A_e^{ref} is solely determined by the state of the protocol at each instant of time, the reference affinities correspond to those of the backward process, $A_e^{\text{ref}}(\pi_{t-\tau}) = A_e^{\text{ref}}(\pi_\tau^\dagger)$. Using the property that $j^e(t-\tau) = j^{\dagger\,-e}(\tau)$ (see eq. 15.36) and $A_e^{\text{ref}} = -A_{-e}^{\text{ref}}$, we finally obtain

$$\Sigma_{\text{nc}}[\boldsymbol{n}_t; \pi_t] = -\int_0^t d\tau \, A_e^{\text{ref}}(\pi_\tau^\dagger) \, j^{\dagger\,e}(\tau) = -\Sigma_{\text{nc}}[\boldsymbol{n}_t^\dagger; \pi_t^\dagger]. \tag{15.127}$$

Concerning the driving contribution of equation (15.30), we obtain

$$\Sigma_{\text{d}}[\boldsymbol{n}_t; \pi_t] = \int_0^t d\tau \left[d_\tau \psi_n^{\text{ref}}(\pi_\tau) \right]\bigg|_{n=n_\tau} \tag{15.128}$$

$$= \int_0^t d\tau \left[-d_\tau \psi_n^{\text{ref}}(\pi_{t-\tau}) \right]\bigg|_{n=n_{t-\tau}}.$$

It is here again crucial that ψ_n^{ref} depends solely on the protocol value so that $\psi_n^{\text{ref}}(\pi_{t-\tau}) = \psi_n^{\text{ref}}(\pi_\tau^\dagger)$. Therefore

$$\Sigma_{\text{d}}[\boldsymbol{n}_t; \pi_t] = -\int_0^t d\tau \left[d_\tau \psi_n^{\text{ref}}(\pi_\tau^\dagger) \right] n = n_\tau^\dagger = -\Sigma_{\text{d}}[\boldsymbol{n}_t^\dagger; \pi_t^\dagger]. \tag{15.129}$$

ALTERNATIVE PROOF 1

Inspired by García-García et al. (2010), we here derive, using an alternative approach, the symmetry of the MGF that underlies our DFT, equation (15.121). In terms of trajectory probabilities, the MGF (eq. 15.105) can be written as $\Lambda_t(q_\text{d}, q_\text{nc}) =$

$$\int \mathfrak{D}\boldsymbol{n}_t \, \mathfrak{P}[\boldsymbol{n}_t; \pi_t] \, p_{n_0}^{\text{ref}}(\pi_0) \exp\left\{ -q_\text{d} \Sigma_\text{d}[\boldsymbol{n}_t; \pi_t] - q_\text{nc} \Sigma_\text{nc}[\boldsymbol{n}_t; \pi_t] \right\}. \tag{15.130}$$

Using the relation between the EP contributions and the stochastic trajectories in forward and backward processes (eq. 15.35), we can recast the MGF into

$$\Lambda_t(q_\mathrm{d}, q_\mathrm{nc}) = \int \mathfrak{D}\boldsymbol{n}_t\, \mathfrak{P}[\boldsymbol{n}_t^\dagger; \pi_t^\dagger]\, p_{n_t}^\mathrm{ref}(\pi_t)$$
$$\exp\{(1-q_\mathrm{d})\Sigma_\mathrm{d}[\boldsymbol{n}_t; \pi_t] + (1-q_\mathrm{nc})\Sigma_\mathrm{nc}[\boldsymbol{n}_t; \pi_t]\}$$
(15.131)

so that using the property of involution (eq. 15.37), we get

$$\Lambda_t(q_\mathrm{d}, q_\mathrm{nc}) = \int \mathfrak{D}\boldsymbol{n}_t\, \mathfrak{P}[\boldsymbol{n}_t^\dagger; \pi_t^\dagger]\, p_{n_t}^\mathrm{ref}(\pi_t)$$
$$\exp\left\{-(1-q_\mathrm{d})\Sigma_\mathrm{d}[\boldsymbol{n}_t^\dagger; \pi_t^\dagger] - (1-q_\mathrm{nc})\Sigma_\mathrm{nc}[\boldsymbol{n}_t^\dagger; \pi_t^\dagger]\right\}.$$
(15.132)

Hence changing and renaming the integration variable, $\boldsymbol{n}_t \to \boldsymbol{n}_t^\dagger$, and using the fact that the Jacobian determinant of this transformation is 1, we finally get

$$\Lambda_t(q_\mathrm{d}, q_\mathrm{nc}) = \int \mathfrak{D}\boldsymbol{n}_t\, \mathfrak{P}[\boldsymbol{n}_t; \pi_t^\dagger]\, p_{n_t}^\mathrm{ref}(\pi_t)$$
$$\exp\left\{-(1-q_\mathrm{d})\Sigma_\mathrm{d}[\boldsymbol{n}_t; \pi_t^\dagger] - (1-q_\mathrm{nc})\Sigma_\mathrm{nc}[\boldsymbol{n}_t; \pi_t^\dagger]\right\}$$
$$= \Lambda_t^\dagger(1-q_\mathrm{d}, 1-q_\mathrm{nc}),$$
(15.133)

which proves equation (15.121). With respect to the previous proof, this one is based on equation (15.35) and on the property of involution, which follow from the specifications of forward and backward processes.

ALTERNATIVE PROOF 2

The joint probability distribution $P_t(\Sigma_\mathrm{d}, \Sigma_\mathrm{nc})$ written in terms of trajectory probabilities (eq. 15.22) reads $P_t(\Sigma_\mathrm{d}, \Sigma_\mathrm{nc}) =$

$$\int \mathfrak{D}\boldsymbol{n}_t\, \mathfrak{P}[\boldsymbol{n}_t; \pi_t]\, p_{n_0}^\mathrm{ref}(\pi_0)\, \delta\left(\Sigma_\mathrm{d}[\boldsymbol{n}_t; \pi_t] - \Sigma_\mathrm{d}\right)$$
$$\delta\left(\Sigma_\mathrm{nc}[\boldsymbol{n}_t; \pi_t] - \Sigma_\mathrm{nc}\right).$$
(15.134)

Using equation (15.35) and then the involution property (eq. 15.37), we finally obtain the DFT (eq. 15.32):

$$P_t(\Sigma_d, \Sigma_{nc}) = \exp\{\Sigma_d + \Sigma_{nc}\} \int \mathfrak{D}\boldsymbol{n}_t \, \mathfrak{P}[\boldsymbol{n}_t^\dagger; \pi_t^\dagger] \, p_{n_t}^{\text{ref}}(\pi_t)$$
$$\delta\left(\Sigma_d[\boldsymbol{n}_t; \pi_t] - \Sigma_d\right) \delta\left(\Sigma_{nc}[\boldsymbol{n}_t; \pi_t] - \Sigma_{nc}\right)$$
$$= \exp\{\Sigma_d + \Sigma_{nc}\} \int \mathfrak{D}\boldsymbol{n}_t \, \mathfrak{P}[\boldsymbol{n}_t^\dagger; \pi_t^\dagger] \, p_{n_t}^{\text{ref}}(\pi_t)$$
$$\delta\left(-\Sigma_d[\boldsymbol{n}_t^\dagger; \pi_t^\dagger] - \Sigma_d\right) \delta\left(-\Sigma_{nc}[\boldsymbol{n}_t^\dagger; \pi_t^\dagger] - \Sigma_{nc}\right)$$
$$= \exp\{\Sigma_d + \Sigma_{nc}\} P_t^\dagger(-\Sigma_d, -\Sigma_{nc}). \tag{15.135}$$

Adiabatic and Nonadiabatic Contributions

We now prove that both the adiabatic and nonadiabatic EP rates are nonnegative. Concerning the adiabatic contribution, using the log-inequality $-\ln x \geq 1 - x$, one obtains

$$\begin{aligned}\langle \dot{\Sigma}_a \rangle &= \sum_e w_e p_{o(e)} \ln \frac{w_e p_{o(e)}^{ss}}{w_{-e} p_{o(-e)}^{ss}} \\ &\geq \sum_e w_e p_{o(e)} \left[1 - \frac{w_{-e} p_{o(-e)}^{ss}}{w_e p_{o(e)}^{ss}}\right] \\ &= \sum_e \left[w_e p_{o(e)}^{ss} - w_{-e} p_{o(-e)}^{ss}\right] \frac{p_{o(e)}}{p_{o(e)}^{ss}} \\ &= \sum_{e,n} D_n^e w_e p_{o(e)}^{ss} \left[-\frac{p_n}{p_n^{ss}}\right] = 0.\end{aligned} \tag{15.136}$$

The last equality follows from the definition of steady-state PMF of equation (15.43). For the nonadiabatic contribution, using the same inequality and similar algebraic steps, one instead obtains

$$\langle \dot{\Sigma}_{\mathrm{na}} \rangle = \sum_e w_e p_{\mathfrak{o}(e)} \ln \frac{p_{\mathfrak{o}(e)} p^{\mathrm{ss}}_{\mathfrak{o}(-e)}}{p^{\mathrm{ss}}_{\mathfrak{o}(e)} p_{\mathfrak{o}(-e)}}$$

$$\geq \sum_e w_e p_{\mathfrak{o}(e)} \left[1 - \frac{p^{\mathrm{ss}}_{\mathfrak{o}(e)} p_{\mathfrak{o}(-e)}}{p_{\mathfrak{o}(e)} p^{\mathrm{ss}}_{\mathfrak{o}(-e)}}\right] \quad (15.137)$$

$$= \sum_e \left[w_e p^{\mathrm{ss}}_{\mathfrak{o}(e)} - w_{-e} p^{\mathrm{ss}}_{\mathfrak{o}(-e)}\right] \frac{p_{\mathfrak{o}(e)}}{p^{\mathrm{ss}}_{\mathfrak{o}(e)}} = 0.$$

Proofs of the DFTs for the Adiabatic and Driving EP Contributions

We here prove the DFTs in equations (15.46) and (15.47) using the same MGF technique described above in the appendix.

PROOF OF THE DFT FOR THE ADIABATIC CONTRIBUTION

The biased generator ruling the sole adiabatic term reads

$$W_{nm}(q_{\mathrm{a}}) = \sum_e w_e \left\{\exp\{-q_{\mathrm{a}} A^{\mathrm{ss}}_e\} \delta_{n,\mathrm{t}(e)} \delta_{m,\mathfrak{o}(e)} - \delta_{n,m} \delta_{m,\mathfrak{o}(e)}\right\}. \quad (15.138)$$

It satisfies the following symmetry:

$$\mathcal{W}(q_{\mathrm{a}}) = \hat{\mathcal{W}}(1 - q_{\mathrm{a}}), \quad (15.139)$$

where $\hat{\mathcal{W}}(q_{\mathrm{a}})$ is the biased generator of the fictitious dynamics ruled by the rates in equation (15.45). Crucially, p^{ss}_n is the steady state of this dynamics, too:

$$\sum_e D^n_e \hat{w}_e p^{\mathrm{ss}}_{\mathfrak{o}(e)} =$$
$$\sum_m \sum_e \hat{w}_e \left\{\delta_{n,\mathrm{t}(e)} \delta_{m,\mathfrak{o}(e)} - \delta_{n,m} \delta_{m,\mathfrak{o}(e)}\right\} p^{\mathrm{ss}}_m = 0, \quad (15.140)$$

for all n.

This fact guarantees that the escape rates of the fictitious dynamics coincide with those of the original ones:

$$-\sum_e \hat{w}_e \delta_{n,m} \delta_{m,\mathfrak{o}(e)} = -\sum_e w_e \delta_{n,m} \delta_{m,\mathfrak{o}(e)}, \quad \text{for all } n. \quad (15.141)$$

We can now proceed to prove the FT (eq. 15.46):

$$\Lambda_t(q_{\mathrm{a}}) = \langle 1|\Lambda_t(q_{\mathrm{a}})\rangle = \langle 1|\mathcal{U}_t(q_{\mathrm{a}})|p\rangle$$
$$= \langle 1|\mathsf{T}_+\exp\left\{\int_0^t d\tau\, \mathcal{W}_\tau(q_{\mathrm{a}})\right\}|p\rangle \qquad (15.142)$$
$$= \langle 1|\mathsf{T}_+\exp\left\{\int_0^t d\tau\, \hat{\mathcal{W}}_\tau(1-q_{\mathrm{a}})\right\}|p\rangle.$$

In the last equality, we made use of the symmetry in equation (15.139). Following the same mathematical steps backward, we readily get

$$\Lambda_t(q_{\mathrm{a}}) = \hat{\Lambda}_t(1-q_{\mathrm{a}}), \qquad (15.143)$$

from which the DFT in equation (15.46) ensues.

PROOF OF THE DFT FOR THE DRIVING CONTRIBUTION

Concerning the DFT of the driving term (eq. 15.47), the generator of the related biased dynamics reads

$$W_{nm}(q_{\mathrm{d}}) = \sum_e w_e \left\{ \delta_{n,\mathfrak{t}(e)}\delta_{m,\mathfrak{o}(e)} - \delta_{n,m}\delta_{m,\mathfrak{o}(e)} \right\} \\ - q_{\mathrm{d}}d_t\psi_m^{\mathrm{ss}}\delta_{n,m}, \qquad (15.144)$$

and it satisfies the following symmetry:

$$\hat{\mathcal{W}}_t^{\mathsf{T}}(q_{\mathrm{d}}, q_{\mathrm{nc}}) = \mathcal{P}_t^{-1}\mathcal{W}_t(q_{\mathrm{d}}, 1-q_{\mathrm{nc}})\mathcal{P}_t, \qquad (15.145)$$

where $\mathcal{P}_t := \mathrm{diag}\{\exp-\psi_m^{\mathrm{ss}}\}$. The finite-time DFT ensues when following the mathematical steps of the main proof and using equation (15.145) at the step at equation (15.114).

REFERENCES

Andrieux, D., and P. Gaspard. 2007. "Fluctuation Theorem for Currents and Schnakenberg Network Theory." *Journal of Statistical Physics* 127 (1): 107–131.

Baiesi, M., and G. Falasco. 2015. "Inflow Rate, a Time-Symmetric Observable Obeying Fluctuation Relations." *Physical Review E* 92 (4): 042162.

Bulnes Cuetara, G., M. Esposito, and A. Imparato. 2014. "Exact Fluctuation Theorem without Ensemble Quantities." *Physical Review E* 89 (May): 052119.

Callen, H. B. 1985. *Thermodynamics and an Introduction to Thermostatistics*. Hoboken, NJ: John Wiley.

Campisi, M., P. Hänggi, and P. Talkner. 2011. "Colloquium: Quantum Fluctuation Relations—Foundations and Applications." *Reviews of Modern Physics* 83, no. 3 (July): 771–791.

Chetrite, R., and S. Gupta. 2011. "Two Refreshing Views of Fluctuation Theorems through Kinematics Elements and Exponential Martingale." *Journal of Statistical Physics* 143, no. 3 (April): 543.

Ciliberto, S. 2017. "Experiments in Stochastic Thermodynamics: Short History and Perspectives." *Physical Review X* 7 (2): 021051.

Crooks, G. E. 1998. "Nonequilibrium Measurements of Free Energy Differences for Microscopically Reversible Markovian Systems." *Journal of Statistical Physics* 90 (5/6): 1481–1487.

———. 1999. "Entropy Production Fluctuation Theorem and the Nonequilibrium Work Relation for Free Energy Differences." *Physical Review E* 60, no. 3 (September): 2721–2726.

———. 2000. "Path-Ensemble Averages in Systems Driven Far from Equilibrium." *Physical Review E* 61, no. 3 (March): 2361–2366.

Esposito, M. 2012. "Stochastic Thermodynamics under Coarse Graining." *Physical Review E* 85 (April): 041125.

Esposito, M., U. Harbola, and S. Mukamel. 2007. "Entropy Fluctuation Theorems in Driven Open Systems: Application to Electron Counting Statistics." *Physical Review E* 76 (September): 031132.

———. 2009. "Nonequilibrium Fluctuations, Fluctuation Theorems, and Counting Statistics in Quantum Systems." *Reviews of Modern Physics* 81 (December): 1665–1702.

Esposito, M., and C. Van den Broeck. 2010a. "Three Detailed Fluctuation Theorems." *Physical Review Letters* 104 (March): 090601.

———. 2010b. "Three Faces of the Second Law. I. Master Equation Formulation." *Physical Review E* 82 (July): 011143.

García-García, R., D. Domínguez, V. Lecomte, and A. B. Kolton. 2010. "Unifying Approach for Fluctuation Theorems from Joint Probability Distributions." *Physical Review E* 82 (September): 030104.

García-García, R., V. Lecomte, A. B. Kolton, and D. Domínguez. 2012. "Joint Probability Distributions and Fluctuation Theorems." *Journal of Statistical Mechanics: Theory and Experiment* 2012 (02): P02009.

Garrahan, J. P. 2016. "Classical Stochastic Dynamics and Continuous Matrix Product States: Gauge Transformations, Conditioned and Driven Processes, and Equivalence of Trajectory Ensembles." *Journal of Statistical Mechanics: Theory and Experiment* 2016 (7): 073208.

Ge, H., and H. Qian. 2010. "Physical Origins of Entropy Production, Free Energy Dissipation, and Their Mathematical Representations." *Physical Review E* 81, no. 5 (May): 051133.

Harris, R. J., and G. M. Schütz. 2007. "Fluctuation Theorems for Stochastic Dynamics." *Journal of Statistical Mechanics: Theory and Experiment*, no. 07 (July): P07020.

Hatano, T., and S. Sasa. 2001. "Steady-State Thermodynamics of Langevin Systems." *Physical Review Letters* 86 (April): 3463–3466.

Jarzynski, C. 1997. "Equilibrium Free-Energy Differences from Nonequilibrium Measurements: A Master-Equation Approach." *Physical Review E* 56 (November): 5018–5035.

———. 2011. "Equalities and Inequalities: Irreversibility and the Second Law of Thermodynamics at the Nanoscale." *Annual Review of Condensed Matter Physics* 2, no. 1 (March): 329–351.

Kelly, F. P. 1979. *Reversibility and Stochastic Networks*. Hoboken, NJ: John Wiley.

Knauer, U. 2011. *Algebraic Graph Theory: Morphisms, Monoids and Matrices*. Vol. 41. Berlin: Walter de Gruyter.

Kolmogoroff, A. 1936. "Zur Theorie der Markoffschen Ketten." *Mathematische Annalen* 112, no. 1 (December): 155–160. https://doi.org/10.1007/bf01565412.

Peliti, L. 2011. *Statistical Mechanics in a Nutshell*. Princeton, NJ: Princeton University Press.

Pérez-Espigares, C., A. B. Kolton, and J. Kurchan. 2012. "Infinite Family of Second-Law-Like Inequalities." *Physical Review E* 85 (3): 031135.

Polettini, M. 2012. "Nonequilibrium Thermodynamics as a Gauge Theory." *Europhysics Letters* 97, no. 3 (March): 30003.

———. 2014. "Cycle/Cocycle Oblique Projections on Oriented Graphs." *Letters in Mathematical Physics* 105, no. 1 (November): 89–107.

Polettini, M., G. Bulnes Cuetara, and M. Esposito. 2016. "Conservation Laws and Symmetries in Stochastic Thermodynamics." *Physical Review E* 94 (5): 052117.

Polettini, M., and M. Esposito. 2014. "Transient Fluctuation Theorems for the Currents and Initial Equilibrium Ensembles." *Journal of Statistical Mechanics: Theory and Experiment* 2014, no. 10 (October): P10033.

———. 2017. "Effective Thermodynamics for a Marginal Observer." *Physical Review Letters* 119 (December): 240601.

———. 2018. "Effective Fluctuation and Response Theory." arXiv 1803.03552v1 (March 9).

Rao, R., and M. Esposito. 2018a. "Conservation Laws and Work Fluctuation Relations in Chemical Reaction Networks." arXiv 1805.12077v1.

———. 2018b. "Conservation Laws Shape Dissipation." *New Journal of Physics* 20 (2): 023007.

Sánchez, R., and M. Büttiker. 2012. "Detection of Single-Electron Heat Transfer Statistics." *Europhysics Letters* 100 (4): 47008.

Schmiedl, T., and U. Seifert. 2007. "Stochastic Thermodynamics of Chemical Reaction Networks." *Journal of Chemical Physics* 126 (4): 044101.

Schnakenberg, J. 1976. "Network Theory of Microscopic and Macroscopic Behavior of Master Equation Systems." *Reviews of Modern Physics* 48 (October): 571–585.

Seifert, U. 2005. "Entropy Production along a Stochastic Trajectory and an Integral Fluctuation Theorem." *Physical Review Letters* 95 (July): 040602.

———. 2012. "Stochastic Thermodynamics, Fluctuation Theorems and Molecular Machines." *Reports on Progress in Physics* 75, no. 12, 126001 (November): 126001.

Speck, T., and U. Seifert. 2005. "Integral Fluctuation Theorem for the Housekeeping Heat." *Journal of Physics A: Mathematical and General* 38 (34): L581.

Strasberg, P., G. Schaller, T. Brandes, and M. Esposito. 2013. "Thermodynamics of a Physical Model Implementing a Maxwell Demon." *Physical Review Letters* 110, no. 4 (January): 040601.

Chapter 15: Detailed Fluctuation Theorems

Thierschmann, H., R. Sánchez, B. Sothmann, F. Arnold, C. Heyn, W. Hansen, H. Buhmann, and L. W. Molenkamp. 2015. "Three-Terminal Energy Harvester with Coupled Quantum Dots." *Nature Nanotechnology* 10, no. 10 (August): 854–858.

Vaikuntanathan, S., and C. Jarzynski. 2009. "Dissipation and Lag in Irreversible Processes." *Europhysics Letters* 87, no. 6 (September): 60005.

Van den Broeck, C., and M. Esposito. 2015. "Ensemble and Trajectory Thermodynamics: A Brief Introduction." *Physica A* 418 (January): 6–16.

Verley, G., R. Chétrite, and D. Lacoste. 2012. "Inequalities Generalizing the Second Law of Thermodynamics for Transitions between Nonstationary States." *Physical Review Letters* 108, no. 12 (March): 120601.

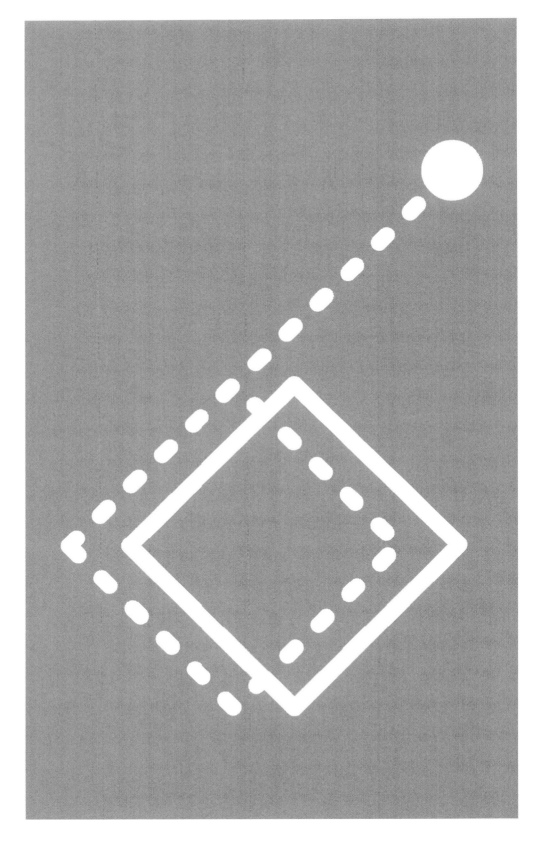

INDEX

A

aggregate 150, 152–160, 163
aging 285, 288, 300, 301
algorithm 18, 91, 93, 95, 98, 216, 219, 243–246, 264, 269, 272, 276, 277
  approximation 219, 248, 249
  evolutionary 385, 388, 390, 391, 399
  genetic (GA) 270, 272, 274
  network-free 93
  polynomial-time 249
  randomized 219, 220, 243, 245, 246
  uniform 250
all-at-once (AO) device 45–55, 230, 231, 234–240, 244
answer-reinitialization 45–52, 55
approximate majority (AM) 63–65, 75–77
atom transition 95, 96
  hypergraph 95, 97
  network 94, 97
automaton 105, 107, 108, 115, 120, 121, 170, 171, *see also* Turing machine (TM)
  chemical 107–111, 114, 120–122
  finite (FA) 11–16, 105, 109–112, 116, 170
  linearly bounded (LBA) 116, 120
  native 121
  probabilistic 15
  pushdown (PDA) 105, 110, 112–116, 118

B

Bayes's theorem 50
Belousov–Zhabotinsky reaction 116, 118, 121

Bennett, Charles H. 191, 193–198, 203–207, 209, 289, 318, 319, 333, 335, 343, 344
Berry's chemical abstract machine 98
bifurcation 287, 295–299
bitstring 286, 293
Boltzmann constant 132, 196, 410
Boltzmann factor 155, 394
Boolean network 173
Brillouin, Léon 317, 318
Brownian
  computer 194
  motion 197

C

catalyst 324, 326, 328–331
category 70–73
  theory 65, 69, 73
cell cycle 138
chain 195, 198, 199, 203, 205, 206
  chain rule 8, 9
Chaitin's Omega 23
chemical bond 85, 86, 88–90, 94, 98, 99
chemical kinetics 106
chemistry 107, 108, 120, 122
  nonbiochemical 106, 120
  one-pot native 110
  oscillatory 116, 121
chemostat 78
chord 423, 424, 427, 439
chromatin 106, 176, 177, 289
Church–Turing
  hypothesis 18, 23
  physical thesis 18
circuit 5, 10–12, 24, 42, 45, 46, 52, 54–56, 219, 222, 226, 228–232, 235–238, 240–243, 245, 247, 250, 251, 405, *see also* reversibility

~457~

extended 44–47, 49–52, 55, 56
family 11, 13, 24
Fredkin 49, 51
irreversible 43, 45, 52, 56
linear 238–241, 250
reversible 5, 42–45, 52
straight-line 10
citric acid (TCA) cycle 95–98
coarse graining 354, 357, 358, 361–363, 366, 368
coin-flipping process 22, 23
compartment 149, 152
compiler 264, 265, 273
complexity 149, 153, 164, 225, 229, 231, 238, 239, 241, 245, 250, 251
average-case 242, 244, 245
Kolmogorov 22, 23, 243
linear 241
nonuniform 229
thermodynamic 219, 229, 242, 243, 250
uniform 230, 242, 245
computation 63, 79, 83–86, 91, 94, 98, 99, 105–118, 120–123, 127, 128, 154, 156, 163, 169–174, 177–181, 191–199, 201–210, 216, 218–226, 228, 238, 240, 242, 245, 246, 248, 249, 263, 264, 266–269, 277, 278, 314, 318, 343, 353–365, 367, 369–374, 377–379
asynchronous 64, 65, 68, 74, 78
biological 180, 288, 289
chemical 105, 107, 112, 115
evolutionary (EC) 265, 270, 278
native 107, 112, 115
noisy, approximate, or inexact 247–249
symmetric 192, 203, 204, 207
thermodynamics of 216, 217, 225, 249, 307, 308, 314
computational machine 3, 13, 17, 23–25, 33, 40, 41
nonuniform 11
uniform 11
computer
chemical 99
"nonuniform" 219
"uniform" 242, 243
computer science theory 3, 10, 18, 217, 218, 232, 242

conservation 216
input–output (IO) 229, 245
law 430, 431, 434–436, 438, 439
continuous-time Markov chain (CTMC) 25, 27, 35, 36, 250, 251
contribution 406, 411, 412, 414, 415, 417, 419, 420, 424–427, 434, 435, 437, 442, 445, 448
adiabatic 421, 449, 450
driving 414–416, 421, 442, 445, 450, 451
nonadiabatic 421, 447, 449
correlation 316, 329, 330, 332–335, 338–340, 366, 367, 369–371, 373, 376, *see also* information
total 372, 373
cost–precision function 287, 293, 294, 296, 300, 301
crossover 270
current 410, 424, 437

D

data 314, 316–318, 326, 328–333, 338, 340
data processing inequality 8, *see also* entropy
KL divergence 9, 36
decomposition 406, 407, 410, 412–414, 418, 419, 421–426, 428, 434, 435, 437, 439
destabilization 150
detailed balance (DB) 28, 31, 39, 406, 410–412, 421, 423, 425, 436
local (LDB) 39, 41, 430, 433
deterioration 298, 299
diffusion 342, 345
directed acyclic graph (DAG) 10, 11
dissipation 224, 353, 354, 358–361, 363, 365, 369–374, 376–379
distribution 3–10, 15, 16, 18, 22, 24, 26, 28–38, 40–43, 45–49, 51, 54–57, 220–224, 227, 230, 232–235, 237–239, 241–243, 252, 310–312, 322, 337, 356, 359–361, 363, 369–372, 375, 376, 378, 386, 387, 391, 394, 397, 400, 406, 407, 426, 448
Boltzmann 23, 39, 311, 386

distributional problem 242, 244, 245
  global-equilibrium 355, 361, 369
  local-equilibrium 356, 357, 361, 363, 364
  nonequilibrium 353, 360, 377
  probability 225, 242, 243, 313, 322, 359
  reference 8
  steady-state 387, 399
  uniform 221, 229, 234–241, 243, 246, 252, 253
  update 15, 16
DNA 106, 123, 289, 294
  DNA strand displacement system (DND) 192, 195–199, 201, 202, 205, 208–210
double-pushout (DPO) 87–89, 93
dual-rail methodology 74
duplication 151, 156, 157, 159, 160, 163, 164
  cycle 163
  spontaneous 156, 160

E

$E.\ coli$ 130, 131, 135, 138, 139
efficiency 13
emulsion 152–154, 158, 164
energy 89, 98, 150, 151, 154, 159, 163, 192, 194–197, 215, 216, 221, 222, 224, 244, 246, 247, 263, 266, 269, 271, 275–278, 309–312, 318, 321, 325, 340, 341, 344, 355, 359, 361, 362, 364, 365, 368, 369, 374, 428, 431–433, 438
  consumption 263–272, 274–278
  cost 307
  efficiency 191, 192, 196, 197, 203, 205, 206, 208, 264, 278
  free 78, 93, 119, 122, 151, 153, 154, 159, 160, 163, 164, 197, 221, 308, 311–313, 321, 322, 324–327, 329–334, 338–340, 343, 344, 356, 359, 364, 365, 368, 376, 390
  local-equilibrium 356, 362, 363, 378
  nonequilibrium 353, 358, 359, 362, 377, 378
  Gibbs free 92, 119, 121, 122, 153

  reduction 264, 268, 270–272, 275
enthalpy 113, 115, 121
  total enthalpy of atomization (TAE) 89, 90, 93
entropy 7–9, 18, 24, 27, 28, 30, 36–39, 43, 46, 48, 78, 155, 164, 197, 221, 222, 230, 233, 240, 246, 247, 252, 308–313, 317, 318, 326, 329, 332, 333, 335, 336, 338–340, 344, 359, 360, 363–368, 376, 377, 384, 387, 390, 391, 407, 411, 412, 421, 430, 431, 433–435
  balance 435
  entropic cost 24
  entropic intensive field 430, 433
  flow (EF) 3, 4, 24, 26–31, 36–40, 43, 46, 48, 52, 54–56, 221
  information-theoretic 216
  physical 216
  production (EP) 6, 24, 27–30, 34, 36–39, 225, 405–407, 410–423, 425, 426, 435, 437–440, 445, 448–450
  residual production 4, 223, 224
  Shannon 4, 7, 29, 162, 359, 366, 377
  thermodynamic 216
equation 93
  chemical master 67
  chemical reaction 198
    tagged 201, 202, 205
  master 407–409, 411, 419, 440, 441
  rate 67
  Schrödinger 84
equilibrium 150–153, 155–157, 159, 160, 163, 164, 194–197, 203, 206, 291, 301, 311, 353, 356, 358, 360, 361, 363, 364, 368, 369, 376, 386, 390, 394, 395, 405, 407, 418, 422, 423, 425, 426, 430, 435, 437, 439
Gibbs 407
out-of-equilibrium process 292
erasure 227, 308, 314, 318, 319, 324–326, 328, 332–339, 343, 396, 397
  bit 31–33, 43, 216–218, 221, 222, 246
  parallel bit 41, 54, 56
  "unerase" 333

error 285–288, 290, 292–294, 296, 298–300
  accumulation 285–287, 290, 294, 298, 300, 301
  fraction 286, 287, 292, 294–298
  threshold 285, 287, 293, 294, 297
evolution 149, 150, 159
  inorganic 150
extremum principle 122

F

feedback cycle 319
finite state machine 11, 16
fission 155, 156, 161
fluctuation 127, 128, 138, 141, 290–292, 295, 298
  stochastic 286, 290
fluctuation theorem (FT) 360, 405, 406, 414, 417, 419, 421, 438, 440, 451
  detailed (DFT) 405, 406, 415–419, 421, 422, 425, 426, 428, 435, 437–440, 442, 445–447, 449–451
  integral (IFT) 405, 406, 416, 417, 421, 422, 425, 435, 437–440, 446
flux 413, 424, 425, 427
  instantaneous 413
  probability 408, 424
forward/reverse ratio 197
functor 71–73
fusion 155, 156, 161

G

gate 5, 24, 33, 37, 42, 44, 45, 49, 51–54, 64, 106
  AND 226, 228, 229, 233, 234, 237, 253
  Fredkin 5, 42, 44
  logic 77, 106, 373, 377
  NAND 354, 373, 374, 376–378
  noisy 247, 248
  NOT 228, 229
  OR 228, 229, 238
  Toffoli 226, 230
  XOR 228, 238–241
gene expression polarity 129
gene regulation 137
generator 388, 400

genetic programming (GP) 271
Gray code 204, 205, 207
growth *see also* model
  rate 134, 135, 137, 139, 140

H

Hamiltonian 23, 28, 29, 39, 46, 226, 355, 356, 360
  Helfrich 154
hardware 263, 264, 268, 276
heat 384, 385, 387, 388, 390, 398, 399
  transfer 336
hypergraph 66, 73 *see also* atom transition
Hückel theory 85

I

information 4, 8–10, 30, 36, 41, 46, 51, 55, 56, 216, 222, 227, 237, 286, 289, 297, 314, 344, 345, *see also* correlation
  processing 169, 285, 286, 288–293, 300
  ratchet 16, 17, 41
  theory 3, 28, 216, 243, 288, 334, 344
input 4, 10–17, 20–23, 33, 35, 37, 42, 44–47, 49, 52, 55–57
instantaneous description (ID) 20
instruction scheduling 265–267, 270
involution property 418, 446, 449
irreversibility 335, 336, 338, 339, *see also* reversibility
  logical 318, 319, 337, 343
  thermodynamic 336–339
isochronous fork 68

K

kinetic proofreading 291, 293
Kolmogorov criterion 425
Kullback–Leibler (KL) divergence 7, 9, 227, *see also* chain rule

L

Landauer, Rolf 192, 217, 221, 222
  Landauer bound 30–32, 36, 221–223

Index

Landauer cost 4, 25, 31, 33, 36, 37, 39–41, 43, 46–51, 54, 55, 222, 224–231, 233, 234, 236–238, 240, 246, 248, 252, 253
Landauer limit 122, 289, 290
Landauer principle 217, 319, 338, 339, 364, 365, 406, 412, 417, 439
language 12, 22, 107, 108, 111, 112, 115–117, 120–122, 273
  assembly 273
  context-free (CFL) 110
  context-sensitive 115, 116, 121
  domain-specific programming 91
  Dyck 110, 112
  high-level 273
  regular 12, 109
lipid 150, 152–158
loop perforation 268, 269

M

Markovian stochastic process 93
mass-action kinetics 93, 136
Massieu potential 436, 437
matrix 386–388, 392, 393, 400, 401, 429, 432, 435, 438
  incidence 408, 409, 422
  rate 388, 408, 441, 443, 445
  transition 390, 392, 395, 397
Maxwell's demon 315–317, 319, 343, 368, 383
Mealy machine 15, 17, *see also* transducer
memory 128, 129, 141, 143, 220, 316–319, 328, 329, 334, 335, 340, 354–357, 359, 362, 363, 365–368, 370–378
  cellular 128
  metastability 353–357, 359, 362–366, 369, 371, 372, 374
minimum circuit size problem 242
mismatch cost 4, 35–37, 39, 46–51, 56, 223, 224, 227, 231, 233–241, 252, 253
model 156, 164, 266, 269, 271, 274, 276, 278
  growth 135–138
  mean-field 136–139
  stochastic 136, 138

molecule 192, 194, 198, 199, 202, 208
  fuel/transformer 199–204, 208, 210
  signal 198, 202
moment generating function (MGF), 440
Moore machine 15, 17, *see also* transducer
morphism 87, 88, 95, 96, 98
  auto- 95, 96, 98
Muller C-element 64, 65, 68, 76
mutation 270

N

neural network 106
node 10
  root 11
noise 290, 291
nonequilibrium statistical physics 3, 42, 216–219, 221, 249

O

operon 129, 131, 133, 137, 140
output 10, 11, 13, 15–17, 20–23, 37, 42, 44–53, 55, 56

P

Pareto
  frontier 271, 272, 275
  optimal 271, 275
patch representation 274
Petri net *see* network, reaction
potential
  chemical 321–323
  difference 132, 139
  energy surface 83, 85
precision 286, 287, 289–294, 300–302
  scaling 268
precursor 152–159, 163
probability mass function (PMF) 405, 440
product 194–201, 203, 208
protocol 355, 357–361, 363, 366, 369, 370, 376, 377
  canonical copy 289

~461~

cellular copy 289
quasistatic 289

Q

quantum field theory (QFT) 83

R

randomness 246, 247
reactant 194–201, 203, 206, 208
reaction 83–95, 97, 98, 194–196, 198–207
  Diels–Alder 89
  diffusion 106
  forward 194–197, 203, 204, 207, 210
    bias 197
  identity 91
  net 90
  network 66–71, 73, 75–77, 91, 93, 94, 97
    chemical (CRN) 64, 65, 68, 78, 85, 92, 94, 95, 98, 99, 198, 199, 201–207, 209, 320
    open 69–72, 78
  reverse 195, 197, 203, 204, 207, 210
reactor 110, 113, 117, 120
  homogeneous one-pot 106, 108, 120
regulator 129–131, 134, 135, 142
  activation of 130
  deactivation of 130
regulon 131, 134–142
relaxation 386, 394–398
repair 286–288, 291, 293–302
reservoir 157, 286, 287, 292, 295, 297–300, 311–313, 322, 344, 345, 385, 387, 407, 428–435, 438, 439
  fuel 332
  heat 153, 158, 309–312, 318, 335, 336
  work 309, 310, 312, 318, 322, 335, 340, 341, 343
reversibility 24, 27, 42, 216, 225–228, 240, 245, 246, 313, 317, 326, 329, 332, 336, 353, 363, 377, 386, 388, 389, 392, 396, 399, 400, *see also* circuit, *see also* irreversibility
  logical 5, 32, 42–46, 49, 51, 52, 54, 56, 191–199, 203, 204, 206, 207, 209, 210, 220, 227, 318, 337, 343
  thermodynamic 32, 33, 36, 41–43, 46, 54, 55, 220, 313, 327, 336–339, 343, 397
running average power limit (RAPL) 276

S

self-replication 149–151, 159, 164
sensor 129–131, 140
side-channel 277
signaling 127–129, 136, 142
  biological 127
simulation 191–193, 195–197, 202–210
software 264, 265, 268–273, 275–278
space 192–194, 196, 197, 202–205, 207–210
  efficiency 191, 192, 197, 203, 204, 208, 209
spanning tree 407, 422–424, 427
stability *see* metastability
stack of trays 112
state 4–6, 9, 11–17, 19–21, 25, 26, 28, 30–32, 34, 37, 40–43, 45, 50, 52, 354–357, 370, 374, 376
  accept(ing) 12, 14–17, 19, 20
  accept/reject 107, 109, 110, 117, 121
  halt 20
  macro- 157, 160, 220, 221, 225, 320–322, 325, 329, 333, 334
  memory 356–358, 361–364, 366, 368, 371–376, 378
  micro- 10, 221, 225, 355, 357–359, 361, 364, 371, 373, 374, 376
  start 12, 14–17, 19
  steady 134, 135, 139, 140, 142
  terminal 15
stoichiometric coefficient 86, 92
stress signal 130
substrate 127
superoptimization 267, 270

system 151, 153–160, 162–164
 autonomous 342, 345
 cellular 149
 closed 121
 protocellular 149, 152
 ternary emulsion 164
 two-component (TCS) 128–142
Szilard, Leo 316, 318, 319, 329, 332, 333, 343, 344, 383
Szilard's engine 308, 315, 317, 319, 320, 324, 329–331, 333, 334, 337–339, 341–343, 383

T

task skipping 268
thermodynamic cost 107, 128
thermodynamic potential 121, 122
thermodynamics 65, 68, 78, 89, 150, 151, 153, 157, 164, 165, 353, 354, 361, 363, 366–369, 371, *see also* computation
 laws of 149
 nonequilibrium 127, 383
 second law of 92, 308–312, 314–316, 318, 319, 332–335, 337, 339, 343
 stochastic 3, 24, 94
time-inhomogeneous Markov jump process, 405, 407
topology 77, 92, 301
 network 409
trade-off 286, 287, 289–293, 300
trajectory 413, 414, 416–418, 425, 435, 437, 440, 442, 447, 448
transcription 128, 129, 131, 132, 136, 138–140, 142
transducer 15–17
transduction 216
transistor 265–267
 dynamic energy 266
 static energy 266
translation 136, 137
Turing machine (TM) 13, 17–24, 105, 107, 115–118, 120, 121, 170, 191–193, 195, 196, 198, 208, 209, 219, 242–244
 prefix-free 22
 universal (UTM) 21–24

V

von Neumann, John 247, 248

W

work 383–385, 387–399

DAVID H. WOLPERT is a professor at the Santa Fe Institute and adjunct professor at Arizona State University. He is the author of three books and more than 200 papers, has three patents, is an associate editor at over half a dozen journals, and is a fellow of the Institute of Electrical and Electronics Engineers (IEEE). His research, spanning thermodynamics of computation and foundations of physics, to machine learning and game theory, has been cited more than 21,000 times. Wolpert has held positions at the Center for Nonlinear Studies, NASA Ames Research Center, Stanford University, and IBM. He holds degrees in physics from Princeton and the University of California.

CHRIS KEMPES is interested in finding theories and principles that apply to the widest range of biological phenomena from the evolution of life, to modern ecology, to astrobiology. Chris often focuses his work on biological architecture—which may include phenomena ranging from explicit biological morphology to metabolic and genetic network structure—as an intermediate between organism physiology and environmental conditions. Mathematical and physical theories lie at the heart of his methodologies to predict how evolution has shaped architecture and how this, in turn, forms a foundation for reliable predictions of environmental response and interaction. His work spans the scales of genetic information architecture to the morphology of microbial individuals and communities to the regional ecological variation. He also uses these principles to understand the major transitions in the evolution of life and what is required of a theory of life such that it can be reliably applied in astrobiological contexts.

PETER F. STADLER studied chemistry, physics, mathematics and astronomy in Vienna, Austria. He received a PhD in chemistry from the University of Vienna in 1990 and then worked as an associate professor for theoretical chemistry at the same school. In 2002 he moved to Leipzig as professor of bioinformatics. Since 1994 he has been an external professor at the Santa Fe Institute. He is a corresponding member abroad of Austrian Academy of Sciences, external scientific member of the Max Planck Society, and honorary professor of the Universidad Nacional de Colombia.

JOSHUA A. GROCHOW is an assistant professor in the departments of computer science and mathematics at the University of Colorado at Boulder, where he is a member of the Computer Science Theory and Complex Systems groups. He was previously an Omidyar Fellow at the Santa Fe Institute and a postdoc in the University of Toronto Computer Science Theory Group, and prior to that he received his PhD in computer science from the University of Chicago.

SFI PRESS BOARD OF ADVISORS

Nihat Ay
Professor, Max Planck Institute for Mathematics in the Sciences; SFI Resident Faculty

Sam Bowles
Professor, University of Siena; SFI Resident Faculty

Jennifer Dunne
SFI Resident Faculty; SFI Vice President for Science

Andrew Feldstein
CEO, Co-CIO & Partner, Blue Mountain Capital Management; SFI Trustee

Jessica Flack
SFI Resident Faculty

Mirta Galesic
SFI Resident Faculty

Murray Gell-Mann
SFI Distinguished Fellow & Life Trustee

Chris Kempes
SFI Resident Faculty

Michael Lachmann
SFI Resident Faculty

Manfred Laublicher
President's Professor, Arizona State University; SFI External Faculty

Michael Mauboussin
Managing Director, Global Financial Strategies; SFI Trustee & Chairman of the Board

Cormac McCarthy
Author; SFI Trustee

Ian McKinnon
Founding Partner, Sandia Holdings LLC; SFI Trustee

John Miller
Professor, Carnegie Mellon University; SFI Science Steering Committee Chair

William H. Miller
Chairman & CEO, Miller Value Partners; SFI Trustee & Chairman Emeritus

Cristopher Moore
SFI Resident Faculty

Mercedes Pascual
Professor, University of Chicago; SFI Science Board Co-Chair

Sidney Redner
SFI Resident Faculty

Dan Rockmore
Professor, Dartmouth College; SFI External Faculty

Jim Rutt
JPR Ventures; SFI Trustee

Daniel Schrag
Professor, Harvard University; SFI External Faculty

Geoffrey West
SFI Resident Faculty; SFI Distinguished Professor & Past President

David Wolpert
SFI Resident Faculty

EDITORIAL

David C. Krakauer
Publisher/Editor-in-Chief

Tim Taylor
Aide-de-Camp

Laura Egley Taylor
Miller Omega Manager for the Press

Sienna Latham
Editorial Coordinator

Katherine Mast
SFI Press Associate

THE SANTA FE INSTITUTE PRESS

The SFI Press endeavors to communicate the best of complexity science and to capture a sense of the diversity, range, breadth, excitement, and ambition of research at the Santa Fe Institute. To provide a distillation of discussions, debates, and meetings across a range of influential and nascent topics.

To change the way we think.

SEMINAR SERIES

New findings emerging from the Institute's ongoing working groups and research projects, for an audience of interdisciplinary scholars and practitioners.

ARCHIVE SERIES

Fresh editions of classic texts from the complexity canon, spanning the Institute's thirty years of advancing the field.

COMPASS SERIES

Provoking, exploratory volumes aiming to build complexity literacy in the humanities, industry, and the curious public.

For forthcoming titles, inquiries, or news about the Press, contact us at
SFIPRESS@SANTAFE.EDU

ABOUT THE SANTA FE INSTITUTE

The Santa Fe Institute is the world headquarters for complexity science, operated as an independent, nonprofit research and education center located in Santa Fe, New Mexico. Our researchers endeavor to understand and unify the underlying, shared patterns in complex physical, biological, social, cultural, technological, and even possible astrobiological worlds. Our global research network of scholars spans borders, departments, and disciplines, bringing together curious minds steeped in rigorous logical, mathematical, and computational reasoning. As we reveal the unseen mechanisms and processes that shape these evolving worlds, we seek to use this understanding to promote the well-being of humankind and of life on Earth.

COLOPHON

The body copy for this book was set in EB Garamond, a typeface designed by Georg Duffner after the Egenolff-Berner type specimen of 1592. Headings are in Kurier, a typeface created by Janusz M. Nowacki, based on typefaces by the Polish typographer Małgorzata Budyta. Additional type is set in Cochin, a typeface based on the engravings of Nicolas Cochin, for whom the typeface is named.

The SFI Press complexity glyphs used throughout this book were designed by Brian Crandall Williams.

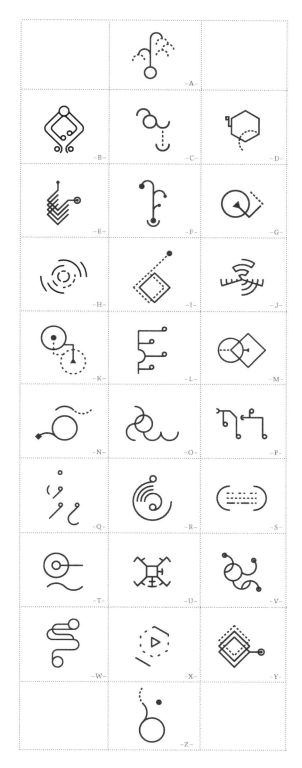

SEMINAR SERIES

Made in the USA
Monee, IL
29 February 2020